Lecture Notes in Mathematics

Edited by A. Dold and B. Eckmann

1286

H. R. Miller D. C. Ravenel (Eds)

Algebraic Topology

Proceedings of a Workshop held at the University of Washington, Seattle, 1985

Springer-Verlag
Berlin Heidelberg New York London Paris Tokyo

Editors

Haynes R. Miller
Room 2-237, Department of Mathematics
Massachusetts Institute of Technology
Cambridge, MA 02139, USA

Douglas C. Ravenel
Department of Mathematics, University of Washington
Seattle, WA 98195, USA

Mathematics Subject Classification (1980): 55-06

ISBN 3-540-18481-3 Springer-Verlag Berlin Heidelberg New York
ISBN 0-387-18481-3 Springer-Verlag New York Berlin Heidelberg

Printing and binding: Druckhaus Beltz, Hemsbach/Bergstr.
2146/3140-543210

Preface

The University of Washington hosted a Workshop in Algebraic Topology during the 1984-85 academic year. The program had several components. Four topics courses were offered:

Doug Ravenel, "Im J and the EHP sequence."

Emmanuel Dror Farjoun, "Homotopy and homology theory for diagrams of spaces."

Fred Cohen, "Some aspects of classical homotopy theory."

Mark Mahowald, "The Kervaire invariant."

Some 29 topologists visited Seattle for varying periods of time, mainly in the Winter and Spring quarters. A very active seminar resulted; a schedule follows on pages v and vi.

We gratefully acknowledge support from several sources. The Department of Mathematics focused its resources on it: Fred Cohen, Emmanuel Dror Farjoun, and Mark Mahowald were given visiting positions; four quarters of topics courses with very low nominal enrollment were offered; and the recently donated Milliman endowment underwrote the visit by J. Frank Adams as well as a number of others. The N.S.F. provided travel and per diem support under grant number DMS-8407234.

In this volume we have collected lecture notes from the courses of Fred Cohen and Emmanuel Dror Farjoun, together with papers submitted by participants in the workshop. All have been refereed; we wish to thank the referees for their assistance.

This volume is dedicated to Seattle's children, Helen Beatrice Whalen Cohen and Meier Amir Farjoun.

Contents

Schedule of Talks

Jan.	11	D.C. Ravenel	What I know about $BP_*\Omega^2S^{2n+1}$
	16	N.J. Kuhn	Stable splittings of BG
	25	A. Pearlman	Bordism with singularities
Feb.	1	B. McQuistan	$bo_*\Omega^2S^3$
	8	M.J. Hopkins	Nilpotence in stable homotopy, I
	8	S.A. Mitchell	Loop groups, I
	12	M.J. Hopkins	II
	15	S.A. Mitchell	II
	15	M.J. Hopkins	III
	20	M.J. Hopkins	IV
	22	D. Sjerve	Actions of finite groups on Riemann surface
	22	J.H. Smith	Splittings constructed from the symmetric group, I
	26	M.J. Hopkins	V
	26	J.H. Smith	II
Mar.	1	M.J. Hopkins	VI
	1	J.H. Smith	III
	4	M.J. Hopkins	VII
	8	M.J. Hopkins	VIII
	12	M.J. Hopkins	IX
	12	J.R. Harper	Cohomology operations and cup products
	15	M.J. Hopkins	X
	15	D. Waggoner	$H_*(\Omega^2SU(n))$
	22	J.R. Harper	Co H-spaces and self-maps
Apr.	3	J.F. Adams	Extensions of the Segal conjecture
	5	J.F. Adams	Equivariant analogues of the Adams spectral sequence
	8	A. Zabrodsky	Fixed points and homotopy fixed points
	10	J.F. Adams	Classifying spaces revisited
	10	N. Yagita	$BP_*(X)$ and $H_*(X;\mathbb{Z}/p)$
	12	J.F. Adams	Problems in the topology of Lie groups and finite H-spaces with classifying spaces.
	15	E. Devinatz	Extended powers in the Adams spectral sequence and the proof of the nilpotence conjecture
	17	J.D.S. Jones	Cyclic homology, I
	19	J.D.S. Jones	II
	22	J.D.S. Jones	III
	24	H. Mui	Homology operations derived from modular coinvariants

	25	R.L. Cohen	The cyclic groups, loop spaces, and K-theory, I
	26	H. Mui	II
	26	R.L. Cohen	II
	29	R.L. Cohen	III
	30	R.L. Cohen	IV
May	8	J.A. Neisendorfer	Moore variations on a theme of Selick
	13	J.A. Neisendorfer	On maps between classifying spaces, I
	15	J.A. Neisendorfer	II
	17	L.G. Lewis	Calculating the $RO(G)$ - graded ordinary homology of a point
	20	C.-F. Bödigheimer	Configuration spaces: general ideas
	22	C.-F. Bödigheimer	Configuration spaces: examples
	22	S.B. Priddy	Invariant theory and characteristic classes
	29	D.J. Pengelley	$H_*(BO)$, I
		V. Giambalvo	$H_*(BO)$, II
	31	M.A. Guest	Loop groups
		C.W. Wilkerson	Aguadé's work on rings of invariants
June	3	M.J. Hopkins	The potence of nilpotence
	4	D.M. Davis	The spectrum $(P\wedge BP\langle 2\rangle)_{-\infty}$
	5	D.M. Davis	The stable homtopy types of stunted real projective spaces
	5	W.M. Singer	The Λ-algebra and the homology of the symmetric groups
	10	R.D. Thompson	Products of unstable elements suspending to ImJ,I
	12	R.D. Thompson	II
	12	J.C. Moore	Lemaire's proof of Halperin's theorem, and the application of McGibbon and Wilkerson
	14	W.G. Dwyer	The algebra of cyclic objects
	14	F. Raymond	Manifolds on which only tori can act
	18	U. Koschorke	Higher order link homotopy invariants

A course in some aspects of classical homotopy theory

F.R. Cohen*

These notes are based on a course in classical homotopy theory given during the 1985 emphasis year in Topology at the University of Washington. The material here is expository and expresses some prejudices of the author; all of it is either in the literature or known to the experts.

The main direction of these notes is based on the Whitehead product and the classical distributivity law. The divisibility of the Whitehead product, the so-called strong form of the Kervaire invariant, comes up in several places where we study the difference between the H-space squaring map and the loopings of the degree 2 map on $\Omega^q S^n$. We study the relation to P. Selick's theorem on the odd primary homotopy groups of S^3 and allied 2-primary decompositions. A smattering of information is given about function spaces together with some remarks about related work of Dickson.

These notes are neither comprehensive nor complete; they are an exposition of some interesting aspects of classical homotopy theory.

We would like to thank Ed Curtis, Haynes Miller, Doug Ravenel, and Steve Mitchell for their kind hospitality; Joe Neisendorfer, Frank Peterson, Paul Selick, Kathleen Whalen, and Helen Beatrice Whalen Cohen for their conversations and help.

---

*Partially supported by an NSF grant.

Table of contents

§1: Whitehead products and Samelson products

Throughout this paper spaces are simply-connected unless otherwise stated. Let X and Y be compactly generated Hausdorff spaces with non-degenerate base-point $*$[St] and let F denote the homotopy theoretic fibre of the inclusion $i: X \vee Y \to X \times Y$. Since $\Omega(i)$ is a principal fibration with a cross-section, it follows that $\Omega(X \vee Y)$ is homotopy equivalent to $\Omega X \times \Omega Y \times \Omega F$. Thus $\pi_q(X \vee Y)$ is isomorphic to $\pi_q X \oplus \pi_q Y \oplus \pi_q F$. If $X = S^k$ and $Y = S^n$, then F is $(n+k-2)$-connected and $\pi_{n+k-1} F$ is isomorphic to $\mathbf{Z}$ by inspection of the Serre spectral sequence for i. The Whitehead product $[\iota_k, \iota_n]$ is the element in $\pi_{n+k-1}(S^k \vee S^n)$ given by the image of a choice of generator for $\pi_{n+k-1} F$[Wh]. Notice that $[\iota_k, \iota_n]$ has been defined up to a sign which will be made precise shortly. Furthermore, since the composite

$$S^{n+k-1} \xrightarrow[{[\iota_k, \iota_n]}]{} S^k \vee S^n \to S^k \times S^n$$

gives a long exact sequence in homology by the above calculation, the following proposition is immediate.

Proposition 1.1. The Whitehead product $[\iota_k, \iota_n]$ is the attaching map of the top cell in $S^k \times S^n$ to $S^k \vee S^n$.

Next notice that there is an induced map

$$[\ ,\ ]: \pi_k X \otimes \pi_n X \to \pi_{n+k-1} X$$

with $[\ ,\ ]$ defined to be the composite

$$S^{n+k-1} \xrightarrow{[\iota_k, \iota_n]} S^k \vee S^n \xrightarrow{\alpha \vee \beta} X \vee X \xrightarrow{\text{fold}} X$$

for α in $\pi_k X$ and β in $\pi_n X$.

Useful properties of the Whitehead product are given below and proven in [H_2,MU,NT,T_4,W_2].

Proposition 1.2. The Whitehead product

$$[\ ,\ ]: \pi_k X \otimes \pi_n X \to \pi_{k+n-1} X$$

satisfies the following properties where degree $(\iota_q)=q$.

(1) It is bilinear.

(2) $[\iota_k,\iota_n] = (-1)^{kn}[\iota_n,\iota_k]$.

(3) $(-1)^{jn}[[\iota_j,\iota_k],\iota_n]+(-1)^{jk}[[\iota_k,\iota_n],\iota_j]+(-1)^{kn}[[\iota_n,\iota_j],\iota_k] = 0.$

Instead of deriving these directly, we use methods of G.W. Whitehead $[W_2]$ and H. Samelson [Sa] after listing the following corollary.

Corollary 1.3.

(1) $[\iota_{2n+1},\iota_{2n+1}]$ has order 2 in $\pi_{4n+1}S^{2n+1}$.

(2) $[\iota_{2n+1}[\iota_{2n+1},\iota_{2n+1}]] = 0$ in $\pi_{6n+1}S^{2n+1}$.

(3) $3[\iota_{2n}[\iota_{2n},\iota_{2n}]] = 0$ in $\pi_{6n-2}S^{2n}$.

(4) $[\iota_{2n}[\iota_{2n}[\iota_{2n},\iota_{2n}]]] = 0$ in $\pi_{8n-3}S^{2n}$.

Remarks 1.4. The element $[\iota_{2n},\iota_{2n}]$ has infinite order in $\pi_{4n-1}S^{2n}$ as we shall see later. That $[\iota_{2n+1},\iota_{2n+1}]$ is non-zero in $\pi_{4n+1}S^{2n+1}$ if $n\neq 0,1,3$ follows from work of J.F. Adams [A1]. The so-called "strong form of the Kervaire invariant" is the question whether $[\iota_{2n+1},\iota_{2n+1}]$ is divisible by 2 when $n=2^k-1$. The element $[\iota_{2n}[\iota_{2n},\iota_{2n}]]$ is discussed by Toda [T1].

Next assume that G is a loop space, ΩA. Consider the group of pointed homotopy classes of maps $[X_1\times X_2,\Omega A]$ which is isomorphic to $[\Sigma(X_1\times X_2),A]$. Since $\Sigma(X_1\times X_2)$ is homotopy equivalent to $\Sigma X_1 \vee \Sigma X_2 \vee \Sigma(X_1\wedge X_2)$, and $X_1\vee X_2 \to X_1\times X_2 \to X_1\wedge X_2$ is a cofibration, there is a short exact sequence of groups which is split as sets,

$$1 \to [X_1\wedge X_2,\Omega A] \to [X_1\times X_2,\Omega A] \to [X_1\vee X_2,\Omega A] \to 1.$$

Next, observe that there is a map $\bar{c}\colon \Omega A\times\Omega A \to \Omega A$ given by $\bar{c}(f,g) = ((f\cdot g)f^{-1})g^{-1}$ which when restricted to $\Omega A\vee\Omega A$ is null-homotopic. Thus there is a map $c\colon \Omega A\wedge\Omega A \to \Omega A$ which is unique up to homotopy and which gives a homotopy commutative diagram

$$\begin{array}{ccc} \Omega A\times\Omega A & \xrightarrow{\ \bar{c}\ } & \Omega A \\ \text{quotient}\downarrow & \nearrow c & \\ \Omega A\wedge\Omega A & & \end{array} .$$

Define the Samelson product

$$\langle\ ,\ \rangle\colon \pi_p\Omega A\otimes\pi_q\Omega A \to \pi_{p+q}\Omega A$$

to be given by the composite

$$S^p\wedge S^q \xrightarrow{\alpha\wedge\beta} \Omega A\wedge\Omega A \xrightarrow{\ c\ } \Omega A.$$

<u>Proposition 1.5</u>. The Samelson product satisfies the following properties.

(1) It is bilinear.

(2) $\langle\alpha,\beta\rangle = (-1)^{pq+1}\langle\beta,\alpha\rangle$ for $\alpha\in\pi_p\Omega A$ and $\beta\in\pi_q\Omega A$.

(3) $\langle\alpha\langle\beta,\gamma\rangle\rangle = \langle\langle\alpha,\beta\rangle,\gamma\rangle + (-1)^{pq}\langle\beta\langle\alpha,\gamma\rangle\rangle$ for $\alpha\in\pi_p\Omega A$, $\beta\in\pi_q\Omega A$, and $\gamma\in\pi_r\Omega A$.

Consider the suspension $E\colon X \to \Omega\Sigma X$ and the induced map $ad\colon X\wedge X \to \Omega\Sigma X$ given by $c\cdot(E\wedge E)$. Inductively define $ad^k\colon \hat{k}X \to \Omega\Sigma X$ by $c\cdot(E\wedge ad^{k-1})$. Recall that with field coefficients, $H_*\Omega\Sigma X$ is isomorphic to the tensor algebra $T(\bar{H}_*X)$ as an algebra; the diagonal is induced by the diagonal for X[BS]. Assume that homology groups are taken with field coefficients in the following lemma.

<u>Lemma 1.6</u>. (1) $ad_*(x\otimes y) = x\otimes y-(-1)^{|x||y|}y\otimes x+\sum_{k>2} z_1\otimes\cdots\otimes z_k$.

(2) If x and y are primitive, $ad_*(x\otimes y) = x\otimes y-(-1)^{|x||y|}y\otimes x$.

Thus consider the fundamental classes ι_k and ι_n in $\pi_*\Omega\Sigma(S^k \vee S^n)$. By Lemma 1.6, the Hurewicz image of $<\iota_k,\iota_n>$ is the primitive $\iota_k \otimes \iota_n - (-1)^{kn}\iota_n \otimes \iota_k$. But by inspection of the Serre spectral sequence, this last element is a generator of $H_{k+n}\Omega F \cong \pi_{k+n}\Omega F$. Since ΩF is $(k+n-1)$-connected, $<\iota_k,\iota_n>$ is the adjoint of the Whitehead product up to a sign. Define the sign of $[\iota_{k+1},\iota_{n+1}]$ by setting $[\iota_{k+1},\iota_{n+1}]$ adjoint to $(-1)^k < \iota_k,\iota_n>$. The details in 1.2 are deleted.

Notice that the Hurewicz image of $<\iota_{2n-1},\iota_{2n-1}>$ is $2(\iota_{2n-1})^2$ and this element has infinite order in $H_{4n-2}(\Omega S^{2n};\mathbb{Z})$.

A good reference for homotopy with coefficients is [N1].

<u>Proof of Proposition 1.5</u>: First check that $<k\alpha,\beta> = k<\alpha,\beta>$. Notice that $k<\alpha,\beta>$ is given by the composite $S^p \wedge S^q \xrightarrow{\alpha\wedge\beta} \Omega A \wedge \Omega A \xrightarrow{c} \Omega A \xrightarrow{k} \Omega A$ where k denotes the kth power map. Thus k is the composite $\Omega A \xrightarrow{\Delta^k} \underset{k}{\times}\Omega A \xrightarrow{\mu_k} \Omega A$ where $\Delta^k(x) = (x,\ldots,x)$ and $\mu_k(x_1,\ldots,x_k) = (x_1(x_2(\cdots(x_{k-1},x_k)\cdots)$. Since $\Delta^k : S^n \to \underset{k}{\times} S^n$ factors through the inclusion of the bouquet $\underset{k}{\vee} S^n$ in $\underset{k}{\times} S^n$, there is a homotopy commutative diagram

$$\begin{array}{ccccc}
 & S^p\wedge S^q & \xrightarrow{c\cdot(\alpha\wedge\beta)} & \Omega A \\
\swarrow & \downarrow \Delta^k & & \downarrow \Delta^k \\
\underset{k}{\vee}(S^p\wedge S^q) \to & \underset{k}{\times} S^p\wedge S^q & \xrightarrow{[c\cdot(\alpha\wedge\beta)]^k} & \underset{k}{\times}\Omega A \\
\text{fold}\downarrow & & & \downarrow \mu_k \\
S^p\wedge S^q & & \xrightarrow[c\cdot(\alpha\wedge\beta)]{} & \Omega A
\end{array}$$

The left hand composite from $S^p\wedge S^q$ to itself is degree k and thus statement (1) follows.

We next prove (2). Let α and β be in elements in $\pi_p\Omega A$ and $\pi_q\Omega A$ respectively. Then $<\alpha,\beta>$ is the composite $S^p\wedge S^q \xrightarrow{\alpha\wedge\beta} \Omega A\wedge\Omega A \xrightarrow{c} \Omega A$. But observe that by the definitions, $\langle\alpha,\beta\rangle$ is homotopic to the composite $S^p\wedge S^q \xrightarrow{\text{switch}} S^q\wedge S^p \xrightarrow{\beta\wedge\alpha} \Omega A\wedge\Omega A \overset{c}{\to} \Omega A \xrightarrow{-1} \Omega A$. Thus (2) follows.

Next recall that if Λ is a group with x,y,z in Λ, then $[x[yz]]\cdot[y[zx]]\cdot[z[xy]] \equiv 1$ modulo commutators of length at least 4[Z]. Consider pointed maps $\alpha\colon S^p \to \Omega A$, $\beta\colon S^q \to \Omega A$ and $\gamma\colon S^t \to \Omega A$. Then the composite $S^p\times S^q\times S^t \to \Omega A$ represented by $[\alpha[\beta\gamma]]+[\beta[\gamma,\alpha]]+[\gamma[\alpha,\beta]]$ is 0 in the group $[S^p\times S^q\times S^t,\Omega A]$ because the diagonal $S^n\to S^n\times S^n$ is null-homotopic in $S^n\wedge S^n$. Now observe that $[\beta[\gamma,\alpha]]$ and $[\gamma[\alpha,\beta]]$ are represented by

$$S^p\wedge S^q\wedge S^t \xrightarrow{\sigma_1} S^q\wedge S^t\wedge S^p \xrightarrow{\beta\wedge\gamma\wedge\alpha} \wedge_3\Omega A \xrightarrow{ad^2} \Omega A, \quad \text{and}$$

$$S^p\wedge S^q\wedge S^t \xrightarrow{\sigma_2} S^t\wedge S^p\wedge S^q \xrightarrow{\gamma\wedge\alpha\wedge\beta} \wedge_3\Omega A \xrightarrow{ad^2} \Omega A$$

respectively where σ_i is the indicated permutation of coordinates. Thus the following equation

$$[\alpha[\beta\gamma]]+(-1)^{p(q+t)}[\beta[\gamma,\alpha]]+(-1)^{t(p+q)}[\gamma[\alpha,\beta]] = 0$$

is satisfied in the group $[S^{p+q+t},\Omega A]$. Proposition 1.5(3) follows.

<u>Proof of Lemma 1.6</u>: By definition, the following diagram homotopy commutes where $\pi\colon X\times X\to X\wedge X$ is the natural projection:

$$\begin{array}{ccccccc}
X\times X & \xrightarrow{E\times E} & (\Omega\Sigma X)^2 & \xrightarrow{\Delta\times\Delta} & (\Omega\Sigma X)^2\times(\Omega\Sigma X)^2 & \xrightarrow{(1\times -1)\times(1\times -1)} & (\Omega\Sigma X)^4 \\
 & & & & & & \downarrow{\scriptstyle 1\times\text{switch}\times 1} \\
\downarrow{\scriptstyle\pi} & & & & & & (\Omega\Sigma X)^4 \\
 & & & & & & \downarrow{\scriptstyle\mu_4} \\
X\wedge X & & & \xrightarrow{\quad ad \quad} & & & \Omega\Sigma X
\end{array}$$

Thus $ad_*(x\otimes y) = \Sigma(-1)^{|x''||y'|}x'\otimes y'\otimes\chi(x'')\otimes\chi(y')$ where $\Delta z = \Sigma z'\otimes z''$ is the coproduct and $\chi=(-1)_*$. Furthermore, it follows from the definition [MM] that $\chi(1)=-1$ and $\Sigma x'\chi(x")=0$ if $|x|>0$. Hence $\chi(x)=-x$ + decomposable elements. The formula in 1.6(1) follows. Notice that if x is primitive, then $\chi(x)=-x$ and so formula (2) follows.

§2: The Hilton-Milnor theorem

Before stating one form of the Hilton-Milnor theorem [H_1,Mr], we point out that it gives a partial description of the group $[\Sigma A, \Sigma X \vee \Sigma Y]$. For example, let $[k]: S^n \to S^n$ denote the degree k map. Since the map $[k]$ is given by $S^n \xrightarrow{\text{pinch}} \underset{k}{\vee} S^n \xrightarrow{\text{fold}} S^n$ if $k \geq 1$, one can use the Hilton-Milnor theorem to study the effect of $[k]$ on the homotopy groups of S^n by factoring the map through the homotopy groups of $\underset{k}{\vee} S^n$.

The Hilton-Milnor theorem gives a specific product decomposition for $\Omega\Sigma(X \vee Y)$. Let $X^{[k]}$ denote the k-fold smash product $\underbrace{X \wedge \cdots \wedge X}_{k}$. Namely there is a homotopy equivalence

$$\theta\colon \Omega\Sigma X \times \Omega\Sigma(Y \underset{k\geq 1}{\vee} (X^{[k]} \wedge Y)) \to \Omega\Sigma(X \vee Y).$$

The usual statement of the Hilton-Milnor theorem is obtained by iterating the above decomposition to exhibit a specific homotopy equivalence between $\Omega\Sigma(X \vee Y)$ and the weak product $\prod_\alpha \Omega\Sigma(Z_\alpha)$ where Z_α is a smash product of copies of X and Y. We will not need this further precision here. However, it is useful to have a precise description of the map θ.

There are canonical maps $E_X: X \to \Omega\Sigma(X \vee Y)$ and $E_Y: Y \to \Omega\Sigma(X \vee Y)$. Recall the map $c: \Omega A \wedge \Omega A \to \Omega A$ of section 1 inducing the Samelson product. Inductively define maps $ad^k: X^{[k]} \wedge Y \to \Omega\Sigma(X \vee Y)$ by setting $ad^1 = c \cdot (E_X \wedge E_Y)$ and $ad^{k+1} = c \cdot (E_X \wedge ad^k)$. Thus there is a map $ad: Y \underset{k\geq 1}{\vee} (X^{[k]} \wedge Y) \to \Omega\Sigma(X \vee Y)$ which is given by E_Y on Y and by ad^k on $X^{[k]} \wedge Y$. Let $\Omega(\lambda): \Omega\Sigma(Y \underset{k\geq 1}{\vee} (X^{[k]} \wedge Y)) \to \Omega\Sigma(X \vee Y)$ denote the canonical multiplicative extension of ad. Define θ to be the composite

$$\Omega\Sigma X \times \Omega\Sigma(Y \underset{k\geq 1}{\vee} (X^{[k]} \wedge Y)) \xrightarrow{\Omega(i_X) \times \Omega(\lambda)} [\Omega\Sigma(X \vee Y)]^2 \xrightarrow{\mu_2} \Omega\Sigma(X \vee Y)$$

where i_X is the natural inclusion $\Sigma X \to \Sigma X \vee \Sigma Y$.

Theorem 2.1[H1,Mr,P]. The map θ is a homotopy equivalence.

An immediate consequence is

Proposition 2.2. The homotopy theoretic fibre of the inclusion $\Sigma X \vee \Sigma Y \subset \Sigma X \times \Sigma Y$ is $\Sigma(\Omega\Sigma X) \wedge (\Omega\Sigma Y)$.

That the homotopy theoretic fibre of the inclusion of $A \vee B$ in $A \times B$ is $\Sigma(\Omega A) \wedge (\Omega B)$ for simply-connected A and B is given in [G].

There are several proofs of this theorem. We reproduce the quick proof in [G1] which does not specifically give the map θ and which is based on the following where we assume that A and B are simply-connected.

Proposition 2.3. The homotopy theoretic fibre of the pinch map $p: A \vee B \to A$ is the half-smash product $B \times \Omega A / * \times \Omega A = B \ltimes \Omega A$.

Proof. Recall that if $f: X \to A$ is any map, then there is a map $\tilde{f}: \tilde{X} \to A$ which is a fibration and $\tilde{X}$ is homotopy equivalent to X; the space $\tilde{X}$ is $\{(x,g) \mid x \in X,\ g: I \to A,\ g(o)=f(x)\}$ and $\tilde{f}(x,g)=g(1)\}$. Apply this to the pinch map $p: A \vee B \to A$ to get $\tilde{p}: \widetilde{A \vee B} \to A$. The fibre of $\tilde{p}$, F, is the space $\{(x,g) \mid x \in A \vee B,\ p(x)=g(o), \text{ and } g(1)=*\}$ where $*$ is the base-point in A.

Write $F_A=\{(x,g) \mid x \in A,\ p(x)=g(o),\ g(1)=*\}$ and $F_B = \{(x,g) \mid x \in B,\ p(x)=g(o),\ g(1)=*\}$. Notice that (1) $F = F_A \cup F_B$, (2) F_B is homeomorphic to $B \times \Omega A$, and (3) F_A is homeomorphic to the path space PA. Next observe that $F_A \cap F_B$ is $\{(x,g) \mid x \in A \cap B,\ p(x)=g(o),\ g(1)=*\}$ which is ΩA. Thus F is homeomorphic to $(B \times \Omega A) \cup_{\Omega A} PA$. Since PA is contractible and (F,PA) is an NDR pair, the quotient map $F \to F/PA$ is a homotopy equivalence. But F/PA is homeomorphic to $(B \times \Omega A) \cup_{\Omega A} PA/PA$ and this latter space is homeomorphic to $B \times \Omega A / * \times \Omega A$. The proposition follows.

Next one has

Lemma 2.4. $\Sigma A \times B / * \times B$ is homotopy equivalent to $A \wedge (\Sigma B \vee S^1)$.

Proof. $\Sigma A\times B/*\times B$ is homeomorphic to $\Sigma A\times(B_+)/(*\times B_+)\cup(\Sigma A\times +)$ where B_+ is the space B with a disjoint base-point +. But this last space is $\Sigma A\wedge(B_+)$ which is homotopy equivalent to $A\wedge(\Sigma B\vee S')$.

Lemma 2.5. If X is connected then $\Sigma\Omega\Sigma X$ is homotopy equivalent to $\Sigma\bigvee_{k\geq 1}X^{[k]}$. Thus there is a map $\tilde{H}_k\colon \Sigma\Omega S^{n+1}\to S^{kn+1}$ which is onto in homology. The maps $\tilde{H}_k$ are natural for maps $\Omega\Sigma(f)\colon \Omega\Sigma X\to\Omega\Sigma Y$.

Proof of Theorem 2.1: Since $\Omega(p)\colon \Omega\Sigma(X\vee Y)\to\Omega\Sigma X$ has a cross-section where $p\colon \Sigma(X\vee Y)\to\Sigma X$ is the pinch map, $\Omega\Sigma(X\vee Y)$ is homotopy equivalent to $\Omega\Sigma X\times\Omega(\Sigma Y\rtimes\Omega\Sigma X)$ by 2.3. Thus Lemma 2.4 gives that $\Omega\Sigma(X\vee Y)$ is homotopy equivalent to $\Omega\Sigma X\times\Omega\Sigma(Y\bigvee_{k\geq 1}X^{[k]}\wedge Y)$. This proof does not explicitly give the equivalence θ.

We outline a proof that θ is an equivalence. It suffices to check that θ_* is an isomorphism with coefficients in a field. Notice that Lemma 1.6 implies the equation

$$ad_*^k(x_1\wedge\cdots\wedge x_k\wedge y)=[x_1[x_2\cdots[x_k,y]\cdots]+\sum_{j>k+1}z_1\otimes\cdots\otimes z_j.$$

Next observe that there is a morphism of fibrations

$$\begin{array}{ccc}
\Omega\Sigma(Y\bigvee_{k\geq 1}X^{[k]}\wedge Y) & \xrightarrow{\Omega g} & \Omega(\Sigma Y\rtimes\Omega\Sigma X)\\
\downarrow & & \downarrow\\
\Omega\Sigma X\times\Omega\Sigma(Y\bigvee_{k\geq 1}X^{[k]}\wedge Y) & \xrightarrow{\theta} & \Omega\Sigma(X\vee Y)\\
\downarrow \text{project} & & \downarrow \Omega(p)\\
\Omega\Sigma X & \xrightarrow[1]{} & \Omega\Sigma X
\end{array}$$

where the left hand fibration is a product. If $(\Omega g)_*$ is a monomorphism, then it must be an isomorphism by 2.2 and 2.3 since one can assume that all spaces are of finite type by passage to limits. If x_i and y run over a basis for $\bar{H}_*X$ and $\bar{H}_*Y$ respectively, then the elements $[x_1[x_2\cdots[x_k,y]\cdots]$ are algebraically independent in $H_*\Omega(\Sigma Y \rtimes \Omega\Sigma X) \cong H_*\Omega\Sigma(Y \vee_{k\geq 1} X^{[k]}\wedge Y)$ by the proof of Proposition 4.5 in $[CMN_1]$. Since $(\Omega g)_*$ is multiplicative and $(\Omega g)_*(x_1\wedge\cdots\wedge x_k\wedge y)=[x_1[x_2\cdots[x_k,y]\cdots]+\sum_{j>k+1} z_1\otimes\cdots\otimes z_j$, $(\Omega g)_*$ is an isomorphism. Thus θ_* is an isomorphism.

<u>Proof of 2.2</u>. Consider the morphism of fibrations

$$
\begin{array}{ccccc}
A & \longrightarrow & \Sigma Y \vee_{k\geq 1} \Sigma(X^{[k]}\wedge Y) & \xrightarrow{\text{pinch}} & \Sigma Y \\
\downarrow & & \downarrow & & \downarrow \\
A & \longrightarrow & \Sigma X \vee \Sigma Y & \longrightarrow & \Sigma X\times\Sigma Y \\
\downarrow & & \downarrow{\scriptstyle\text{pinch}} & & \downarrow{\scriptstyle\text{project}} \\
* & \longrightarrow & \Sigma X & \xrightarrow[1]{} & \Sigma X
\end{array}
$$

and apply 2.1 to get that A is homotopy equivalent to $\vee_{k\geq 1}\Sigma(X^{[k]}\wedge Y) \vee_{j,k\geq 1} \Sigma Y^{[j]}\wedge X^{[k]}\wedge Y$. By Lemma 2.5, A is homotopy equivalent to $\Sigma(\Omega\Sigma X)\wedge(\Omega\Sigma Y)$.

<u>Proof of 2.5</u>: It suffices to give a map $\Sigma\vee_{k\geq 1} X^{[k]} \to \Omega\Sigma X$ which is an isomorphism in homology with any field coefficients. Define $\theta_k\colon X^k \to \Omega\Sigma X$ to be the composite

$$X^k \xrightarrow{E^k} (\Omega\Sigma X)^k \xrightarrow{\text{multiply}} \Omega\Sigma X.$$

Thus $\theta_{k*}(x_1\otimes\cdots\otimes x_k) = x_1\otimes\cdots\otimes x_k$. Suspending, we obtain $\Sigma X^{[k]} \to \Sigma(X^k) \to \Sigma\Omega\Sigma X$ which gives an isomorphism from $\bar{H}_*\Sigma X^{[k]}$ to the submodule of $H_*\Sigma\Omega\Sigma X$ spanned by the suspensions of $x_1\otimes\cdots\otimes x_k$, $x_i\in\bar{H}_*X$. The lemma follows.

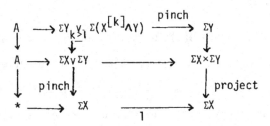

§3: James' EHP sequence

I.M. James [J1,J3] showed that there is a long exact sequence on the 2-primary components of homotopy groups given by

$$\cdots \to \pi_q S^n \xrightarrow{E} \pi_{q+1} S^{n+1} \xrightarrow{H} \pi_{q+1} S^{2n+1} \xrightarrow{P} \pi_{q-1} S^n \to \cdots.$$

Throughout the rest of this paper, we assume that all spaces are 2-local unless otherwise stated. This long exact sequence is obtained from the following lemma 2.5 which gives maps $\tilde{H}_k : \Sigma\Omega S^{n+1} \to S^{nk+1}$ inducing epimorphisms on H_{nk+1}.

Theorem 3.1 [J3]. Let $H: \Omega S^{n+1} \to \Omega S^{2n+1}$ be any map which induces an epimorphism on $H_{2n}(\ ; \mathbb{Z})$. Then there is a (2-local) fibration

$$S^n \xrightarrow{E} \Omega S^{n+1} \xrightarrow{H} \Omega S^{2n+1}$$

where E is the Freudenthal suspension. The EHP is the long exact homotopy sequence for this fibration.

Remark 3.2. In case spaces are localized at an odd prime p, Toda gave similar fibrations [T2]: Namely, ΩS^{2n} is homotopy equivalent to $S^{2n-1} \times \Omega S^{4n-1}$ and there are p-local fibrations

(1) $J_{p-1} S^{2n} \to \Omega S^{2n+1} \xrightarrow{H_p} \Omega S^{2np+1}$, and

(2) $S^{2n-1} \to \Omega J_{p-1} S^{2n} \to \Omega S^{2np-1}$

where $J_{p-1} S^{2n}$ is the 2n(p-1)-skeleton of ΩS^{2n+1}.

Proof of 3.1: The proof which we give is due to J.C. Moore. There are 2 cases to check, namely the cases where $n \equiv 1(2)$ and $n \equiv 0(2)$. We do the case where $n \equiv 0(2)$ and remark that the case $n \equiv 1(2)$ is quite similar.

Recall that $H_*(\Omega S^{2n+1}; \mathbb{Z})$ is isomorphic to $T[x_{2n}]$ as a Hopf algebra where T[V] denotes the tensor algebra on V by the Bott-Samelson theorem

[BS]. Notce that x_{2n} is primitive. Thus $H^*(\Omega S^{2n+1};\mathbb{Z})$ is isomorphic to $\Gamma[y_{2n}]$, the divided polynomial algebra on y_{2n}.

A map $H: \Omega S^{2n+1} \to \Omega S^{4n+1}$ induces a map of algebras $H^*: H^*(\Omega S^{4n+1};\mathbb{Z}) \to H^*(\Omega S^{2n+1};\mathbb{Z})$ with $H^*(y_{4n}) = \gamma_2(y_{2n})$. Recall that $\gamma_p(y_{2n})\gamma_q(y_{2n}) = (p,q)\gamma_{p+q}(y_{2n})$ where $(p,q) = \frac{(p+q)!}{p!q!}$ if $p,q \geq 0$. Thus $H^*(\gamma_q(y_{4n})) = H^*(\frac{1}{q!}\gamma_1(y_{4n})^q) = \frac{1}{q!}\gamma_2(y_{2n})^q = \frac{(2q)!}{q!2^q}\gamma_{2q}(y_{2n})$. Since

$\frac{(2q)!}{q!2^q} \equiv 1(2)$, it follows that $H^*: H^*(\Omega S^{4n+1};\mathbb{Z}_{(2)}) \to H^*(\Omega S^{2n+1};\mathbb{Z}_{(2)})$ satisifes $H^*(\gamma_q(y_{4n})) = u\cdot\gamma_{2q}(y_{2n})$ where u is a unit in $\mathbb{Z}_{(2)}$. Thus $H^*(\Omega S^{2n+1};\mathbb{Z}_{(2)})$ is isomorphic to $H^*(\Omega S^{4n+1};\mathbb{Z}_{(2)}) \oplus \gamma_1(y_{2n})\cdot H^*(\Omega S^{4n+1};\mathbb{Z}_{(2)})$ as a $H^*(\Omega S^{4n+1};\mathbb{Z}_{(2)})$-module and is thus a free $H^*(\Omega S^{4n+1};\mathbb{Z}_{(2)})$-module on 2 generators, 1 and $\gamma_1(y_{2n})$).

Next, consider the Eilenberg-Moore spectral sequence for the 2-local fibration $Y \to \Omega S^{2n+1} \xrightarrow{H} \Omega S^{4n+1}$. The E_2-term is $\mathrm{Tor}_{H^*(\Omega S^{4n+1};\mathbb{Z}_{(2)})}(\mathbb{Z}_{(2)}; H^*(\Omega S^{2n+1};\mathbb{Z}_{(2)}))$ which by the previous paragraph is isomorphic to an exterior algebra on a (2n)-dimensional class. Evidently, $E_2 = E_\infty$ and the lift g given by

$$\begin{array}{ccc} & & Y \\ & \overset{g}{\nearrow} & \downarrow \\ S^{2n} & \xrightarrow{E} & \Omega S^{2n+1} \end{array}$$

induces an isomorphism in $H^*(\quad;\mathbb{Z}_{(2)})$. Thus g is a 2-local equivalence and Theorem 3.1 follows.

There is an exceptional fibration obtained from the EHP sequence. Let X<k> denoted the k-connected cover of X. Write H for the composite $\Omega(S^3\langle 3\rangle) \to \Omega S^3 \xrightarrow{H} \Omega S^5$.

Theorem 3.3[T_1,T_2]. There is a 2-local fibration $S^3 \xrightarrow{\eta} \Omega S^3\langle 3\rangle \xrightarrow{H} \Omega S^5$ where η represents the generator of $\pi_3 \Omega S^3\langle 3\rangle$.

Proof. Consider the morphism of fibrations

$$\begin{array}{ccccc} X & \xrightarrow{\eta} & \Omega S^3\langle 3\rangle & \xrightarrow{H} & \Omega S^5 \\ \downarrow & & \downarrow & & \downarrow 1 \\ S^2 & \longrightarrow & \Omega S^3 & \longrightarrow & \Omega S^5 \\ j\downarrow & & \downarrow & & \downarrow \\ K(\mathbb{Z},2) & \longrightarrow & K(\mathbb{Z},2) & \longrightarrow & * \end{array}$$

where $j\colon S^2 \to K(\mathbb{Z},2)$ is degree one. Thus X is S^3. Since ΩS^5 is 3-connected η induces an epimorphism on π_3.

Finally, notice that the composite $S^{2n-1} \xrightarrow{E^2} \Omega^2 S^{2n+1} \xrightarrow{P} S^n$ is the Whitehead product.

§4: The distributivity law; order of the Hopf invariant

Consider a suspension ΣX and "the" degree k map $[k]: \Sigma X \to \Sigma X$ given as the composite $\Sigma X \xrightarrow{\text{pinch}} \bigvee_k \Sigma X \xrightarrow{\text{fold}} \Sigma X$. Of course one must consider the order in which the pinching is done, but assume that one fixed choice has been made; generally our spaces X will be suspensions and so this ordering will be irrelevant.

The classical method for studying [k] on homotopy groups is to loop [k] and then to apply the Hilton-Milnor theorem: There is a homotopy commutative diagram

$$\begin{array}{ccccc} \Omega\Sigma X & \xrightarrow{\Omega(\text{pinch})} & \Omega\Sigma(\bigvee_k X) & \xrightarrow{\Omega(\text{fold})} & \Omega\Sigma X \\ & \searrow & 21 \big\downarrow \phi & \nearrow \lambda & \\ & & \prod_\alpha \Omega\Sigma(X_\alpha) & & \end{array}$$

where X_α is given by certain smash products of X [Theorem 2.1], and $\prod_\alpha \Omega\Sigma(X_\alpha)$ is the weak product which is the colimit of finite products of $\Omega\Sigma(X_\alpha)$ with $X_\alpha = X^{[n_\alpha]}$. There are projection maps $\pi_\alpha: \prod_\alpha \Omega\Sigma(X_\alpha) \to \Omega\Sigma X_\alpha$ and the composites $\pi_\alpha \cdot \phi \cdot \Omega(\text{pinch})$ are the Hilton-Hopf invariants [Hl,Mr]. They are natural for maps which are suspensions. The map λ is obtained by multiplication (in some fixed order) of loop maps $\Omega f_\alpha: \Omega\Sigma(X_\alpha) \to \Omega\Sigma X$ which themselves are induced by the canonical multiplicative extensions of iterated Samelson products $X_\alpha \to \Omega\Sigma X$. M.G. Barratt has studied this situation in detail [B].

Specialize to $X = S^n$ and recall Corollary 1.3 which states that $\underbrace{[\iota_n[\iota_n[\cdot\cdot[\iota_n,\iota_n]]}_{k} = 0$ if either $n \equiv 1(2)$ and $k \geq 3$ or $n \equiv 0(2)$ and $k \geq 4$. We use this to study the order of a Hopf invariant.

Let $h: \Omega S^{2n+1} \to \Omega S^{4n+1}$ be any collection of maps, one for each n, satisfying (1) h_* is onto $H_{4n}(\Omega S^{4n+1}; \mathbb{Z})$ and (2) there are homotopy commutative diagrams

$$\begin{array}{ccc} \Omega S^{2n+1} & \xrightarrow{\Omega\Sigma(f)} & \Omega S^{2n+1} \\ h\downarrow & & \downarrow h \\ \Omega S^{4n+1} & \xrightarrow[\Omega\Sigma(f\wedge f)]{} & \Omega S^{4n+1} \end{array} .$$

Such maps exist by the Hilton-Milnor theorem or by Lemma 2.5.

The following lemma is an unpublished result due to M.G. Barratt. We gave a different proof in [C_2]. Let $(\Omega X)\{k\}$ denote the homotopy theoretic fibre of the kth power map on ΩX.

<u>Lemma 4.1</u>. The map Ωh has order 2 in the abelian group $[\Omega^2 S^{2n+1}, \Omega^2 S^{4n+1}]$. Thus there is a lift of Ωh to $\bar{h}\colon \Omega^2 S^{2n+1} \to (\Omega^2 S^{4n+1})\{2\}$.

<u>Remark 4.2</u>. The map h has infinite order in the group $[\Omega S^{2n+1}, \Omega S^{4n+1}]$ since h_* is onto $H_{4n}(\Omega S^{4n+1};\mathbb{Z})$.

Next, consider the group $[X, \Omega S^{2n+1}]$ of homotopy classes of pointed maps and write + for the operation in this (generally non-abelian) group. Since this last group is non-abelian, we keep track of the order of addition. Let $H\colon \Omega S^{2n+1} \to \Omega S^{4n+1}$ denote the Hilton-Hopf invariant.

<u>Proposition 4.3</u>. (1) The element $\Omega[2]$ in $[\Omega S^{2n+1}, \Omega S^{2n+1}]$ is equal to $2+(\Omega[\iota_{2n+1}, \iota_{2n+1}]\cdot H)$. (2) The element $\Omega[-1]$ in $[\Omega S^{2n+1}, \Omega S^{2n+1}]$ is equal to $(\Omega[\iota_{2n+1}, \iota_{2n+1}]\cdot H)-1$.

Similar formulae apply to $\Omega[k]$ in $[\Omega S^{2n}, \Omega S^{2n}]$. We shall not need these formulae here.

To prove 4.3, consider

<u>Observation 4.4</u>. If $f_i\colon X_i \to \Omega Y$ are continuous maps with f_2 null-homotopic, then the composite $X_1 \times X_2 \xrightarrow{f_1 \times f_2} \Omega Y \times \Omega Y \xrightarrow{\text{multiply}} \Omega Y$ factors through the

map $X_1 \times X_2 \xrightarrow{\text{project}} X_1 \xrightarrow{f_1} \Omega Y$ up to homotopy.

Since the proof of 4.1 depends on 4.3, Proposition 4.3 is proven first. The reader should compare proofs here to those of James [J4].

Proof of Proposition 4.2: Consider the map [2] which is the composition $S^{2n+1} \xrightarrow{\text{pinch}} S^{2n+1} \vee S^{2n+1} \xrightarrow{\text{fold}} S^{2n+1}$. Loop this composite and expand $\Omega(S^{2n+1} \vee S^{2n+1})$ by the Hilton-Milnor theorem as in section 2. By iterating Theorem 2.1, we obtain a homotopy equivalence

$$\Omega S^{2n+1} \times \Omega S^{2n+1} \times \Omega\Sigma(\bigvee_{i,j\geq 1} S^{2n(i+j)}) \to \Omega(S^{2n+1} \vee S^{2n+1}).$$

By the identification of the maps given in the Hilton-Milnor theorem, the map $\Omega\Sigma(\bigvee_{i,j\geq 1} S^{2n(i+j)}) \to \Omega(S^{2n+1} \vee S^{2n+1})$ is the canonical multiplicative extension of a map $\gamma\colon \bigvee_{i,j\geq 1} S^{2n(i+j)} \to \Omega(S^{2n+1} \vee S^{2n+1})$ given by Samelson products of length $i+j$. Since such maps are null-homotopic in ΩS^{2n+1} if $i+j \geq 3$ by Corollary 1.3, the map $\Omega(\text{fold})\colon \Omega(S^{2n+1} \vee S^{2n+1}) \to \Omega S^{2n+1}$ factors through $\Omega S^{2n+1} \times \Omega S^{2n+1} \times \Omega S^{4n+1}$.

Notice that the composite

$$\Omega S^{2n+1} \xrightarrow{\Omega(\text{pinch})} \Omega(S^{2n+1} \vee S^{2n+1}) \to (\Omega S^{2n+1})^2 \times \Omega S^{4n+1}$$

is given by $\Delta^3\colon \Omega S^{2n+1} \to (\Omega S^{2n+1})^3 \xrightarrow{1\times 1\times H} (\Omega S^{2n+1})^2 \times \Omega S^{4n+1}$. Since the map $(\Omega S^{2n+1})^2 \times \Omega S^{4n+1} \to \Omega S^{2n+1}$ above is the composite

$(\Omega S^{2n+1})^2 \times \Omega S^{4n+1} \xrightarrow{1\times 1\times \Omega[\iota_{2n+1},\iota_{2n+1}]} (\Omega S^{2n+1})^3 \xrightarrow{\text{multiply}} \Omega S^{2n+1}$, it follows immediately that $\Omega[2] = 2 + (\Omega[\iota_{2n+1},\iota_{2n+1}] \cdot H)$ in $[\Omega S^{2n+1}, \Omega S^{2n+1}]$.

To prove 4.2(2), first observe that the composite $S^n \xrightarrow{\text{pinch}} S^n \vee S^n \xrightarrow{[-1]\vee[1]} S^n \vee S^n \xrightarrow{\text{fold}} S^n$ is null-homotopic. Again, loop to apply the Hilton-Milnor theorem and naturality to get a homotopy commutative diagram

$$\begin{array}{ccccccc}
\Omega S^{2n+1} & \xrightarrow{\Omega(\text{pinch})} & \Omega(S^{2n+1}\vee S^{2n+1}) & \xrightarrow{\Omega([-1]\vee[1])} & \Omega(S^{2n+1}\vee S^{2n+1}) & \xrightarrow{\Omega(\text{fold})} & \Omega S^{2n+1} \\
 & \searrow^{\Delta\times H} & \downarrow & & \downarrow & \nearrow & \\
 & & (\Omega S^{2n+1})^2\times\Omega S^{4n+1} & \xrightarrow[\Omega[-1]\times 1\times\Omega([-1]\wedge[1])]{} & (\Omega S^{2n+1})^2\times\Omega S^{4n+1} & & .
\end{array}$$

Since $[-1]\wedge[1]$ is of degree -1 and $-[\iota_{2n+1},\iota_{2n+1}]=[\iota_{2n+1},\iota_{2n+1}]$ by Corollary 1.3, the formula in 4.2(2) follows.

Proof of Lemma 4.1: By naturality, there is a commutative diagram

$$\begin{array}{ccc}
\Omega S^{2n+1} & \xrightarrow{\Omega\Sigma[-1]} & \Omega S^{2n+1} \\
h\downarrow & & \downarrow h \\
\Omega S^{4n+1} & \xrightarrow[\Omega\Sigma([-1]\wedge[-1])]{} & \Omega S^{4n+1}
\end{array}$$

and thus $h-(h\cdot\Omega\Sigma[-1])$ is null-homotopic. Next, use Proposition 4.2 to give $\Omega\Sigma[-1]=-1+\Omega[\iota_{2n+1},\iota_{2n+1}]\cdot H$. We would like to expand $h\cdot(-1+\Omega[\iota_{2n+1},\iota_{2n+1}]\cdot H)$. Since h is not homotopy multiplicative, we can't compare $h\cdot[-1]$ and $-h$. However after looping, one has

$$\Omega h\cdot(-1+\Omega^2[\iota_{2n+1},\iota_{2n+1}]\cdot\Omega H) = -\Omega h+\Omega h\cdot\Omega^2[\iota_{2n+1},\iota_{2n+1}]\cdot\Omega H$$

in $[\Omega^2S^{2n+1},\Omega^2S^{4n+1}]$. If $\Omega h\cdot\Omega^2[\iota_{2n+1},\iota_{2n+1}]$ is null-homotopic, then it follows at once that $2\Omega h=0$ in $[\Omega^2S^{2n+1},\Omega^2S^{4n+1}]$. But this last statement follows from the following lemma.

Lemma 4.5: The composite $\Omega S^{4n+1}\xrightarrow{\Omega[\iota_{2n+1},\iota_{2n+1}]}\Omega S^{2n+1}\xrightarrow{h}\Omega S^{4n+1}$ is null-homotopic.

Proof: Notice that $[\iota_{2n+1},\iota_{2n+1}]$ desuspends by the EHP sequence because it is of finite order and thus has trivial Hopf invariant. Let v be a choice of desuspension for $[\iota_{2n+1},\iota_{2n+1}]$. By naturality, there is a homotopy commutative diagram

$$\begin{array}{ccc} \Omega S^{4n+1} & \xrightarrow{\Omega\Sigma v} & \Omega S^{2n+1} \\ \downarrow h & & \downarrow h \\ \Omega S^{8n+1} & \xrightarrow[\Omega\Sigma(v\wedge v)]{} & \Omega S^{4n+1} \end{array} .$$

Since $\Sigma(v\wedge v)$ is homotopic to the composite

$$S^{4n+1}\wedge S^{4n} \xrightarrow{(\Sigma v)\wedge 1} S^{2n+1}\wedge S^{4n} \xrightarrow{1\wedge v} S^{2n+1}\wedge S^{2n},$$

and $\Sigma^2 v$ is null-homotopic, the lemma follows.

§5. Further properties of the EHP sequence

I.M. James' fibrations (Theorem 3.1) directly apply to give global information about the image of maps $f\colon S^n \to X$ in homotopy. For example, the EHP sequence gives that if $i\colon S^3 \to SU(3)$ is the canonical inclusion, then $2i_*(\alpha)=0$ for $\alpha\in\pi_q S^3$ and $q>3$. As before, $X<k>$ denotes the k-connected cover of X.

Proposition 5.1. (1) Let $f\colon \Omega S^3<3> \to X$ be any map which is trivial on π_3. Then $2(\Omega f)=0$ in $[\Omega^2 S^3<3>,\Omega X]$.

(2) Let $g\colon \Omega S^{2n+1} \to X$ be any map which is trivial on π_{2n}. Then $2(\Omega g)=0$ in $[\Omega^2 S^{2n+1},\Omega X]$.

As a corollary, one gets an unpublished result proven by J.C. Moore and the implications on homotopy groups first proven by James [J4].

Corollary 5.2. The 2^{2n}-th power map on $\Omega^{2n}S^{2n+1}<2n+1>$ is null-homotopic. Thus 2^{2n} annihilates $\pi_q S^{2n+1}$ if $q>2n+1$.

Some mild improvements of 5.2 will be given in section 6. The following related conjecture has been made by Barratt and Mahowald.

Conjecture 5.3. The $2^{\phi(2n)}$-th power map on $\Omega^{2n+1}S^{2n+1}<2n+1>$ is null-homotopic where $2^{\phi(2n)}$ is the order of the canonical line bundle η over $\mathbb{R}P^{2n}$.

It is easy to see that 5.3 is best possible since there is a factorization

$$\begin{array}{ccc} \mathbb{R}P^{2n} & \xrightarrow{\ \eta\ } & BO \\ \downarrow & \nearrow_{g} & \\ \Omega^{2n+1}S^{2n+1}<2n+1> & & \end{array}$$

where g is a loop map and η classifies the canonical line bundle over $\mathbb{R}P^{2n}$.

Other examples are

Corollary 5.4. (1) The 8th power map on $\Omega^4 S^5\langle 5\rangle$ is null-homotopic.

(2) Let $i\colon S^3 \to SU(3)$ denote the canonical inclusion. Then $2\Omega^2(i)=0$ in $[\Omega^2 S^3\langle 3\rangle, \Omega^2 SU(3)\langle 3\rangle]$ and the 16th power map on $\Omega^4(SU(3)\langle 5\rangle)$ is null-homotopic.

Question 5.5. We do not know whether there are elements of order 16 or even order 8 in $\pi_* SU(3)$.

Corollary 5.6. $2\Omega^2[-1]=-2$ in $[\Omega^2 S^{2n+1}, \Omega^2 S^{2n+1}]$ and so $2([-1]_*(x))=-2(x)$ for $x \in \pi_* S^{2n+1}$.

An interesting question is given by

Question 5.6. Is $\Omega^2[-1]$ homotopic to -1? Partial information is given in section 11.

Proof of Proposition 5.1: To prove 5.1(1), recall Theorem 3.3 which gives a fibration $S^3 \xrightarrow{\eta} \Omega S^3\langle 3\rangle \xrightarrow{H} \Omega S^5$ with η the generator of $\pi_3 \Omega S^3\langle 3\rangle = \mathbb{Z}/2$. Let $f\colon \Omega S^3\langle 3\rangle \to X$ be any map which is trivial on π_3. Since f is trivial on π_3, there is a homotopy commutative diagram

$$(*)\qquad \begin{array}{ccc} S^3 & \longrightarrow & * \\ \downarrow & & \downarrow \\ \Omega S^3\langle 3\rangle & \longrightarrow & X \end{array} .$$

Thus there is a map of fibrations

$$\begin{array}{ccc} \Omega^2 S^3\langle 3\rangle & \xrightarrow{\Omega f} & \Omega X \\ {\scriptstyle \Omega H}\downarrow & & \| \\ \Omega^2 S^5 & \xrightarrow{\ell} & \Omega X \\ \downarrow & & \downarrow \\ S^3 & \longrightarrow & * \end{array}$$

for some lift ℓ given by (*). But $2(\Omega f)$ is given by the composite $\Omega^2S^3\langle 3\rangle \xrightarrow{2} \Omega^2S^3\langle 3\rangle \xrightarrow{\Omega f} \Omega X$ since Ωf is multiplicative. There is a homotopy commutative diagram

$$\begin{array}{ccccc} \Omega^2S^3\langle 3\rangle & \xrightarrow{2} & \Omega^2S^3\langle 3\rangle & \xrightarrow{\Omega f} & \Omega X \\ & \searrow^{\simeq *} & \downarrow{\scriptstyle \Omega H} & \nearrow_{\ell} & \\ & & \Omega^2S^5 & & \end{array}$$

and since $2\Omega H$ is null by 4.1, the result follows.

A similar proof applies to $g\colon \Omega S^{2n+1} \to X$. There is a homotopy commutative diagram

$$\begin{array}{ccc} S^{2n} & \longrightarrow & * \\ \downarrow & & \downarrow \\ \Omega S^{2n+1} & \xrightarrow{g} & X \end{array}$$

by hypothesis. Hence $\Omega^2 g$ factors through $\Omega^2S^{4n+1} \xrightarrow{\ell} \Omega X$ for some lift ℓ. Since $2\Omega H$ is null by 4.1, 5.1(2) follows

<u>Proof of 5.2</u>: To prove 5.2, induct on n starting with n=1. Here recall that $2\Omega H=0$ in $[\Omega^2S^3\langle 3\rangle, \Omega^2S^5]$. Thus there is a lift λ,

$$\begin{array}{ccccc} & & \Omega^2S^3\langle 3\rangle & & \\ & \swarrow^{\lambda} & \downarrow{\scriptstyle 2} & \searrow^{\simeq *} & \\ \Omega S^3 & \longrightarrow & \Omega^2S^3\langle 3\rangle & \xrightarrow[\Omega H]{} & \Omega^2S^5 . \end{array}$$

Hence $2\colon \Omega^2S^3\langle 3\rangle \to \Omega^2S^3\langle 3\rangle$ factors through $\Omega\eta\colon \Omega S^3 \to \Omega^2S^3\langle 3\rangle$. Notice that $\eta\colon S^3 \to \Omega S^3\langle 3\rangle$ has order 2 and since S^3 is an H-space $\Omega[2]$ is homotopic to 2 on S^3. Thus there is another homotopy commutative diagram

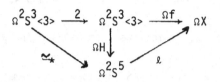

$$\begin{array}{ccccc} \Omega S^3 & \xrightarrow{1} & \Omega S^3 & & \\ \downarrow{\scriptstyle 2} & & \downarrow{\scriptstyle \Omega[2]} & \searrow{\scriptstyle \simeq *} & \\ \Omega S^3 & \xrightarrow[1]{} & \Omega S^3 & \xrightarrow{\Omega\eta} & \Omega^2 S^3\langle 3\rangle \end{array} .$$

Since $\Omega\eta$ is multiplicative, it follows that $2(\Omega\eta)=0$ in $[\Omega S^3, \Omega^2 S^3\langle 3\rangle]$.

Remark 5.8. That the map $\eta: S^3 \to \Omega S^3\langle 3\rangle$ is not an H-map, follows by computing the H-deviation which is the generator of $\pi_6 \Omega S^3 \cong \mathbb{Z}/2$ as in Lemma 21.5.

Next inductively assume that the 2^{2k}-th power map on $\Omega^{2k} S^{2k+1}\langle 2k+1\rangle$ is null-homotopic for $k<n$ and consider the fibration $\Omega H: \Omega^2 S^{2n+1} \to \Omega^2 S^{4n+1}$. By 4.1, $2\Omega H$ lifts to ΩS^{2n}.

Next consider $[-1]: S^{2n} \to S^{2n}$. Since there is a homotopy commutative diagram

$$\begin{array}{ccc} \Omega S^{2n} & \xrightarrow{\Omega[-1]} & \Omega S^{2n} \\ \downarrow{\scriptstyle H} & & \downarrow{\scriptstyle H} \\ \Omega S^{4n-1} & \xrightarrow[\Omega\Sigma[-1]\wedge[-1]]{} & \Omega S^{4n-1} \end{array} ,$$

$\Omega H \cdot \Omega^2[-1] - \Omega H$ factors through ΩS^{2n-1}. Also, notice that the diagram

$$\begin{array}{ccc} \Omega S^{2n} & \xrightarrow{\Omega E} & \Omega^2 S^{2n+1} \\ \downarrow{\scriptstyle 1-\Omega[-1]} & & \downarrow{\scriptstyle 2} \\ \Omega S^{2n} & \xrightarrow[\Omega E]{} & \Omega^2 S^{2n+1} \end{array}$$

homotopy commutes. Thus there is a large homotopy commutative diagram by Lemma 4.1

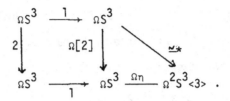

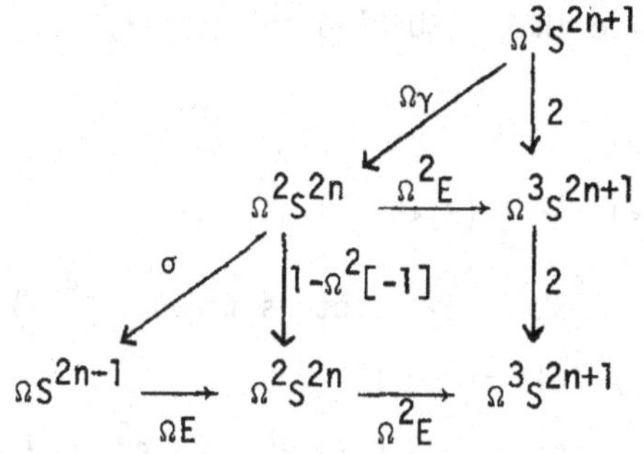
Ω^3S^{2n+1}
$\Omega\gamma$
2
Ω^2S^{2n}
Ω^2E
Ω^3S^{2n+1}
σ
$1-\Omega^2[-1]$
2
ΩS^{2n-1}
Ω^2S^{2n}
Ω^3S^{2n+1}
ΩE
Ω^2E

which implies the following lemma.

Lemma 5.9. There is a homotopy commutative diagram

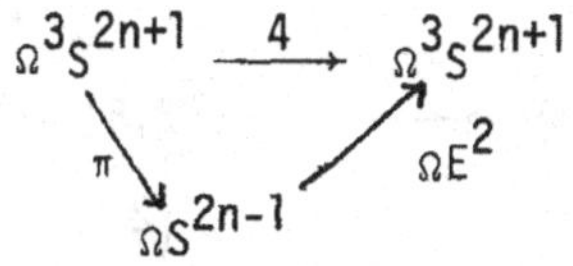
Ω^3S^{2n+1}
4
Ω^3S^{2n+1}
π
ΩE^2
ΩS^{2n-1}

By inspection, this diagram lifts to connected covers so that the inductive hypothesis implies that 2^{2n} is null on $\Omega^{2n}S^{2n+1}\langle 2n+1\rangle$. Proposition 5.2 follows.

Proof of Theorem 5.4: By Theorem 3.3, there is a fibration $S^3 \xrightarrow{\eta} \Omega S^3\langle 3\rangle \xrightarrow{H} \Omega S^5$ and thus there is a map $\partial\colon \Omega^2S^5 \to S^3$ which is degree 2 on the bottom cell. By Lemma 5.9, there is a homotopy commutative diagram

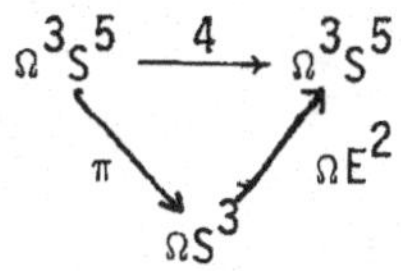
Ω^3S^5
4
Ω^3S^5
π
ΩE^2
ΩS^3

Thus $4=\Omega^2E^2\cdot\Omega\pi$ in $[\Omega^4S^5,\Omega^4S^5]$. Notice that because S^3 is an H-space, $4\cdot\Omega\partial=\Omega\partial\cdot\Omega^2E^2\cdot\Omega\pi=2\Omega\pi$ in $[\Omega^4S^5,\Omega^2S^3]$. Lifting to simply-connected covers, $8=2(\Omega^2E^2\cdot\Omega\pi)=(\Omega^2E^2)(2\Omega\pi)=(\Omega^2E^2)(4\cdot\Omega\partial)$ in $[\Omega^4S^5\langle 5\rangle,\Omega^4S^5\langle 5\rangle]$. Since the identity map of $\Omega^2S^3\langle 3\rangle$ has order 4, $4\Omega\partial=0$ in $[\Omega^4S^5\langle 5\rangle,\Omega^2S^3\langle 3\rangle]$ and so 5.4(1) follows.

To prove 5.4(2), notice that $\pi_4 SU(3)=0$ and that $\pi_5 SU(3) \cong \mathbf{Z}$. Thus there is a fibration

$$S^3\langle 4\rangle \xrightarrow{i} SU(3)\langle 5\rangle \xrightarrow{\pi} S^5\langle 5\rangle .$$

Since the 8th power map on $\Omega^4 SU(3)\langle 5\rangle$ factors through $\Omega^4(i)$ by 5.4(1), it suffices to check that $2\Omega^4(i)$ is null.

Consider the fibration $S^3\langle 3\rangle \longrightarrow SU(3)\langle 3\rangle \longrightarrow S^5$ and replace $SU(3)\langle 3\rangle$ by $SU(3)\langle 4\rangle$ since $\pi_4 SU(3)=0$. By inspection, there is a map of fibrations

$$\begin{array}{ccc} S^3\langle 3\rangle & \longrightarrow & K(\mathbb{Z}/2,4) \\ \downarrow j & & \downarrow \\ SU(3)\langle 4\rangle & \longrightarrow & K(\mathbb{Z},5) \\ \downarrow & & \downarrow 2 \\ S^5 & \longrightarrow & K(\mathbb{Z},5) \end{array}$$

and so there is a map $i\colon S^3\langle 4\rangle \longrightarrow SU(3)\langle 3\rangle$ lifting j. But $\Omega^4(S^3\langle 3\rangle)$ is a disjoint union of 2 copies of $\Omega^4 S^3\langle 4\rangle$. Thus $2\Omega^4(i)$ restricted to the component of the base-point in $\Omega^4(S^3\langle 3\rangle)$ is trivial by 5.1(1) and 5.4(2) follows.

§6. Improvements

In [C_2], it was observed that Corollary 5.2 can be improved slightly in case n=2 or 4. P. Selick then proved the following in a restated form [S_5].

Theorem 6.1. There is a factorization

$$\begin{array}{ccc} \Omega^4 S^{4n+1} & \xrightarrow{\;2\;} & \Omega^4 S^{4n+1} \\ & {}_{f}\searrow \quad \nearrow_{g} & \\ & \Omega^2 S^{4n-1} & \end{array}$$

for some maps f and g.

Our proof is based on the observation in 5.1(2) that if $g:\Omega S^{2n+1} \to X$ is any map which is zero on π_{2n}, then $2(\Omega g)$ is null-homotopic.

Proof: The proof here is analogous to that given in [S_5]. Consider the pull-back diagram

$$\begin{array}{ccc} Q & \longrightarrow & \Omega S^{2n+1} \\ \downarrow & & \downarrow H \\ (\Omega S^{4n+1})\{2\} & \longrightarrow & \Omega S^{4n+1} \end{array}$$

to obtain a morphism of fibrations

$$\begin{array}{ccccccc} * & \longrightarrow & \Omega S^{2n} & \longrightarrow & \Omega S^{2n} & \longrightarrow & * \\ \downarrow & & \downarrow & & \downarrow & & \downarrow \\ \Omega^3 S^{4n+1} & \xrightarrow{\Omega(f)} & \Omega Q & \longrightarrow & \Omega^2 S^{2n+1} & \longrightarrow & \Omega^2 S^{4n+1} \\ 1 \downarrow & & \downarrow \Omega(i) & & \downarrow \Omega H & & \| \\ \Omega^3 S^{4n+1} & \longrightarrow & (\Omega^2 S^{4n+1})\{2\} & \longrightarrow & \Omega^2 S^{4n+1} & \xrightarrow{\;2\;} & \Omega^2 S^{4n+1} \end{array} .$$

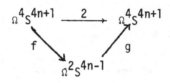

Since $2\Omega H$ is null by Lemma 4.1, the principal fibration $\Omega Q \to \Omega^2 S^{2n+1}$ has a section and is thus trivial. Thus ΩQ is homotopy equivalent to $\Omega^2 S^{2n+1} \times \Omega^3 S^{4n+1}$ and there is a map $s: \Omega Q \to \Omega^3 S^{4n+1}$ such that $s \cdot \Omega(f)$ is homotopic to the identity.

Notice that $2\Omega(if)$ factors through the map $2\colon \Omega^3 S^{4n+1} \to \Omega^3 S^{4n+1}$ and is thus null-homotopic. Hence $2\Omega(f)$ factors through ΩS^{2n}.

Since $\Omega^3 S^{4n+1}$ is $(4n-3)$-connected and $4n-2 \geq 2n-1$, Proposition 5.1(2) applies to give that $\Omega[s\cdot 2\Omega(f)]$ factors through $\Omega^2 S^{4n-1}$. But $\Omega[s\cdot 2(\Omega(f))]$ is homotopic to $2(\Omega s)\cdot(\Omega^2 f)$. Since $s\cdot\Omega(f)$ is homotopic to the identity, the theorem follows.

As Selick proves [S_5], this gives

<u>Corollary 6.3</u>. $2^{(3/2)n+\varepsilon}\pi_q S^{2n+1}=0$ for $q>2n+1$ where

$$\varepsilon = \begin{cases} 0 & \text{if } n \equiv 0(2) \\ 1 & \text{if } n \equiv 1(2)\ . \end{cases}$$

Combining Theorem 6.1 and Lemma 5.9, one gets

<u>Corollary 6.4</u>. The identity map of $\Omega^{2n}S^{2n+1}\langle 2n+1\rangle$ has order bounded by $2^{(3/2)n+\varepsilon}$. If n=1 or 2, this is best possible (since the 2-primary components of $\pi_6 S^3$ and $\pi_8 S^5$ are $\mathbb{Z}/4$ and $\mathbb{Z}/8$ respectively).

§7: Homology of loop spaces

In much of what we do later, we need to know the mod-2 homology of $\Omega^k S^n$, $k\leq n$ and shall record that information here. Throughout the rest of this paper, all homology groups are taken with $\mathbb{Z}/2$-coefficients unless otherwise specified. The following theorems are based on [[AK], [Br], [DL], [Ni], [CLM]].

There are operations which are natural for n-fold loop maps

$$Q_i: H_q\Omega^n X \to H_{2q+i}\Omega^n X, \quad 0\leq i\leq n-1,$$

which are linear if $0\leq i<n-1$. The Freudenthal suspension $E^k: S^n \to \Omega^k S^{n+k}$ gives a map in homology which is a monomorphism because the composite

$$S^k\wedge S^n \xrightarrow{1\wedge E^k} S^k\wedge\Omega^k S^{n+k} \xrightarrow{\text{evaluate}} S^{n+k} \quad \text{is an equivalence.}$$

Write x_n for E^k_* (fundamental class) and $\Omega^n_0 S^n$ for the component of the base-point. Let $Q_I x_n$ denote $Q_{i_1}Q_{i_2}\cdots Q_{i_j}x_n$ if $I=(i_1,\ldots,i_j)$. The sequence I admissible if $0<i_1\leq i_2\leq\cdots\leq i_j$; write $\lambda(I)\leq q$ if $i_j\leq q$ and $\ell(I)=j$.

Theorem 7.1

(1) There is an isomorphism of Hopf algebras

$$\mathbb{Z}/2[Q_I x_n] \to H_*\Omega^k S^{n+k}, \quad n\geq 1,$$

for admissible I with $\lambda(I)\leq k-1$ and $Q_I x_n$ is primitive.

(2) There is an isomorphism of algebras

$$\mathbb{Z}/2[Q_I[1]]\to H_*\Omega^n_0 S^n$$

where I is admissble with $\lambda(I)\leq n-1$, $\ell(I)\geq 1$, and [1] is of degree o.

(3) The Steenrod operations are obtained inductively by the Nishida relations

$$Sq^t_* Q_r x_q = \sum_i (t-2i, r+q-2t+2i) Q_{r-t+2i} Sq^i_* x_q \,.$$

Remark 7.2. The coalgebra structure for $H_*\Omega^n_0 S^n$ is given more precisely in [CLM]. Namely, the class [1] is the class of the non-base-point in $H_*\Omega^n S^n$ and the element $Q_I[1]$ is shorthand for $Q_I[1]*[-2^{\ell(I)}]$. The diagonal is then given by the Cartan formula.

Recall that $QX=\varinjlim \Omega^n\Sigma^n X$.

Theorem 7.3. If X is path-connected, there is an isomorphism of algebras

$$\mathbb{Z}/2[Q_I x] \to H_*QX$$

for admissible I and x runs over a basis for $\bar{H}_*X$. If X is a suspension, then the elements $Q_I x$ are primitive. The Steenrod operations are given by the Nishida relations

$$Sq^t_* Q_r x_q = \sum_i (t-2i, r+q-2t+2i) Q_{r-t+2i} Sq^i_* x_q .$$

We will also have occasion to use the mod-p homology of $\Omega^2 S^3$ for p odd and state the requisite result here.

Theorem 7.4. Let p denote an odd prime. Then the mod-p homology of $\Omega^2 S^3$ is isomorphic as a Hopf algebra to

$$\bigotimes_{i\geq 0} \Lambda[y_{2p^i-1}] \bigotimes_{j\geq 1} \mathbb{Z}/p[x_{2p^i-2}] \text{ for } |y_i|=i \text{ and } |x_j|=j.$$

The homology Bockstein, β, is given by $\beta y_{2p^j-1}=x_{2p^j-2}$. The Steenrod operations are given by

(1) $P^{p^k}_* y_{2p^i-1}=0$ for all i and k.

(2) $P^1_* x_{2p^{j+1}-2} = -(x_{2p^j-2})^p$ if $j\geq 1$, and

(3) $P^{p^k}_* x_{2p^{j+1}-2}=0$ if $k\geq 1$.

§8. Selick's theorem on the odd primary homotopy groups of S^3

In [S1], P. Selick proved

Theorem 8.1. If p is an odd prime, then p annihilates the p-primary component of π_*S^3.

James had already proven that 4 annihilates the 2-primary component of π_*S^3. (See Corollary 5.2.) This is best possible since π_6S^3 is isomorphic to $\mathbb{Z}/12$.

The method of proof is to show that $\Omega^2S^3\langle 3\rangle$ is a retract of a space whose homotopy groups are a priori annihilated by p. Consider $map_*(A,X)$ the space of pointed map from A to X. If $A=\Sigma B$, then $map_*(A,X)=map_*(\Sigma B,X) \cong map_*(B,\Omega X)$. The degree k map $[k]: \Sigma B \to \Sigma B$ thus induces the kth power map on the loop space $map_*(\Sigma B,X)$. Write $k: \Omega X \to \Omega X$ for the kth power map as before. The starting point of Selick's method is

Lemma 8.2. If $[p]: \Sigma B \to \Sigma B$ is null-homotopic, then the pth power map in $map_*(\Sigma B,X)$ is null-homotopic and $p\pi_q map_*(\Sigma B,X)=0$.

Proof: From the above remarks, it suffices to recall that the pth power map on an H-space Y induces multiplication by p on π_*Y.

Write $P^n(k)$ for the mod-k Moore space which is the cofibre of the degree k map, $[k]: S^{n-1} \to S^{n-1}$. If k is odd, then the suspension order of the identity for $P^3(k)$ is k[N1]. Thus Theorem 8.1 follows immediately from

Theorem 8.3. Localized at the prime p, p>2, $\Omega^2S^3\langle 3\rangle$ is a retract of $map_*(P^3(p),S^{2p+1})$.

To prove 8.3, it suffices to give maps $h: \Omega^2S^3\langle 3\rangle \to map_*(P^3(p),S^{2p+1})$ and $\alpha: map_*(P^3(p),S^{2p+1}) \to \Omega^2S^3\langle 3\rangle$ such that $(\alpha h)_*$ is $\neq 0$ on $H_{2p-2}\Omega^2S^3\langle 3\rangle$ by

Lemma 8.4. Let $f\colon \Omega^2S^3\langle 3\rangle \to \Omega^2S^3\langle 3\rangle$ be any map which induces a non-zero map on H_{2p-2}. Then f is a p-local equivalence.

We start by constructing h. Since there is a cofibre sequence $S^2 \xrightarrow{[p]} S^2 \to P^3(p)$, there is a fibration $\Omega^2X \xleftarrow{p} \Omega^2X \leftarrow \mathrm{map}_*(P^3(p),X)$ where p denotes the pth power map on Ω^2X. Thus we consider the looped Hopf invariant $\Omega h_p\colon \Omega^2S^3 \to \Omega^2S^{2p+1}$ where h_p is the adjoint of $\tilde{H}_p$ in Lemma 2.5 and show that $p\Omega h_p$ is null. Hence there is a map $h\colon \Omega^2S^3 \to \mathrm{map}_*(P^3(p),S^{2p+1})$ lifting Ωh_p. We claim that h is non-zero on H_{2p-2}. Notice that a choice of generator for $H_{2p-1}\Omega^2S^3$ suspends to the p-th power in $H_{2p}\Omega S^3$. Thus $(\Omega h_p)_*$ is onto H_{2p-1}. Since $H_{2p-2}(\mathrm{map}_*(P^3(p),S^{2p+1});\mathbb{Z})$ is $\mathbb{Z}/p$ by the Hurewicz theorem, it follows that h_* is an isomorphism on H_{2p-2} by commutation with the homology Bockstein. Thus to construct h, it suffices to prove the following lemma.

Lemma 8.5. Ωh_p has order p in $[\Omega^2S^{2n+1},\Omega^2S^{2np+1}]$.

Proof: There is a commutative diagram

$$\begin{array}{ccc} \Omega S^{2n+1} & \xrightarrow{h_p} & \Omega S^{2np+1} \\ {\scriptstyle \Omega[p]}\big\downarrow & & \big\downarrow{\scriptstyle \Omega[p^p]} \\ \Omega S^{2n+1} & \xrightarrow[h_p]{} & \Omega S^{2np+1} \end{array}$$

by naturality of h_p. Since S^{2k+1} is a p-local H-space if $p>2$, $\Omega[p^j]$ is homotopic to the p^j-th power map on ΩS^{2k+1}. Looping gives $(p^p-p)\Omega h_p=0$ in $[\Omega^2S^{2n+1},\Omega^2S^{2np+1}]$. But $p^{p-1}-1$ is a unit in $\mathbb{Z}_{(p)}$ and thus $p\Omega h_p=0$ when spaces are localized at p.

Next, Selick constructs $\alpha\colon \mathrm{map}_*(P^3(p),S^{2p+1}) \to \Omega^2S^3\langle 3\rangle$ as follows: There is a homotopy commutative diagram

$$\begin{array}{ccc} S^{2p+1} & \longrightarrow & * \\ \downarrow{[p]} & & \downarrow \\ S^{2p+1} & \xrightarrow[\bar{\alpha}]{} & BS^3\langle 3\rangle \end{array}$$

where $\bar{\alpha}$ generates the p-primary component of $\pi_{2p+1}BS^3$. Thus there is a homotopy commutative diagram

$$\begin{array}{ccc} \Omega S^{2p+1} & \longrightarrow & * \\ p\downarrow & & \downarrow \\ \Omega S^{2p+1} & \xrightarrow[\Omega\bar{\alpha}]{} & \Omega S^3\langle 3\rangle \end{array}$$

because S^{2p+1} is a p-local H-space and thus $\Omega[p]$ is homotopic to the pth power map. Since there is a lift $\tilde{\alpha}$: $map_*(P^2(p),S^{2p+1}) \to \Omega S^3\langle 3\rangle$ there is a loop map α: $map_*(P^3(p),S^{2p+1}) \to \Omega^2S^3\langle 3\rangle$. Since $\Omega^2\bar{\alpha}$: $\Omega^2S^{2p+1} \to \Omega S^3\langle 3\rangle$ induces an epimorphism on H_{2p-1}, α induces an epimorphism on H_{2p-2} by inspection.

Thus there exist maps α and h inducing isomorphisms on $H_{2p-2} \cong \mathbb{Z}/p$. To finish, it suffices to give

Lemma 8.6. Let B denote the acyclic differential Hopf algebra $\Lambda[y]\otimes\mathbb{Z}/p[dy]$ with p prime and let f: B → B be any map of differential coalgebras with f(dy)≠0. Then f is an isomorphism.

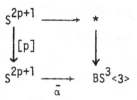

Proof: It suffices to check that f is a monomorphism. Since f is a morphism of coalgebras, it suffices to check that f is a monomorphism on the module of primitives [MM]. By inspection, there is at most one primitive in any fixed degree and $\{y,(dy)^{p^i}\}$ is a basis for the module of primitives. If f has a kernel, then $(dy)^{p^k}$ is the element of least degree in ker(f). Since $f(y(dy)^{p^k-1})=\lambda y(dy)^{p^k-1}$, $\lambda\neq 0$, by commutation with the coproduct, $f((dy)^{p^k})\neq 0$ and the lemma follows.

Proof of Theorem 8.4. Recall from Theorem 7.2 that the mod-p homology of $\Omega^2S^3\langle 3\rangle$ is isomorphic to $\underset{k\geq 1}{\otimes}\Lambda[y_{2p^k-1}]\underset{k\geq 1}{\otimes}\mathbb{Z}/p[x_{2p^k-2}]$ and $\beta y_{2p^k-1}=x_{2p^k-2}$. By inspection there is a most one primitive in any degree and a basis for the module of primitives is $\{y_{2p^k-1},(x_{2p^k-2})^{p^j}\mid k\geq 1,\ j\geq 0\}$. Write $f=\alpha\cdot h$. It suffices to check that f_* (primitive) is non-zero. Since $P_*^j(x_{2p^{k+1}-2})^{p^j}=-(x_{2p^k-2})^{p^{j+1}}$ and f_* commutes with P_*^j, it suffices to check that $f_*(x)\neq 0$ for $x=y_{2p-1}$ or $x_{2p-2}^{p^j}$.

Consider the differential Hopf algebra $B=\Lambda[y_{2p-1}]\otimes\mathbb{Z}/p[x_{2p-2}]$ with differential given by the homology Bockstein. The composite

$$B\subset H_*\Omega^2S^3\langle 3\rangle \xrightarrow{f_*} H_*\Omega^2S^3\langle 3\rangle \xrightarrow{\text{project}} B$$

is a map of differential coalgebras which is non-zero on x_{2p-2}. Thus the composite is an isomorphism by lemma 8.6 and the theorem follows.

Remark 8.7. That p^n annihilates the p-primary component of π_*S^{2n+1}, p odd, is proven in [CMN2] and [N3] using apparently different techniques. We shall try to find 2 primary analogues of Selick's retract theorem rather than pursuing odd primary information.

§9: Remarks on Selick's theorem for $\pi_* S^3$; desuspensions

One consequence of the proof of Selick's theorem is

Theorem 9.1. Let p be an odd prime and let $f\colon \Omega P^{2p+2}(p) \to S^3$ be any map which is onto π_{2p}. Then f_* is a split epimorphism on the p-primary component of $\pi_q S^3$, $q>3$.

This theorem can be thought of as an analogue of the Kahn-Priddy theorem for S^3. It is proven by looping f twice, constructing a map $h\colon \Omega^2 S^3 \to \Omega^3 P^{2p+2}(p)$ which is onto H_{2p-2} and then repeating the proof of Theorem 8.4. Thus $\Omega^2 S^3\langle 3\rangle$ is a p-local retract of $\Omega^3 P^{2p+2}(p)$. By adjointness, one has

Observation 9.2. (1) $\Omega^2 S^3\langle 3\rangle$ is a retract of $\Omega^3 P^{2p+2}(p)$ if and only if there is a map $\Sigma^3 \Omega^2 S^3\langle 3\rangle \to P^{2p+2}(p)$ which is onto in homology.

(2) $\Omega^2 S^3\langle 3\rangle$ is a retract of $\Omega^2 P^{2p+1}(p)$ if and only if there is a map $\Sigma^2 \Omega^2 S^3\langle 3\rangle \to P^{2p+1}(p)$ which is onto in homology.

We do not know whether there exists a map $\Sigma^2 \Omega^2 S^3\langle 3\rangle \to P^{2p+1}(p)$ which is onto in homology. The existence of such a map immediately implies that any map $P^{2p+1}(p) \to S^3\langle 3\rangle$ which is onto π_{2p} gives a split epimorphism on homotopy groups.

A map $\tilde{h}\colon \Sigma^3 \Omega^2 S^3\langle 3\rangle \to P^{2p+2}(p)$ which is onto in homology if $p>2$, is obtained as follows. We've given a map $\Omega^2 S^{2n+1} \to \mathrm{map}_*(P^3(p), S^{2np+1})$ since $p\Omega h_p = 0$ in $[\Omega^2 S^{2n+1}, \Omega^2 S^{2np+1}]$. Since there is a homotopy commutative diagram

$$\begin{array}{ccc} \Omega S^n & \longrightarrow & \Omega^2 S^{n+1} \\ {\scriptstyle p}\big\downarrow & & \big\downarrow{\scriptstyle \Omega^2[p]} \\ \Omega S^n & \longrightarrow & \Omega^2 S^{n+1} \end{array},$$

the composite $\Omega S^n \xrightarrow{p} \Omega S^n \xrightarrow{\Omega E} \Omega^2 S^{n+1} \xrightarrow{\Omega^2(\text{inclusion})} \Omega^2 P^{n+2}(p)$ is null homotopic, and there is a degree one map j: $map_*(P^2(p),S^n) \to \Omega^2 P^{n+2}(p)$. The map $\tilde{h}$ may be chosen to be the adjoint of the composite

$$\Omega^2 S^3\langle 3\rangle \to map_*(P^3(p),S^{2p+1}) \xrightarrow{\Omega j} \Omega^3 P^{2p+2}(p) .$$

The 2-primary analogue fails as we shall see in the next section. Namely, there does not exist a map $\Sigma^3\Omega^2 S^3\langle 3\rangle \to P^6(2)$ which is onto in homology. Thus we are unable to prove a retract theorem such as 9.1 if p=2. By other means, we shall show that $\Omega^2 S^3\langle 3\rangle$ is a 2-local retract of $map_*(P^3(2),S^5)$.

<u>Question 9.3</u>. Does there exist a map $\Sigma^2\Omega^2 S^3 \to P^{2p+1}(p)$ which is onto in homology (where p is an odd prime)?

§10. Desuspension and non-desuspension theorems

In the last section we saw that Selick's retraction theorem is equivalent ot the existence of a map $\bar{h}\colon \Sigma^3\Omega^2S^3 \to \Sigma^3P^{2p-1}_{(p)}$ which is onto in homology if p is an odd prime. Thus we are interested in the same question at the prime 2.

Lemma 10.1. There is a map $\tilde{h}\colon \Sigma^4\Omega^2S^{2n+1} \to \Sigma^4P^{4n-1}(2)$ which is onto in homology.

Proof: Since $2\Omega H = 0$ in $[\Omega^2S^{2n+1},\Omega^2S^{4n+1}]$, there is a lift, $\bar{h}$, of ΩH to $(\Omega^2S^{4n+1})\{2\}$, the fibre of the squaring map. Since the suspension of the Whitehead product is zero, there is a homotopy commutative diagram

$$\begin{array}{ccccc}
\Omega S^{n+1} & \xrightarrow{\Omega E} & \Omega^2S^{n+2} & \longrightarrow & * \\
{\scriptstyle 2}\big\downarrow & & \big\downarrow{\scriptstyle \Omega^2[2]} & & \big\downarrow \\
\Omega S^{n+1} & \xrightarrow[E]{} & \Omega^{2p+2} & \longrightarrow & \Omega^2P^{n+2}(2) .
\end{array}$$

Thus there is a degree one map, g, from $(\Omega S^{n+1})\{2\}$ to $\Omega^3P^{n+2}(2)$. The map in Lemma 10.1 is the adjoint of

$$\Omega^2S^{2n+1} \xrightarrow{\bar{h}} (\Omega^2S^{4n+1})\{2\} \xrightarrow{\Omega g} \Omega^4P^{4n+1}(2).$$

We remark that a related map can be obtained by desuspending Snaith's stable decomposition. A quick proof is given in the appendix of [C_2].

Next, we check that the map in 10.1 does not desuspend. Let $\tilde{f}\colon \Sigma^\infty X \to \Sigma^\infty Y$ denote a stable map with adjoint $f\colon X \to QY$. By definition one has

Lemma 10.2. The map $\tilde{f}$ is the stablization of a map which exists after q-suspensions if and only if there is a homotopy commutative diagram

$$\begin{array}{ccc} & & \Omega^q\Sigma^q Y \\ & \overset{g}{\nearrow} & \downarrow \\ X & \xrightarrow[f]{} & QY \end{array} .$$

Direct primary computations then give non-desuspension results; some examples will be given here.

<u>Proposition 10.3</u>. If $\Sigma^q\Omega^2S^{2n+1} \to \Sigma^q P^{4n-1}(2)$ is onto in homology, then $q\geq 4$.

<u>Proof</u>: This was proven in [CM] with the same method which we give below.

Consider $f\colon \Omega^2S^{2n+1} \to QP^{4n-1}(2)$ with f_* onto H_{4n-2}. Write v as the generator of $H_{4n-1}QP^{4n-1}(2)$ and $u=Sq^1_*v$. Write x_{2^in-1} for the unique primitive in $H_{2^in-1}\Omega^2S^{2n+1}$.

Then by the Nishida relations $Sq^1_*x_{8n-1}=x^2_{4n-1}$ and $Sq^{2^k}_*x_{8n-1}=0$ if $k>0$. Since x_{8n-1} an odd degree primitive,

$$f_*x_{8n-1} = AQ_3u+BQ_1v$$

for some A and B. Next observe that $Sq^2_*x_{8n-1}=0$ and $Sq^2_*Q_3u=Sq^2_*Q_1v=Q_1u$ because the degree of u is 4n-2. Thus by naturality, A=B. Apply Sq^1_* to get

$$f_*(x^2_{4n-1}) = B(v^2).$$

Apply Sq^2_* to this last equation to get

$$f_*(x^4_{2n-1}) = Bu^2.$$

Notice that $f_*(x^2_{2n-1})=u$ and $f_*(x_{4n-1})=v$. Thus $f_*(x^2_{2n-1}\cdot x_{4n-1})=uv$ + primitive by commutation with the coproduct. There is exactly one (8n-3)-dimensional primitive in $H_*QP^{4n-1}(2)$, namely Q_1u. Since $Sq^1_*Q_1u=0$ and $Sq^1_*x_{4n-1}=x^2_{2n-1}$, we get $f_*(x^4_{2n-1})=u^2$. Thus A=B=1 and the lemma follows.

<u>Example 10.4</u>. I.M. James proved that $\Sigma\mathbb{C}P^{n-1}$ is a stable retract of SU(n) in [J5]. Thus SU(3) is stably equivalent to $\Sigma\mathbb{C}P^2\vee S^8$. The group $\pi_{7+k}\Sigma^{k+2}\mathbb{C}P^2$

stabilizes when k=3; it follows that $\Sigma^3 SU(3)$ is equivalent to $\Sigma^4 \mathbb{C}P^2 \vee S^{11}$. We shall check that this is best possible in the following proposition which is a special case of the results in [CP].

Proposition 10.5. Let $\Sigma^q SU(3) \to \Sigma^{q+1}\mathbb{C}P^2$ be any map which is onto in homology. Then $q \geq 3$.

Using the same methods, we check the following which was given in [CP].

Proposition 10.6. The Lie group G_2 is stably equivalent to $X \vee S^{14}$ where a degree one stable self-map of X is a homotopy equivalence.

Proof of 10.5: Let f: $SU(3) \to Q\Sigma\mathbb{C}P^2$ be degree one. We shall check that if f lifts to $\Omega^q\Sigma^{q+1}\mathbb{C}P^2$, then $q \geq 3$.

Write v for the generator of $H_4\mathbb{C}P^2$ and $u = Sq^2_* v$. Write $H_*SU(3) = \Lambda[x,y]$ with $|x|=3$. Thus $f_*(x)=u$ and $f_*(y)=v$ and by inspection

$$f_*(xy) = Auv + BQ_2u \quad \text{for some A and B.}$$

Since u is primitive, the reduced coproduct of $f_*(xy)$ is $A(u \otimes v + v \otimes u)$. But by naturality of the coproduct, it is also equal to $(f_* \otimes f_*)(x \otimes y + y \otimes x) = u \otimes v + v \otimes u$. Thus A=1.

Applying Sq^2_*, we get

$$o = f_*(x^2) = Au^2 + Bu^2.$$

Thus A=B. Since A=1, we get $f_*(xy) = ux + Q_2u$ and so $q \geq 3$. The proposition follows.

Proof of Proposition 10.6: Recall that

$$H^*G_2 \cong \mathbb{Z}/2[x_3]/_{x_3^4=o} \otimes \Lambda[Sq^2x_3]$$

by [W1]. We use this to check that a degree one stable self-map of X is an equivalence.

Since G_2 is a Lie group, the top cell is stably spherical. Thus G_2 is stably equivalent to a complex $X \vee S^{14}$ with partial cell diagram for X given by

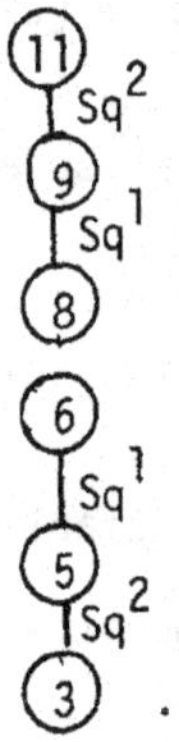

Let $f\colon \Sigma^q X \to \Sigma^q X$ be a map inducing an isomorphism on H_{q+3}. Let Tel_f denote the mapping telescope of f, $i\colon X \to \mathrm{Tel}_f$ the natural map with fibre F.

Assume that f is not an equivalence. Then f^* is zero on H^{8+q}. We show that this is a contradiction. Consider the Serre spectral sequence for i and observe that a cell decomposition for F with $q \geq 2$ is given by

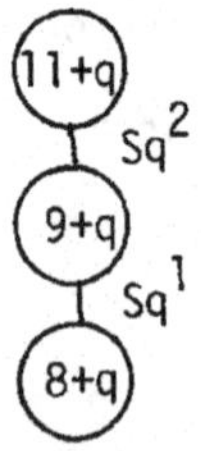

in dimensions at most 11+q. Thus it follows that X is stably a bouquet $A \vee B$ with $\bar{H}_i X \cong \bar{H}_i A$ for $i \leq 6$ and $\bar{H}_i X \cong \bar{H}_i B$ for $i \geq 8$.

Thus there is a map $g\colon G_2 \to QA$ which induces an isomorphism on H_3. Let x_i denote a generator for $H_i G_2$ and y_i a generator for $H_i A$. Then $g_*(x_i)=y_i$ for $i \leq 6$. By a check of degrees, $g_*(x_8)=Ay_3y_5+BQ_2y_3$, $g_*(x_9)=Cy_3y_6+DQ_3y_3$ and $g_*(x_{11})=Ey_5y_6+FQ_5y_3+GQ_1y_5$. Note that A=C=E=1 by commutation with the coproduct. Further note that (1) B=1 by commutation with Sq_*^2, (2) D=1 by commutation with Sq_*^1, and (3) F=0 and G=1 by commutation with Sq^1_*. But then $Sq^2_* g_*(x_{11}) \neq g_*(x_9)$ which is a contradiction and the proposition follows.

The group G_2 has some interesting properties implied by its cohomology and we make a slight digression here to discuss it. There is a fibration [W1]

$$SU(3) \to G_2 \to S^6$$

and we shall consider the composition $\Omega G_2 \to \Omega S^6 \xrightarrow{h} \Omega S^{11}$. Notice there is a morphism of fibrations

$$\begin{array}{ccccc}
X & \longrightarrow & \Omega G_2\langle 3\rangle & \xrightarrow{f} & \Omega S^{11} \\
\downarrow & & \downarrow & & \downarrow = \\
S^5 & \longrightarrow & \Omega S^6 & \xrightarrow{h} & \Omega S^{11} \\
\downarrow g & & \downarrow & & \downarrow \\
SU(3)\langle 3\rangle & \longrightarrow & SU(3)\langle 3\rangle & \longrightarrow & *
\end{array}$$

where X is the fibre of f (or of g). By direct calculation, the mod-2 homology of X is isomorphic to the underlying coalgebra of $\mathbb{Z}/2[x_6, x_{2^i-1} \mid i \geq 3]$ with $|x_i| = i$. It would be interesting to know whether X is homotopy equivalent to the space BW_2 recently constructed by B. Gray [G2].

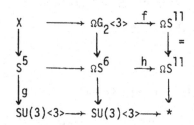

§11. Divisibility of the Whitehead product and the distributivity formula

Let w_n denote the Whitehead product $[\iota_n,\iota_n]$ in $\pi_{2n-1}S^n$. In section 4 we gave the following formulas for expanding $\Omega[k]$ where $[k]$ is the degree k map on S^n. Namely, it was checked that

(i) $\Omega[-1] = \Omega w_{2n+1}\cdot H-1$, and

(ii) $\Omega[2] = 2+\Omega w_{2n+1}\cdot H$

in $[\Omega S^{2n+1},\Omega S^{2n+1}]$ where H is the second Hilton-Hopf invariant.

Thus consider the self-map of ΩS^{2n+1} given by $\Omega w_{2n+1}\cdot H=\phi$. We study null-homotopies of $\Omega^q\phi$ in this section.

Proposition 11.1.

(1) ϕ is null-homotopic if and only if n=0,1, or 3.

(2) $\Omega\phi$ is null-homotopic if and only if w_{2n+1} is divisible by 2.

The proof of 11.1 gives a bit more

Proposition 11.2.

1. ϕ is null homotopic on the 4n-skeleton of ΩS^{2n+1} if and only if n=0,1, or 3.

2. $\Omega\phi$ is null homotopic on the (4n-1)-skeleton of Ω^2S^{2n+1} if and only if w_{2n+1} is divisible by 2.

3. $\Omega^2\phi$ is null-homotopic on the (4n-2)-skeleton of Ω^3S^{2n+1} if and only if

a. $n\equiv 0(2)$ and $w_{2n+1}=2x+y\eta_{4n}$ or

b. $n\equiv 1(2)$ and $w_{2n+1}=2x$.

Write P^n_j for $\mathbf{R}P^n/\mathbf{R}P^{j-1}$.

Proposition 11.3. The map $\Omega^q\phi$ is null-homotopic on the (4n-q+1)-skeleton of Ω^qS^{2n+1}, $n\geq q$, if and only if there is a homotopy commutative diagram

$$\begin{array}{ccc} S^{4n+1} & \xrightarrow{w_{2n+1}} & S^{2n+1} \\ \Big\downarrow{\scriptstyle\alpha} & \nearrow{\scriptstyle g} & \\ \Sigma^{2n+2}P^{2n-1}_{2n-q-1} & & \end{array}$$

where the cofibre of α is $\Sigma^{2n}P^{2n}_{2n-q-1}$.

There are several well-known equivalent formulations of this last question and these can be found in work of Barratt, Mahowald, Jones, and Selick [M,BJM,S_4].

To start consider the long sequence given by the EHP sequence

$$\cdots \to \pi_{4n+3}S^{4n+3} \xrightarrow{P} \pi_{4n+1}S^{n+1} \xrightarrow{E} \pi_{4n+2}S^{2n+2} \xrightarrow{H} \pi_{4n+2}S^{4n+3} \to \cdots .$$

Since there exists an element of Hopf invariant 2, there is a short exact sequence

$$0 \to \mathbb{Z}/2 \to \pi_{4n+1}S^{2n+1} \xrightarrow{E} \pi_{4n+2}S^{2n+2} \to 0$$

if $n \neq 0,1,3$. In the following proposition assume that $n \neq 0,1,3$ in order to avoid special arguments. Thus $w_{2n+1}=0$ if $n=0,1,3$, and $w_{2n+1} \neq 0$ otherwise [A1]. Recall that all cohomology groups are taken with $\mathbb{Z}/2$-coefficients.

Proposition 11.4. Let $n \neq 0,1,3$. The following are equivalent.

(1) The Whitehead product $[\iota_{2n+1},\iota_{2n+1}]$ is divisible by 2.

(2) The short exact sequence

$$0 \to \mathbb{Z}/2 \to \pi_{4n+1}S^{2n+1} \to \pi_{4n+2}S^{2n+2} \to 0$$

is not split.

(3) There is a map $P^{4n+2}(2) \to \Omega S^{2n+2}$ which is non-zero in homology.

(4) There exists a space X with $H^i X \cong \mathbb{Z}/2$ for i=2n+2, 4n+3, and 4n+4 and zero otherwise with $Sq^{2n+2}: H^{2n+2}X \to H^{4n+4}X$ and $Sq^1: H^{4n+3}X \to H^{4n+4}X$ isomorphisms.

(5) $\Omega^2[-1]$ is homotopic to -1 in $\Omega^2 S^{2n+1}$.

We will give the proof of 11.4 in the next section. Observe that the Adem relations imply that X in 11.4 fails to exist whenever $2n+1 \neq 2^k-1$. It has been conjectured that w_{2n+1} is divisible by 2 whenever $2n+1=2^k-1$. For example Toda [T3] proves this when 2n+1=15 and Barratt and Mahowald prove this when 2n+1=31.

We record the following equivalent formulation in

Proposition 11.5. The degree -1 map on S^{2n+1} acts by multiplication by -1 on the abelian group $[\Sigma^2 A, S^{2n+1}]$ for all A if and only if $2n+1=2^k-1$ and w_{2n+1} is divisible by 2.

Remark 11.6. By propositions 11.2 and 11.3, $\Omega\phi$ fails to be null-homotopic on the 8-skeleton of $\Omega^2 S^5$. Since $[\iota_5, \iota_5] = \sigma_5 \eta_8$ [T1], 11.2(3) gives that $\Omega^2\phi$ is null-homotopic on the 7-skeleton of $\Omega^3 S^5$. This leads one to wonder whether $\Omega^2\phi$ is null-homotopic on $\Omega^3 S^5$. It would be interesting to see whether $\Omega^3[2]$ and 2 are homotopic on $\Omega^3 S^5$ or more generally $\Omega^3 S^{4n+1}$.

Proof of Proposition 11.1: We start by proving (1). Notice that if n=0,1, or 3, $w_{2n+1}=0$ since S^{2n+1} is an H-space. Thus ϕ is null-homotopic. Assume that ϕ is null-homotopic and let $J_2 S^{2n}$ denote the 4n-skeleton of ΩS^{2n+1}.

There is a cofibration sequence $S^{2n} \to J_2 S^{2n} \to S^{4n}$ which gives a long exact sequence

$$[S^{4n-1}, \Omega S^{2n+1}] \leftarrow [S^{2n}, \Omega S^{2n+1}] \leftarrow [J_2 S^{2n}, \Omega S^{2n+1}] \leftarrow [S^{4n}, \Omega S^{2n+1}] \leftarrow [S^{2n+1}, \Omega S^{2n+1}] \leftarrow .$$

The map $[S^{2n+1},\Omega S^{2n+1}] \to [S^{4n},\Omega S^{2n+1}]$ is trivial because ΣJ_2S^{2n} is homotopy equivalent to $S^{2n+1}\vee S^{4n+1}$ and thus $[J_2S^{2n},\Omega S^{2n+1}] \to [S^{2n},\Omega S^{2n+1}]$ is split as sets.

Since ϕ is null-homotopic, so is the composite

$$J_2S^{2n} \xrightarrow{\text{pinch}} S^{4n} \xrightarrow{E} \Omega S^{4n+1} \xrightarrow{\Omega w_{2n+1}} \Omega S^{2n+1}.$$

But as observed above, the homomorphism $[S^{4n},\Omega S^{2n+1}] \to [J_2S^{2n},\Omega S^{2n+1}]$ is a monomorphism. Hence $w_{2n+1}=0$ and so $n=0,1$, or 3 by [A1].

We next prove 11.1(2). First assume that w_{2n+1} is divisible by 2 and write $w_{2n+1}=2x$. Since x is of finite order, $H(x)$ is zero in $\pi_{4n}\Omega S^{4n+1}$. Thus x desuspends; write y for a desuspension of x. We want to prove that the composite $\Omega^2S^{2n+1} \xrightarrow{\Omega H} \Omega^2S^{4n+1} \xrightarrow{\Omega^2(2x)} \Omega^2S^{2n+1}$ is null-homotopic. Since x desuspends, we may write $2x$ as the composite $S^{4n+1} \xrightarrow{x} S^{2n+1} \xrightarrow{[2]} S^{2n+1}$. Thus we must check that $\Omega^2[2]\cdot\Omega^2(x)\cdot\Omega H$ is null-homotopic.

Recall that $\Omega[2]=2+\Omega w_{2n+1}\cdot H$ in $[\Omega S^{2n+1},\Omega S^{2n+1}]$ and so $\Omega[2]\cdot\Omega(x)\cdot H = 2\Omega(x)\cdot H+\Omega w_{2n+1}\cdot H\cdot\Omega(x)\cdot H$. We claim that $\Omega w_{2n+1}\cdot H\cdot\Omega(x)$ is zero in $[\Omega S^{4n+1},\Omega S^{2n+1}]$ [and we remark that $2\Omega(x)\cdot H$ is non-zero because it represents w_{2n+1} when restricted to the bottom cell of ΩS^{4n+1}]. Since $2\Omega^2(x)\cdot\Omega H$ is zero by Lemma 4.1, the result follows from the claim. By naturality of H, there is a homotopy commutative diagram

$$\begin{array}{ccccc}
\Omega S^{4n+1} & \xrightarrow{\Omega\Sigma(y)} & \Omega S^{2n+1} & & \\
\downarrow H & & \downarrow H & \overset{\phi}{\searrow} & \\
\Omega S^{8n+1} & \xrightarrow[\Omega\Sigma(y\wedge y)]{} & \Omega S^{4n+1} & \xrightarrow[\Omega(2x)]{} & \Omega S^{2n+1}
\end{array}.$$

But $(2x)\cdot\Sigma(y\wedge y)=x\cdot(\Sigma(2y)\wedge y)=x\cdot(w_{2n+1}\wedge y)=0$ in $\pi_{8n+1}S^{2n+1}$, and thus the claim follows.

To finish, we must check that if $\Omega\phi$ is null-homotopic, then w_{2n+1} is divisible by 2. By Theorem 7.1, the (4n-1)-skeleton of Ω^2S^{2n+1}, X, has partial cell diagram

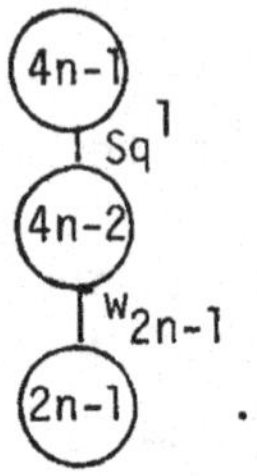

By inspection of the Serre Spectral sequence, we see that the composite, π, $X \hookrightarrow \Omega^2S^{2n+1} \xrightarrow{\Omega H} \Omega^2S^{4n+1}$ is homotopic to the composite $X \xrightarrow{\text{pinch}} S^{4n-1} \xrightarrow{E^2} \Omega^2S^{4n+1}$. Since there is an isomorphism of groups $[X,\Omega^2S^{4n+1}] \cong [P^{4n-1}(2),\Omega^2S^{4n+1}]$ and $\Omega w_{2n+1}\cdot\pi$ is assumed to be null, we get a map $P^{4n-1}(2) \to \Omega^2S^{2n+1}$ which is also null and which is the adjoint of w_{2n+1} on $P^{4n-1}(2)/S^{4n-2}$. Thus w_{2n+1} is divisible by 2.

<u>Proof of Proposition 11.2</u>: Notice that the proof of 11.1 gives 11.2(1) and 11.2(2). Consider the (4n-2)-skeleton of Ω^3S^{2n+1} and observe that it has the following partial cell diagram:

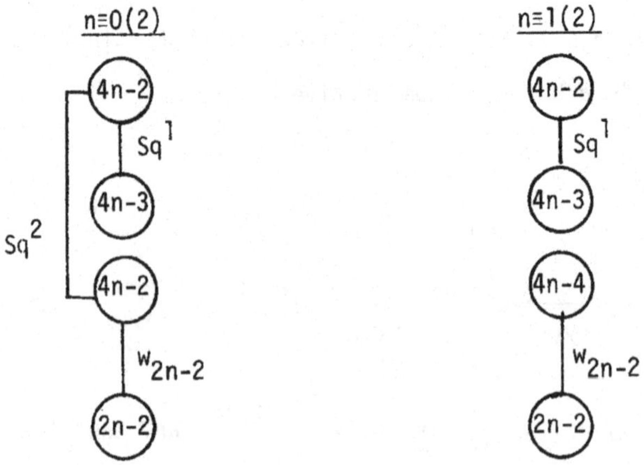

Thus 11.2 follows immediately from 11.3

<u>Proof of Proposition 11.3</u>: Let Z denote the (4n-q+1)-skeleton of $\Omega^q S^{2n+1}$, $n \geq q$. Recall that Z/S^{2n+1-q} is homotopy equivalent to

$$\Sigma^{2n+1-q}(P^{2n}_{2n-q+1})(=\Sigma^{2n+1-q}\,\mathbb{R}P^{2n}/\mathbb{R}P^{2n-q}).$$

Write $\pi\colon Z \to Z/S^{2n+1-q}$ for the collapse map and observe that the map π induces an isomorphism $\pi^*\colon [Z/S^{2n+1-q},\Omega^q S^{4n+1}] \to [Z,\Omega^q S^{4n+1}]$. Thus $\Omega^{q-1}\phi$ is null-homotopic on the (4n-q+1)-skeleton of $\Omega^q S^{2n+1}$ if and only if the induced composite $Z/S^{2n+1-q} \xrightarrow{p} \Omega^q S^{4n+1} \xrightarrow{\Omega^q w_{2n+1}} \Omega^q S^{2n+1}$ is null-homotopic where p is the restriction of $\Omega^{q-1}H$. Since $(\Omega^{q-1}H)_*$ is onto in mod-2 homology if $q \leq 4n+1$, the map p is given by the collapse map $Z/S^{2n+1-q} \to S^{4n+1-q}$. Thus the adjoint of the map $(\Omega^q w_{2n+1})\cdot p$ is given by the composite $\Sigma^{2n+1}P^{2n}_{2n-q+1} \xrightarrow{f} S^{4n+1} \xrightarrow{w_{2n+1}} S^{2n+1}$ where f is the collapse map. This composite is null-homotopic if and only if there is a homotopy commutative diagram

$$\begin{array}{ccc} S^{4n+1} & \xrightarrow{w_{2n+1}} & S^{2n+1} \\ {\scriptstyle\alpha}\big\downarrow & \nearrow{\scriptstyle g} & \\ \Sigma^{2n+2}P^{2n-1}_{2n-q+1} & & \end{array}$$

where α is an attaching map whose cofibre is $\Sigma^{2n+2}P^{2n}_{2n-q+1}$. Thus Proposition 11.3 follows.

§12: Proofs of classical results on the divisibility of $[\iota_{2n+1},\iota_{2n+1}]$

In this section we prove 11.4 which gives well-known and equivalent formulations on the divisibility of $w_{2n+1}=[\iota_{2n+1},\iota_{2n+1}]$. Good references are [M,BJM,S4]. Since $w_{2n+1}\neq 0$ if $n\neq 0,1$, or 3, suspends to zero and has order 2 (by 1.3), the equivalence of 11.4(1) and 11.4(2) is immediate. The equivalence of 11.4(1) and 11.4(5) was checked in section 11.

<u>(1) is equivalent to (3)</u>: Assume that $w_{2n+1}=2x$; we construct a map $f\colon P^{4n+2}(2)\to\Omega S^{2n+2}$ which is non-zero in mod-2 homology. Since $E(2x)$ is null, there is a map $f\colon P^{4n+2}(2)\to\Omega S^{2n+2}$ which is $E(x)$ on the bottom cell. To show that f_* is non-zero on H_{4n+2}, it suffices to check that the Hopf invariant of f in $[P^{4n+2}(2),\Omega S^{4n+2}]$ is essential. If $h(f)=0$, then there is a map $g\colon P^{4n+2}(2)\to S^{2n+1}$ with $E(g)=f$. Let y denote the restriction of g to the bottom cell of $P^{4n+2}(2)$. Thus $E(x-y)=0$ and so $x-y$ is a multiple of w_{2n+1}. Since $2w_{2n+1}=2y=0$, it follows that $w_{2n+1}=2x$ is zero. This is a contradiction.

Assume that there is a map $f\colon P^{4n+2}(2)\to\Omega S^{2n+2}$ which is non-zero on H_{4n+2}. We check that w_{2n+1} is divisible by 2. Let g denote a desuspension of f restricted to S^{4n+1} (which exists because ΩS^{4n+3} is $(4n+1)$-connected). Thus there is a homotopy commutative diagram

$$\begin{array}{ccc} S^{4n+1} & \xrightarrow{i} & P^{4n+2}(2) \\ g\downarrow & & \downarrow f \\ S^{2n+1} & \xrightarrow[E]{} & \Omega S^{2n+2} \end{array} .$$

Thus $E(2g)=0$ and so $2g$ is a multiple of w_{2n+1}. If $2g=0$, then there is a map $\alpha\colon P^{4n+2}(2)\to S^{2n+1}$ with $\alpha i=g$ in $\pi_{4n+1}S^{2n+1}$. But then $f-E(\alpha)$ factors through S^{4n+2} which forces the Pontrjagin square in $H_*\Omega S^{2n+2}$ to be spherical. Thus ΩS^{2n+2} splits as $S^{2n+1}\times\Omega S^{4n+1}$ which contradicts [A1] since S^{2n+1} is an H-space.

(3) is equivalent to (4): We check that there is a map $f\colon P^{4n+2}(2) \to \Omega S^{2n+2}$ which is non-zero in homology, if and only if there exists a space X with

$$H^i(X) \cong \begin{cases} \mathbb{Z}/2 \text{ if } i=0,\ 2n+2,\ 4n+3,\ 4n+4 \\ 0 \quad \text{otherwise.} \end{cases}$$

with $Sq^{2n+2} \neq 0$. The space X is the mapping cone of $g\colon P^{4n+3}(2) \to S^{2n+2}$ which follows by inspection of

Lemma 12.1 . Let B be n-connected. There is a map $g\colon \Sigma B \to S^{n+1}$ with mapping cone X satisfying Sq^{n+1} is non-zero on $H^{n+1}(X)$ if and only if there is a map $f\colon B \to \Omega S^{n+1}$ which gives an epimorphism on H_{2n}.

Proof: Assume that f exists. Let $g\colon \Sigma B \to S^{n+1}$ denote the adjoint of f and write X for the mapping cone of g. Then there is a morphism of fibrations

$$\begin{array}{ccccc} \Omega\Sigma B & \xrightarrow{\Omega(g)} & \Omega S^{n+1} & \xrightarrow{\Omega(i)} & \Omega X \\ \downarrow & & \downarrow & & \downarrow \\ * & \longrightarrow & * & \longrightarrow & * \\ \downarrow & & \downarrow & & \downarrow \\ \Sigma B & \xrightarrow{g} & S^{n+1} & \xrightarrow{i} & X \end{array} .$$

Suppose that $Sq^{n+1}(\iota)=0$ for $\iota \in H^{n+1}X$ with $i^*(\iota) \neq 0$. Consider the Serre spectral sequence for these fibrations in mod-2 cohomology. In the right-hand fibration, there is a unique element x_n transgressing to the class ι in $H^{n+1}X$. Since $Sq^{n+1}\iota=0$, $\iota \otimes x_n$ is an infinite cycle which must be hit by an element z from $H^{2n}\Omega X$. By naturality, z pulls back to the generator of $H^{2n}\Omega S^{2n+1}$ which itself pulls back non-trivially to $H^{2n}\Omega\Sigma B$. Thus is a contradiction because ig is null-homotopic.

Assume that g exists and let $f\colon B \to \Omega S^{n+1}$ be the adjoint of g. Consider the homotopy theoretic fibre of the inclusion $i\colon S^{n+1} \to X$ to obtain a homotopy commutative diagram

$$\begin{array}{ccccc} \Omega\Sigma B & \xrightarrow{\Omega g} & \Omega S^{n+1} & \xrightarrow{\Omega(i)} & \Omega X \\ {\scriptstyle \Omega\alpha}\downarrow & & \downarrow{\scriptstyle 1} & & \downarrow{\scriptstyle 1} \\ \Omega F & \xrightarrow[\Omega(j)]{} & \Omega S^{n+1} & \xrightarrow[\Omega(i)]{} & \Omega X \end{array}$$

where α_* induces an isomorphism in homology through dimensions 2n+1 by the Serre exact sequence. But the Pontrjagin square $\iota_n^2=0$ in $H_{2n}\Omega X$ by inspection of the Serre spectral sequence and the fact that $Sq^{n+1}: H^{n+1}X \to H^{2n+2}X$ is monic. Inspection of the Serre spectral sequence for $\Omega(i)$ gives that ι_n^2 is in the image of $\Omega(j)_*$. Thus ι_n^2 is in the image of $\Omega(j\alpha)_*$. But $B \to \Omega F$ induces an epimorphism in homology through dimension 2n and the lemma follows.

§13. The order of a map

Let α be an element of finite order in the group $[\Sigma A,B]$. One could ask about the order of $\Omega^n\alpha$ in the group $[\Omega^n\Sigma A,\Omega^n B]$, $n\geq 1$. The example which we have in mind is a 2-primary analogue of Selick's retraction theorem. Namely, if there were a map $f\colon \Omega^2S^5 \to \Omega S^3$ which is onto π_3 with the composite $\Omega^2S^5 \xrightarrow{2} \Omega^2S^5 \xrightarrow{f} \Omega S^3$ null-homotopic then one could mimic the proof of Theorem 8.3 to prove that $\Omega^2S^3\langle 3\rangle$ is a 2-local retract of $\mathrm{map}_*(P^3(2),S^5)$. This retraction result is correct and will be proven in section 19 but one has

Lemma 13.1. The element $\Omega^2\eta_4$ in $[\Omega^2S^5,\Omega^2S^4]$ has order 4. Furthermore, an H-map $f\colon \Omega^2S^5 \to \Omega S^3$ which induces an epimorphism on π_3 has order at least 4.

Another consequence of James' fibrations is

Lemma 13.2: Let $\alpha\colon \underset{I}{\vee}S^n \to X$ be of order 2^r. Then (1) $2^{r+1}\Omega^2(\alpha)=0$ in $[\Omega^2(\underset{I}{\vee}S^n),\Omega^2X]$ if $r\equiv 1(2)$ and (2) $2^r\Omega^2\alpha=0$ if $r\equiv 0(2)$.

A related conjecture due to M.G. Barratt is as follows. Suppose that the suspension order of the identity in $[\Sigma^2A,\Sigma^2A]$ is p^r. Then Barratt conjectures that $p^{r+1}\Omega^2(1)=0$ in $[\Omega^2\Sigma^2A,\Omega^2\Sigma^2A]$. It is essential to loop twice here. Indeed if Σ^2A has non-vanishing homology groups, then $\Omega(1)$ in $[\Omega\Sigma^2A,\Omega\Sigma^2A]$ has infinite order [CMN3]. There is one known example of Barratt's conjecture which is given by $\Sigma^2A=P^n(p^r)$ with $n>3$ and p an odd prime [N2]. In a similar vein, one might observe that if $\alpha\in[\Sigma A,BS^3]$ has finite order, then $\Omega^3(\alpha)$ has finite order by Corollary 5.2.

Let $\alpha\in[\Sigma^2B,X]$ with $2^r\alpha=0$. The following is a slight modification of a theorem due to M.G. Barratt [B].

Proposition 13.3. If $\Sigma^2 B$ is n-connected and $H_i(\Sigma^2 A)=0$ for $i\geq n2^q-1$, then $2^{rq}(af)=0$ for and any $f\colon \Sigma^2 A \to \Sigma^2 B$.

Proof of 13.2. Let $\alpha\colon \underset{I}{\vee} S^n \to X$ be of order 2^r. Loop and apply the Hilton-Milnor theorem (2.1) to get

$$\Omega(\underset{I}{\vee} S^n) \xrightarrow{\simeq} \prod_\lambda \Omega S^{m_\lambda} \xrightarrow{\Pi\Omega(\alpha_\lambda)} \Omega X$$

where α_λ has order 2^r by bilinearity of the Samelson product. Thus it suffices to prove 13.2 if $\alpha\colon S^m \to X$.

Recall that $\Omega[2]=2+\Omega[\iota_m,\iota_m]\cdot H$ in $[\Omega S^m,\Omega S^m]$ if $m\equiv 1(2)$. If $m\equiv 0(2)$, then $\Omega[2]=2+\Omega[\iota_m,\iota_m]\cdot H_2+\Omega[\iota_m[\iota_m,\iota_m]]\cdot H_3$ as in Proposition 4.3 where $H_i\colon \Omega S^m \to \Omega\Sigma S^{i(m-1)}$, $i=2,3$, is given by the composite

$$\Omega S^m \xrightarrow{\Omega(\text{pinch})} \Omega(S^m \vee S^m)\simeq(\Omega S^m)^2\times\Omega S^{2m-1}\times\Omega S^{3m-2}\times\Omega(\underset{\alpha>3m-2}{\vee} S^\alpha)_2 \xrightarrow{\text{project}} \Omega S^{i(m-1)+1}.$$

Since $3[\iota_m[\iota_m,\iota_m]]=0$ by 1.3, the formula $\Omega[2]=2+\Omega[\iota_m,\iota_m]\cdot H_2$ is correct in $[\Omega S^m,\Omega S^m]$ when S^m is localized at 2. As a consequence, we compute $\Omega(2\alpha\cdot[\iota_m,\iota_m])$. Using bilinearity, $2\alpha\cdot[\iota_m,\iota_m]$ is the composite $S^{2m-1} \xrightarrow{[2]} S^{2m-1} \xrightarrow{[\iota_m,\iota_m]} S^m \xrightarrow{\alpha} X$. Loop and use the formula $\Omega[2]=2+\Omega[\iota_m,\iota_m]\cdot H$ in $[\Omega S^m,\Omega S^m]$ to get

$$\Omega(2\alpha\cdot[\iota_m,\iota_m])=2\Omega(\alpha\cdot[\iota_m,\iota_m])+\Omega(\alpha\cdot[\iota_m,\iota_m]\cdot[\iota_{2m-1},\iota_{2m-1}])\cdot H.$$

Since $[\iota_m,\iota_m]\cdot[\iota_{2m-1},\iota_{2m-1}]=[[\iota_m,\iota_m],[\iota_m,\iota_m]]=0$, one has

Lemma 13.4. $\Omega(2\alpha\cdot[\iota_m,\iota_m])=2\Omega(\alpha\cdot[\iota_m,\iota_m])$ in $[\Omega S^m,\Omega X]$.

To finish the proof of 13.2, we shall need to loop to get the following where $r\geq 2$:

$$\begin{aligned}\Omega^2(2^r_\alpha) &= 2\Omega^2(2^{r-1}\alpha)+\Omega^2(2^{r-1}\alpha)\Omega^2([\iota_m,\iota_m])\cdot\Omega H\\ &= 2[2\Omega^2(2^{r-2}\alpha)+\Omega^2(2^{r-2}\alpha)\cdot\Omega^2[\iota_m,\iota_m]\cdot\Omega H]+\Omega^2(2^{r-1}\alpha)\Omega^2[\iota_m,\iota_m]\cdot\Omega H\\ &= 4\Omega^2(2^{r-2}\alpha)+\Omega^2(2^r\alpha)\cdot\Omega^2([\iota_m,\iota_m])\cdot\Omega H \quad \text{by 13.4.}\end{aligned}$$

Thus write $2^r=4^s\cdot 2^\varepsilon$, $\varepsilon=0,1$ and observe that if $2^r\alpha=0$, then

$$\Omega^2(4^s\cdot 2^\varepsilon\alpha) = 4^s\Omega^2(2^\varepsilon\alpha).$$

If $\varepsilon=0$ and $4^s\alpha=0$, then $4^s\Omega^2(\alpha)=0$ and 13.2(2) follows. If $\varepsilon=1$ and $2\cdot 4^s\alpha=0$, then $0=2\Omega^2(2\cdot 4^s\alpha)=2\cdot 4^s\Omega^2(2\alpha)$. Now $\Omega^2(2\alpha)=2\Omega^2(\alpha)+\Omega^2(\alpha)\Omega^2[\iota_m,\iota_m]\cdot\Omega H$ and $2\cdot 4^s\Omega^2(2\alpha)=4^{s+1}\Omega^2(\alpha)+\Omega^2(2\cdot 4^s\alpha)\Omega^2[\iota_m,\iota_m]\cdot\Omega H$ by 13.4; thus 13.2(1) follows.

Proof of 13.3. Let ∇_y: $\Sigma^2Y \to \Sigma^2Y\vee\Sigma^2Y$ denote the pinch map. The homotopy theoretic fibre, F_1, of the inclusion i_Y: $\Sigma^2Y\vee\Sigma^2Y \to \Sigma^2Y\times\Sigma^2Y$ is homotopy equivalent to $\Sigma(\vee_{i,j\geq 1}(\Sigma Y)^{[i+j]})$ by section 2. Let $f\varepsilon[\Sigma^2A,\Sigma^2B]$ and $\alpha\varepsilon[\Sigma^2B,X]$ with $2\alpha=0$. Define $\bar{D}(f)=\nabla_B\cdot f-(f\vee f)\cdot\nabla_A$ and notice that $i_A\cdot\bar{D}(f)$ is null. Hence $\bar{D}(f)$ lifts to F_1 by a map $D(f)$. This lift is not unique. Inductively, define F_{q+1} to be the fibre of the inclusion $F_q\vee F_q \to F_q\times F_q$ and $D_{q+1}(f)=D(D_q(f))$. Inductively define α_{q+1}: $F_{q+1} \to X$ by setting α_{q+1} equal to the composite $F_{q+1} \to F_q\vee F_q \xrightarrow{\alpha_q\vee\alpha_q} X$ where α_1 is the composite $F_1 \to \Sigma^2B\vee\Sigma^2B \xrightarrow{\alpha\vee\alpha} X$. Notice that α_{q+1} has order 2 because it is given in terms of iterated Whitehead products. By definition

$\alpha_{s+1}\cdot D_{s+1}(f)=(2\alpha_s)\cdot D_s(f)-2(\alpha_s\cdot D_s(f))$.

If $H_i(\Sigma^2A)=0$ for $i\geq n2^q-1$ and Σ^2B is n-connected, then $D_q(f)$ is null and thus $2^q(\alpha f)=0$. The other cases are quite similar and are deleted.

Proof of 13.1. Since $2\eta_4=0$, $4\Omega^2\eta_4=0$ in $[\Omega^2S^5,\Omega^2S^4]$ by 13.2. Let X denote the 6-skeleton of Ω^2S^5. We shall check that the composite

$$X \xrightarrow{\text{inclusion}} \Omega^2S^5 \xrightarrow{f} \Omega S^3$$

has order 4 in $[X,\Omega S^3]$ where f is an H-map which is onto π_3.

We claim that X is the cofibre of the composite

$S^6 \xrightarrow{\text{pinch}} S^6\vee S^6 \xrightarrow{3\nu'\vee[2]} S^3\vee S^6$ where ν' is a generator for $\pi_6S^3[T_1]$.

To prove this, first notice that ΩS^4 splits as $S^3 \times \Omega S^7$ forcing the class in $H_6(\Omega^2 S^5)$ to be spherical. By inspection of the Serre exact sequence, X is the cofibre of some map $g\colon S^6 \to S^3 \vee S^6$. By the Hilton-Milnor theorem, $\pi_6(S^3 \vee S^6)$ is isomorphic to $\pi_6 S^3 \oplus \pi_6 S^6$. Since $H_6(\Omega^2 S^5;\mathbb{Z})$ is $\mathbb{Z}/2$, g is given by $(k\nu')\vee[2]$ for some integer k. In addition the double suspension of ν' is 2ν [T1]. Since $S^6 \xrightarrow{g} S^3 \vee S^6 \to \Omega^2 S^5$ is a fibration through dimension 7, it follows that $2k\nu + 2\nu = 0$. Thus $2(k+1) \equiv 0(8)$ and so $k \equiv 3(4)$.

Assume that $f\colon \Omega^2 S^5 \to \Omega S^3$ is an H-map inducing an epimorphism on π_3. We shall check that $2f \neq 0$ when restricted to X. First consider $\bar{\eta}\colon S^5 \to BS^3$ the generator of $\pi_5 BS^3$. Thus $\Omega^2\bar{\eta}\colon \Omega^2 S^5 \to \Omega S^3$ induces an epimorphism on π_3 and so $(\Omega^2\bar{\eta})-f$ is trivial on π_3. Since $(\Omega^2\bar{\eta})-f$ is again an H-map, it is null-homotopic on the 6-skeleton of $\Omega^2 S^5$. Thus restricting $(\Omega^2\bar{\eta})-f$ to X, there is a factorization

$$\begin{array}{ccc} X & \longrightarrow & \Omega^2 S^5 \\ {\scriptstyle \text{pinch}}\downarrow & & \downarrow {\scriptstyle (\Omega^2\bar{\eta})-f} \\ S^7 & \xrightarrow[g]{} & \Omega S^3 \end{array} .$$

Since $\pi_7 \Omega S^3 \cong \mathbb{Z}/2$ [T1], $2g = 0$. Hence $2[\Omega^2(\bar{\eta})-f] = 2\Omega^2\bar{\eta} - 2f$ restricted to X is zero. If $2f = 0$, then $2\Omega^2\bar{\eta}$ is zero when restricted to X. This is a contradiction by

Lemma 13.5. $\Omega^2\bar{\eta}$ restricted to X has order 4.

Proof: Since $\Omega^2[2] = 2 + \Omega^2 w_5 \cdot \Omega H$ by Proposition 4.3, $0 = \Omega^2(2\bar{\eta}) = (\Omega^2\bar{\eta})\cdot(2 + \Omega^2 w_5 \cdot \Omega H)$. Thus if $2\Omega^2\bar{\eta} = 0$, it follows that $\Omega^2\bar{\eta}\cdot\Omega^2 w_5 \cdot \Omega H = 0$. Restricting $\Omega^2\bar{\eta}\cdot\Omega^2 w_5\cdot\Omega H$ to X, we obtain the composite

$$\begin{array}{ccccccc} X & \xrightarrow{\text{pinch}} & S^7 & \xrightarrow{E^2} & \Omega^2 S^9 & \xrightarrow{\Omega^2 w_5} & \Omega^2 S^5 \\ & \searrow^{\beta} & & & & & \downarrow {\scriptstyle \Omega^2\bar{\eta}} \\ & & & & & & \Omega S^3 \end{array}$$

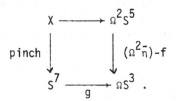

with $\beta=0$ in $[X,\Omega S^3]$.

Since $S^6 \xrightarrow{3\nu' v[2]} S^3 \vee S^6 \to X$ is a cofibration and the adjoint of β is $\Sigma^2 X \to S^9 \xrightarrow{w_5} S^5 \xrightarrow{\bar{\eta}} BS^3$, it follows that there is a homotopy commutative diagram

$$\begin{array}{ccc} S^9 & \xrightarrow{E^3(\nu')v[2]} & S^6 \vee S^9 \\ {\scriptstyle \bar{\eta}\cdot w_5}\downarrow & \swarrow {\scriptstyle \gamma} & \\ BS^3 & & \end{array}$$

We claim that $\bar{\eta}\cdot w_5$ is essential while $\gamma\cdot(E^3(\nu')v[2])$ is not and this gives the lemma.

By [T1], $[S^6 \vee S^9, BS^3]$ is isomorphic to $\pi_5 S^3 \oplus \pi_8 S^3$ with generators η_3^2 and $\nu'\cdot\eta_6^2$. But $2\nu'\eta_6=0$ and $\eta_3\cdot E(\nu')=0$ [T1]. Thus $\gamma\cdot(E^3(\nu')v[2])$ is null.

To check that $\bar{\eta}\cdot w_5$ is essential, recall that $w_5=\nu_5\eta_8$ and so $\eta_4\cdot w_5=\eta_4\nu_5\eta_8=(E\nu')\eta_7\eta_8$ [T1]. Thus there is a homotopy commutative diagram

$$\begin{array}{ccccc} S^8 & \xrightarrow{\nu'\eta_6\eta_7} & S^3 & & \\ {\scriptstyle E}\downarrow & & {\scriptstyle E}\downarrow & \searrow {\scriptstyle 1} & \\ \Omega S^9 & \xrightarrow[\Omega(E\nu'\cdot\eta_7\eta_8)]{} & \Omega S^4 & \longrightarrow & S^3 \end{array}.$$

Since $\nu'\eta_6\eta_7$ generates $\pi_8 S^3 \cong \mathbb{Z}/2$ [T1], the claim and the lemma follow.

We remark that Lemma 13.5 has another implication. By Theorem 3.4, there is a fibration $S^3 \xrightarrow{\eta} \Omega S^3\langle 3\rangle \xrightarrow{H} \Omega S^5$ and the resulting map $\partial\colon \Omega^2 S^5 \to S^3$.

<u>Lemma 13.6</u>. The diagram

$$\begin{array}{ccc} \Omega^2 S^5 & \xrightarrow{2} & \Omega^2 S^5 \\ & {\scriptstyle \partial}\searrow \quad \nearrow {\scriptstyle E^2} & \\ & S^3 & \end{array}$$

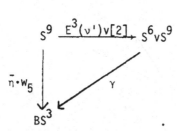

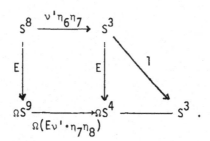

does not homotopy commute.

<u>Proof</u>: Suppose that the diagram in the lemma does homotopy commute, then by inspection there is another homotopy commutative diagram

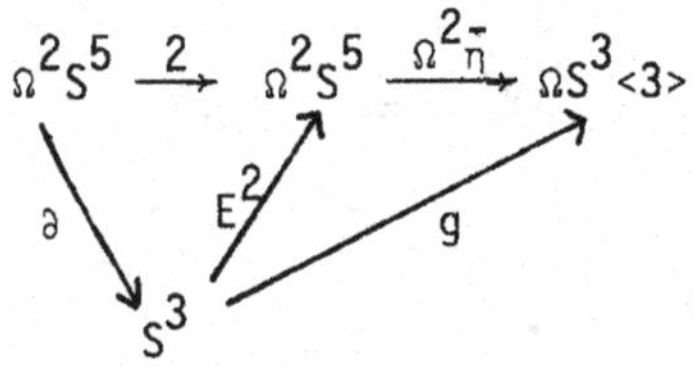

where $g=\Omega^2\bar{\eta}\cdot E^2$. But then $g=\eta$. Also $g\partial$ is null because it is the composition of two successive maps in a fibration. However $2\Omega^2\bar{\eta}\neq 0$ by 13.5 and the lemma follows.

The reader should compare Lemma 13.6 and Theorem 6.1.

§14. A filtration of $Q\mathbb{R}P^\infty$

In this section we give a natural filtration of $Q\mathbb{R}P^\infty$. The point of this is to consider the Kahn-Priddy theorem, one form of which states that the component of the base-point $\Omega_0^\infty S^\infty$ is a 2-local retract of $Q\mathbb{R}P^\infty$[KP].

Let $map_*(A,X)$ denote the space of pointed maps from A to X. Let $f_n: \mathbb{R}P^{2n} \to S^{4n}$ be a fixed smooth embedding and write $A_{2n+1}=S^{4n}-f_n(\mathbb{R}P^{2n})$. Thus consider the two embeddings $\mathbb{R}P^{2n} \xrightarrow{f_n} S^{4n} \overset{j}{\subset} S^{4n+4}$ and $\mathbb{R}P^{2n} \subset \mathbb{R}P^{2n+2} \xrightarrow{f_{n+1}} S^{4n+4}$ induced by equatorial inclusions of spheres. By the Whitney embedding theorem, these two embeddings are isotopic [Wt]. Thus $S^{4n+4}-jf(\mathbb{R}P^{2n})$ is homotopy equivalent to $\Sigma^4 A_{2n+1}$. Hence there is a map $\alpha: A_{2n+3} \to \Sigma^4 A_{2n+1}$ which is a specific destabilization of map induced on the S-duals of projective space given by the inclusions of $\mathbb{R}P^{2n}$ in $\mathbb{R}P^{2n+2}$.

Consider the composition

$$map_*(A_{2n+1},X) \xrightarrow{map_*(1,\Omega^4\Sigma^4)} map_*(A_{2n+1},\Omega^4\Sigma^4 X) \xrightarrow{\alpha^*} map_*(A_{2n+3},\Sigma^4 X)$$

to obtain

$$\sigma_n: map_*(A_{2n+1},X) \to map_*(A_{2n+3},\Sigma^4 X).$$

<u>Lemma 14.1</u>: If $k \geq 4n-1$, there is a homotopy equivalence

$$j: \underset{\sigma_{n+j}}{\underrightarrow{colim}}\, map_*(A_{2(n+j)+1},S^{k+4j}) \to Q\Sigma^{k-4n+1}\mathbb{R}P^\infty.$$

Thus the spaces $map_*(A_{2n+1},S^{4n-1})$ give a filtration of $Q\mathbb{R}P^\infty$. In case n=1, this filtration corresponds to the map $\Omega_0^3 S^3 \to \Omega_0^\infty S^\infty$. Namely Theorem 19.1 gives a homotopy commutative diagram

$$\begin{array}{ccc} map_*(\Sigma^2 A_1, S^5) & \xrightarrow{\theta'} & \Omega_0^3 S^3 \\ \downarrow & & \downarrow \\ Q\mathbb{R}P^\infty & \xrightarrow[\theta]{} & \Omega_0^\infty S^\infty \end{array}$$

where θ and θ' induce split epimorphisms on homotopy groups. One might wonder if a similar statement is correct for $\Omega_0^5 S^5$.

The existence of a map from $map_*(\Sigma^2 A_3, S^9)$ to $\Omega_0^5 S^5$ implies the existence of a map $\Omega^8 S^9 \to \Omega_0^5 S^5$ which is an epimorphism on π_1. Such a map exists naturally from James' fibrations given in section 3.

Lemma 14.2. There is a map $f: \Omega^5 S^9 \to \Omega^3 S^5$ which induces an epimorphism on π_4 and $2f=0$ in $[\Omega^5 S^9, \Omega^3 S^5]$.

Proof: Consider the composite f given by $\Omega^5 S^9 \xrightarrow{\Omega^3 P} \Omega^3 S^4 \xrightarrow{\Omega^3 i} \Omega^2 S^3 \xrightarrow{\Omega H} \Omega^2 S^5$ where $P: \Omega^2 S^9 \to S^4$ is the boundary induced by the fibration $S^4 \to \Omega S^5 \xrightarrow{H} \Omega S^9$ and $i: S^4 \to BS^3$ is the inclusion of the bottom cell. Since all maps are multiplicative, $2f=0$ by Lemma 4.1. By [T1, chap.5], f induces an epimorphism on π_4: $P(\iota_q)=\pm(2\nu_4 - E\nu')$, $i(2\nu_4 - E\nu')=2k\nu'-\nu'$ for some k, and $H(\nu')=\eta_5$. The lemma follows.

Next observe that there is a cofibration $P^5(2)\vee S^3 \xrightarrow{g} S^3 \xrightarrow{i} \Sigma A_3$ where the map g is $\bar{\eta}_3 \vee [2]$ and $\bar{\eta}_3: P^5(2) \to S^3$ is η_3 on the bottom cell. Lemma 14.2 implies that there is a map $g: map_*(P^6(2), S^9) \to \Omega^3 S^5$ inducing an isomorphism on π_3 where $map_*(X,Y)$ is the space of pointed maps from X to Y. In Lemma 19.4 a map $k: \Omega S^9 \to S^5$ is given which induces an epimorphism on $\pi_8 S^5$. Thus there is a map

$$\lambda: map_*(P^6(2), S^9) \times \Omega^4 S^9 \to \Omega^3 S^5.$$

Since $map_*(P^6(2), S^9) \times \Omega^4 S^9$ is homeomorphic to $map_*(P^6(2)\vee S^4, S^9)$ one might ask if the composite $map_*(S^4, S^9) \xrightarrow{map_*(\Sigma g, 1)} map_*(P^6(2)\vee S^4, S^9) \xrightarrow{\lambda} \Omega^3 S^5$ is null-homotopic. If this map is trivial, then there exists $map_*(\Sigma A_3, S^9) \to \Omega^4 S^5$ inducing an isomorphism on π_2. By a verification which we delete, this map exists through a range and $\Omega_0^5 S^5$ <u>through a range</u>, is a retract of $map_*(\Sigma^2 A_3, S^9)$.

§15: The space of maps of Moore spaces to spheres

In the next few sections, we shall study $map_*(P^k(2),S^n)$ where $P^k(2)$ is the cofibre of $[2]: S^{k-1} \to S^{k-1}$. It will be shown that $map_*(P^k(2),S^n)$ fails to split as a non-trivial product for most values of n and that there are product decompositions for some values of n. These sections are based completely on work in [[CCPS],[C1]].

In order to study $map_*(P^k(2),S^n)$, it is necessary to compute it's homology as a Hopf algebra over the Steenrod algebra. Notice that the H-space squaring map $2: \Omega^{k-1}S^n \to \Omega^{k-1}S^n$ has homotopy theoretic fibre $map_*(P^k(2),S^n)$ because $S^{k-1} \xrightarrow{[2]} S^{k-1} \to P^k(2)$ is a cofibration. Since there is a map of fibrations

$$\begin{array}{ccc} \Omega^k S^n & \xrightarrow{2} & \Omega^k S^n \\ \downarrow & & \downarrow \\ * & \longrightarrow & map_*(P^k(2),S^n) \\ \downarrow & & \downarrow\pi \\ \Omega^{k-1}S^n & \xrightarrow[1]{} & \Omega^{k-1}S^n \end{array} ,$$

2 is a loop map if $k\geq 2$, and $H_*\Omega^k S^n$ is primitively generated if $k<n$, the Serre spectral sequence collapses for the right-hand fibration. Since $map_*(P^k(2),S^n)$ is a double loop space if $k\geq 4$, the Pontrjagin ring is commutative. However, since the diagram

$$\begin{array}{ccc} \Omega S^n & \xrightarrow{\Omega E^\infty} & QS^{n-1} \\ 2\downarrow & & \downarrow Q[2] \\ \Omega S^n & \xrightarrow[\Omega E^\infty]{} & QS^{n-1} \end{array}$$

homotopy commutes where $E^\infty: S^n \to QS^n$ is the stabilization map, we get a map

$$\phi: map_*(P^2(2),S^n) \to QP^{n-1}(2)$$

with $(\Omega^q\phi)_*$ a monomorphism if $q+2<n$. Thus one has

Lemma 15.1. If $2\leq k<n$, then $H_*\mathrm{map}_*(P^k(2),S^n)$ is isomorphic to $H_*\Omega^{k-1}S^n \otimes H_*\Omega^k S^n$ as a coalgebra. If $3\leq k<n$, this is an isomorphism of algebras and if $3\leq k<n-1$, this is an isomorphism of Hopf algebras.

Since $\pi\colon \mathrm{map}_*(P^k(2),S^n) \to \Omega^{k-1}S^n$ is a $(k-2)$-fold loop map, π_* commutes with Q_i for $0\leq i\leq k-2$. [See chapter 7]. Write x_j for the fundamental class in $H_*\Omega^n S^{n+j}$. By Lemma 15.1, there is a choice of primitive element $\bar{Q}^a_{k-2}x_{n-k+1}$, $3<k<n$, which projects to $Q^a_{k-2}x_{n-k+1}=\underbrace{Q_{k-2}\cdots Q_{k-2}}_{a}x_{n-k+1}$ in $H_*\Omega^{k-1}S^n$ if $3\leq k<n-1$. Again by Lemma 15.1, we have

$$H_*\mathrm{map}_*(P^k(2),S^n) \cong \mathbb{Z}/2[Q_I x_{n-k}] \otimes \mathbb{Z}/2[Q_J \bar{Q}^a_{k-2}x_{n-k+1}]$$

where J is admissible with $\lambda(J)<k-3$, $a\geq 0$, and I is admissible with $\lambda(I)\leq k-1$. Thus to compute Steenrod operations in $H_*\mathrm{map}_*(P^k(2),S^n)$, it suffices to give their values on $\bar{Q}^a_{k-2}x_{n-k+1}$. Notice that there is another homotopy commutative diagram

$$\begin{array}{ccc}
\Omega S^n & \xrightarrow{\Omega E} & \Omega^2 S^{n+1} \\
{\scriptstyle 2}\big\downarrow & & \big\downarrow{\scriptstyle \Omega^2[2]} \\
\Omega S^n & \xrightarrow[\Omega E]{} & \Omega^2 S^{n+1}
\end{array}$$

and thus we get a lift

$$g\colon \mathrm{map}_*(P^2(2),S^n) \to \Omega^2(S^{n+1}\{2\})$$

where $S^{n+1}\{2\}$ is the fibre of the degree 2 map $[2]\colon S^{n+1} \to S^{n+1}$. Notice that $(\Omega^k g)_*$ is a monomorphism if $k+2<n$ by a comparison of Serre spectral sequences.

Since $\Omega^2(S^{n+1}\{2\})$ is a double loop space, we will be able to compute Steenrod operations in its homology. By identifying the embedding $(\Omega^k g)_*$, we will be able to compute Steenrod operations in $H_*\mathrm{map}_*(P^{k+2}(2),S^n)$. Observe that

$$H_*\Omega^k(S^{n+1}\{2\}) \cong H_*\Omega^{k+1}S^{n+1}\otimes H_*\Omega^{k+2}S^{n+1}$$

if $k<n-1$. Write y_i for the generator of $H_i\Omega^k(S^{n+1}\{2\})$, $i=n-k,n-k-1$.

<u>Theorem 15.2</u>.

(1) $(\Omega g)_*(Q_1^a x_{n-2}) = Q_1^a y_{n-2}+Q_1^{a-1}Q_3 y_{n-3}$ if $n\not\equiv 3(4)$.

(2) $(\Omega^k g)_*(\bar{Q}_k^a x_{n-k-2}) = Q_k^a y_{n-k-2}+Q_k^{a-1}Q_{k+2}y_{n-k+3}$+others if $n\not\equiv 3(4)$.

(3) $Sq_*^1\bar{Q}_1^a x_{n-2} = (\bar{Q}_1^{a-1}x_{n-2})^2$, $(\Omega g)_*(\bar{Q}_1^a x_{n-2}) = Q_1^a y_{n-2}+\varepsilon Q_1^{a-1}Q_3 y_{n-3}$

for some fixed $\varepsilon=0,1$ and $Sq_*^{2^j}Q_1^{a-1}Q_3 y_{n-3}=0$ for $j>0$ if $n\equiv 3(4)$.

The proof of 15.2 is very much like the calculations in section 10; we postpone the proof until section 16.

Some applications follow. Recall that an $(r-1)$-connected space X is atomic if a self-map $f\colon X\to X$ which induces an isomorphism on $\pi_r X$ is an equivalence. We add the remark given by

<u>Lemma 15.2</u>. If X is atomic, then it does not admit a product decomposition in the homotopy category.

<u>Theorem 15.3</u>. If $2n+1\neq 3,5,9$, or 17, then $\mathrm{map}_*(P^3(2),S^{2n+1})$ is atomic.

In [CCPS] where this theorem was proven, analogous results for $\mathrm{map}_*(P^3(2),S^{2n})$ are also proven, but we shall not need these results here.

A theorem which we shall use later and which is proven in [CCPS] is

<u>Theorem 15.4</u>. Let $f\colon X\to \mathrm{map}_*(P^3(2),S^{2n+1})$ be a map which induces an isomorphism on the module of primitives in dimensions $2n-2$ and $4n-3$. If the mod-2 homology of X is isomorphic to that of $\mathrm{map}_*(P^3(2),S^{2n+1})$ as

a coalgebra over the Steenrod algebra, then f_* is an isomorphism.

In section 19, we shall obtain some product decompositions of $map_*(P^k(2),S^n)$ for n=5 or 9. These decompositions along with a decomposition of $map_*(P^4(2),S^{17})$ are given in [CS].

It is worthwhile mentioning that $map_*(P^k(2),S^{2n})$, n=2,4 splits if $k\geq 3$ because ΩS^{2n} is homotopy equivalent to $S^{2n-1}\times\Omega S^{4n-1}$. Thus changing the source from $A=S^1$ to $A=P^k(2)$ gives other decompositions of $map_*(A,S^n)$. It would be interesting to know what happens in case A is the S-dual of $\mathbb{R}P^k$ as in section 14.

§16. Further homological calculations

In this section, we give the proof of Theorem 15.2. Recall that there is a homotopy commutative diagram

$$\begin{array}{ccc} \mathrm{map}_*(P^{k+2}(2),S^n) & \xrightarrow{\Omega^k g} & \Omega^{k+2}(S^{n+1}\{2\}) \\ \downarrow & & \downarrow \\ \Omega^{k+1}S^n & \xrightarrow[\Omega^{k+1}E]{} & \Omega^{k+2}S^{n+1} \end{array} .$$

Thus $(\Omega^k g)_*\bar{Q}_k^a x_{n-k-1} = Q_k^a y_{n-k-1}+\Delta$ where Δ is primitive with trivial image in $H_*\Omega^{k+2}S^{n+1}$. Hence Δ is a linear combination of elements $(Q_I x_{n-k-2})^{2j}$. In case k=1, degree considerations give

$$(\Omega g)_*\bar{Q}_1^a x_{n-2} = Q_1^a y_{n-2}+\varepsilon_a Q_1^{a-1}Q_3 y_{n-3} .$$

By commutation with Sq_*^1, $(\Omega g)_*\ Sq_*^1\bar{Q}_1^a x_{n-2}=(Q_1^{a-1}y_{n-2})^2+\varepsilon_a(Q_1^{a-2}Q_3 y_{n-3})^2$ if $a\geq 2$. By a check of degrees,

$$Sq_*^1\bar{Q}_1^a x_{n-2} = (\bar{Q}_1^{a-1}x_{n-2})^2+\delta Q_2^a x_{n-3},\ \delta=0,1.$$

Since $Sq_*^1 Q_2^a x_{n-3}=Q_1Q_2^{a-1}x_{n-3}$, δ must be zero and so $\varepsilon_1=\varepsilon_a$ for $a>1$.

To check that $\varepsilon_1=1$ when $n\not\equiv 3(4)$, first let $n=4j+1$. By Lemma 4.1, there is a homotopy commutative diagram

$$\begin{array}{ccc} & & \mathrm{map}_*(P^3(2),S^{4j+1}) \\ & \overset{\bar{h}}{\nearrow} & \downarrow \pi \\ \Omega^2S^{2j+1} & \xrightarrow[\Omega H]{} & \Omega^2S^{4j+1} \end{array} .$$

Thus we may choose $\bar{Q}_1x_{4j-1}$ to be $\bar{h}_*(Q_1^2x_{2j-1})$ since $(\Omega H)_*Q_1^2x_{2j-1}=Q_1x_{4j-1}$. By Theorem 7.1, $Sq_*^1Q_1^2x_{2j-1}=(Q_1x_{2j-1})^2$ and $Sq_*^2Q_1^2x_{2j-1}=0$. Thus $Sq_*^2(Q_1y_{4j-1}+\varepsilon_1Q_3y_{4j-2})=0$. But $Sq_*^2Q_1y_{4j-1}=Q_1y_{4j-2}$ and $Sq_*^2Q_3y_{4j-2}=Q_1y_{4j-2}$.

Thus $\varepsilon_1=1$.

Assume that $n\equiv 0(2)$. Consider the adjoint of the Whitehead product $f\colon S^{2n-2} \to \Omega S^n$. Since $f_*(\iota_{2n-2})=2x^2$ by section 1, it follows that $g_*(\iota_{2n-3})=2(\text{generator})$ where $g\colon S^{2n-3} \to \Omega^2 S^n$ is the double adjoint of $[\iota_n,\iota_n]$. But $H_q(\Omega^2 S^n;\mathbb{Z})$ is isomorphic $\mathbb{Z}$ when $q=n-2$, $2n-4$, or $2n-3$. Since twice a generator of $H_{2n-3}(\Omega^2 S^n;\mathbb{Z})$ is spherical, the map $2\colon \Omega^2 S^n \to \Omega^2 S^n$ induces multiplication by 2 on $H_{2n-3}(\ ;\mathbb{Z})$. Thus an inspection of the Serre spectral sequence gives $H_q(\mathrm{map}_*(P^3(2),S^n);\mathbb{Z})$ is $\mathbb{Z}/2$ if $q=n-3$, $2n-6$, $2n-4$ and is $\mathbb{Z}/2^r$ for some r if $q=2n-5$. Thus $Sq^1_*\bar{Q}_1 x_{n-2}\neq 0$ and so $Sq^1_*(\Omega g)_*(\bar{Q}_1 x_{n-2})\neq 0$. But $Sq^1_*(Q_1 y_{n-2}+\varepsilon_1 Q_3 y_{n-3})=\varepsilon_1 Q_3 y_{n-3}$ and thus $\varepsilon_1=1$.

The above gives Theorem 15.2(1). By commutation with the homology suspension, 15.2(2) follows from 15.2(1).

Next assume that $n\equiv 3(4)$ and recall that $(\Omega g)_*(\bar{Q}_1^a x_{n-2})=Q_1^a y_{n-2}+\varepsilon_a Q_1^{a-1} Q_3 y_{n-3}$. By a previous argument with Sq^1_*, we get $\varepsilon_a=\varepsilon_1$ for $a>1$. Thus $(\Omega g)_*(\bar{Q}_1^a x_{n-2})=Q_1^a y_{n-2}+\varepsilon Q_1^{a-1} Q_3 y_{n-3}$ for a fixed ε. Notice that this forces $Sq^1_*\bar{Q}_1^a x_{n-2}=(\bar{Q}_1^{a-1} x_{n-2})^2$ if $a\geq 1$. We now check that $Sq^{2^j}_* Q_1^{a-1} Q_3 y_{n-3}=0$ if $j>0$. But $Sq^{2^a}_* Q_1^{a-1} Q_3 y_{n-3}=0$ because $Sq^2_* Q_3 y_{n-3}=0$, $Sq^{2^{a-1}}_* Q_1^{a-1} Q_3 y_{n-3}=0$ because $Sq^1_* Q_3 y_{n-3}=0$ and $Sq^{2^i}_* Q_1^{a-1} Q_3 y_{n-3}=0$ if $i<a-1$ because $Q_1(z^2)=0$. This suffices.

§17: Endomorphisms of $H_*map_*(P^3(2),S^{2n+1})$, $n\geq 2$

In this section we prove Theorem 15.4. Thus assume that X is a space whose mod-2 homology as a coalgebra over the Steenrod algebra is isomorphic to $H_*map_*(P^3(2),S^n)$; write K_n for this coalgebra. To prove Theorem 15.4 it suffices to show that if $f\colon K_{2n+1}\to K_{2n+1}$ is any morphism of coalgebras over the Steenrod algebra which induces an isomorphism on the module of primitives in dimensions 2n-2 and 4n-3, then f is an isomorphism.

Since K_n is of finite type, it suffices to prove that f is a monomorphism when restricted to the module of primitives PK_n. By Lemma 15.1 and degree considerations, one has

Lemma 17.1. If $n\geq 5$, $PK_n=PH_*\Omega^2S^n\oplus PH_*\Omega^3S^n$. Furthermore, exactly one of the following holds in any fixed degree: (i) PK_n is of dimension at most one with basis

(a) $(Q_1^aQ_2^bx_{n-3})^{2^j}$, $j\geq 0$, $a\geq 1$, $b\geq 0$.

(b) $x_{n-3}^{2^j}$, $j\geq 0$

(c) x_{n-2}

(d) $\bar{Q}_1^a x_{n-2}$, $a\geq 1$, or

(ii) PK_n is of dimension 2 with basis $(\bar{Q}_1^a x_{n-2})^{2^{j+1}}$ and $(Q_2^{a+1}x_{n-3})^{2^j}$, $a,j\geq 0$.

Proof of Theorem 15.4: Let $f\colon K_{2n+1}\to K_{2n+1}$. The following equations which will be checked using the hypotheses of 15.4 imply that f is an isomorphism by 17.1:

(1) $f(x_{2n-1}^{2^j})=x_{2n-1}^{2^j}$, $f(x_{2n-2}^{2^j})=x_{2n-2}^{2^j}$,

(2) $f((\bar{Q}_1^a x_{2n-2})^{2^j})=(\bar{Q}_1^a x_{2n-2})^{2^j}$,

(3) $f((Q_1^aQ_2^bx_{2n-2})^{2^j})=(Q_1^aQ_2^bx_{2n-2})^{2^j}$, and

(4) $f(Q_2^a x_{2n-2})^{2^j} = (Q_2^a x_{2n-2})^{2^j} + \lambda(\bar{Q}_1^{a-1} x_{2n-1})^{2^{j+1}}$.

First notice that $f(x_{2n-2}) = x_{2n-2}$ and $f(Q_1 x_{2n-3}) = Q_1 x_{2n-3}$ by hypothesis. Next consider $f(x_{2n-1}^{2^j}) = Ax_{2n-1}^{2^j} + B(Q_2 x_{2n-2})^{2^{j-1}}$ by 17.1. Apply $Sq_*^{2^{j-1}}$ to get B=0. Since $f(x_{2n-1}) = x_{2n-1}$ by commutation with Sq_*^1, we prove the formula

(1) $$f(x_{2n-1}^{2^j}) = x_{2n-1}^{2^j} \text{ and } f_*(x_{2n-2}^{2^j}) = x_{2n-2}^{2^j}$$

by induction on j. Assume this for $j \leq k$ and notice that $f(x_{2n-1}^{2^k} \cdot x_{2n-2}^{2^{k+1}}) = x_{2n-1}^{2^k} \cdot x_{2n-2}^{2^k} + \varepsilon(Q_1 x_{2n-2})^{2^k}$ by commutation with the coproduct. Since $Sq_*^1 Q_1 x_{2n-2} = 0$, we apply $Sq_*^{2^k}$ to get $f(x_{2n-2}^{2^{k+1}}) = x_{2n-2}^{2^{k+1}}$. Since $f(x_{2n-1}^{2^{k+1}}) = Ax_{2n-1}^{2^k}$, applying $Sq_*^{2^{k+1}}$ gives A=1 and formula (1) follows.

Next, we claim that

(2) $$f(\bar{Q}_1^a x_{2n-1})^{2^k} = (\bar{Q}_1^a x_{2n-1})^{2^k}.$$

Clearly $f_*(\bar{Q}_1^b x_{2n-1}) = C(\bar{Q}_1^b x_{2n-1})$. Apply $Sq_*^{2^{b-1}} \cdots Sq_*^2 Sq_*^1$ to this last equation to get $f_*(x_{2n-1}^{2^b}) = Cx_{2n-1}^{2^b}$. Thus C=1 by equation (1). Thus $f_*(\bar{Q}_1^b x_{2n-1}) = \bar{Q}_1^b x_{2n-1}$ for all b. Equation (2) results by applying $Sq_*^{2^{k-1}} \cdots Sq_*^2 Sq_*^1$ to $f_*(\bar{Q}_1^{a+k} x_{2n-1})$.

Next, notice that $f_*(Q_1^a Q_2^b x_{2n-2})^{2^k} = D(Q_1^a Q_2^b x_{2n-2})^{2^k}$ if $a \geq 1$. Since $Sq_*^1 Q_2^b x_{2n-2} = Q_1 Q_2^{b-1} x_{2n-2}$, $Sq_*^1 Q_1 Q_1 z = (Q_1 z)^2$, and $Sq_*^{2j} Q_1 z = Q_1 Sq_*^j z$, there is an operation Sq^I such that $Sq_*^I (Q_1^a Q_2^b x_{2n-2})^{2^k} = (Q_1 x_{2n-2})^{2^{k+a+b-1}}$. Thus $f_*(Q_1 x_{2n-2})^{2^{k+a+b-1}} = D(Q_1 x_{2n-2})^{k+a+b-1}$. If $n \equiv 1(2)$, then $Sq_*^2 \bar{Q}_1 x_{2n-1} = Q_1 x_{2n-2}$; that D=1 follows by applying a Steenrod operation to equation (2). Assume that $n \equiv 0(2)$. By the hypotheses in 15.4, $f(Q_1 x_{2n-2}) = Q_1 x_{2n-2}$. Observe that

$f((Q_2x_{2n-2})^{2^j})=E_j(Q_2x_{2n-2})^{2^j}+F_jx_{2n-1}^{2^{j+1}}$ by 17.1. Apply $Sq_*^{2^j}$ to get $f(Q_1x_{2n-2})^j=E_j(Q_1x_{2n-2})^{2^j}$. Apply $Sq_*^{2^{j+1}}$ to $f(Q_2x_{2n-2})^{2^j}$ and equation (1) to get $E_j+F_j=1$. Since $E_1=1$, inductively assume that $E_j=1$. By commutation with the coproduct $f((Q_1x_{2n-2})^{2^j}(Q_2x_{2n-2})^{2^j})=(Q_1x_{2n-2}\ Q_2x_{2n-2})^{2^j}+\Delta$ where Δ is primitive. By inspection $\Delta=G(\bar{Q}_1^2x_{2n-1})^{2^j}+K(Q_2^3x_{2n-2})^{2^{j-1}}$. Apply $Sq_*^{2^{j-1}}$ to give K=0. Apply $Sq_*^{2^{j+1}}\ Sq_*^{2^j}$ to get G=0. Finally apply $Sq_*^{2^j}$ to the equation for $f(Q_1x_{2n-2}\cdot Q_2x_{2n-2})^{2^j}$ to get $E_{j+1}=1$. By commutation with $Sq_*^{2^j}$, D=1. Thus

(3) $$f(Q_1^aQ_2^bx_{2n-2})^{2^k}=(Q_1^aQ_2^bx_{2n-2})^{2^k}.$$

Apply $Sq_*^{2^k}$ to $f(Q_2^bx_{2n-2})^{2^k}=L(Q_2^bx_{2n-2})^{2^k}+\lambda(\bar{Q}_1^{b-1}x_{2n-1})^{2^{k+1}}$ to get

(4) $$f(Q_2^bx_{2n-2})^{2^k}=(Q_2^bx_{2n-2})^{2^k}+\lambda(\bar{Q}_1^{b-1}x_{2n-1})^{2^{k+1}}.$$

An inspection of the primitives listed in 17.1 together with equations (1)-(4), gives f that is an isomorphism. Thus Theorem 15.4 follows.

§18: The Hurewicz image for $map_*(P^3(2),S^{2n+1})$ and Theorem 15.3

In this section we shall show that certain low dimensional elements cannot be in the mod-2 reduction of the Hurewicz homomorphism. There are several choices here. During the course we proved this theorem by constructing unstable higher order cohomology operations detecting the elements in question [BP]. Clearly, looping down a secondary operation "often enough" converts the secondary operation to a primary one. We shall do this here, but will not elaborate on this for other operations. A direct consequence is Theorem 15.3.

A basis for the module of primitives $PH_*map_*(P^3(2),S^n)$ is given by 2^i-th powers of $\bar{Q}_1^a x_{n-2}$ and $Q_1^a Q_2^b x_{n-3}$. We need to consider the module of primitives in dimensions at most 2n-3. A basis is

$\{x_{n-3},x_{n-2},x_{n-3}^2,x_{n-2}^2,Q_1x_{n-3},Q_2x_{n-3},\bar{Q}_1x_{n-2}\}$. Let $\phi\colon \pi_qX \to H_q(X;\mathbb{Z}/2)$ denote the mod-2 reduction of the Hurewicz homomorphism. Recall that H_*X means homology with $\mathbb{Z}/2$-coefficients.

Proposition 18.1: The map $\phi\colon \pi_{4n-3}map_*(P^3(2),S^{2n+1}) \to H_{4n-3}map_*(P^3(2),S^{2n+1})$ is zero if $n\neq 2^k$. More precisely, the element Q_1x_{2n-2} is not in the image of ϕ if $n\neq 2^k$.

Proof: Assume that Q_1x_{2n-2} is in the image of ϕ. Now consider $map_*(P^{n-1}(2),S^n)$ and recall that from 15.1 that $H_*map_*(P^{n-1}(2),S^n)$ is isomorphic to $H_*\Omega^{n-2}S^n \otimes H_*\Omega^{n-1}S^n$ as an algebra.

Write $n=(2j+1)2^k$ for $j\geq 1$. If there is a map $f\colon S^{4n-3} \to map_*(P^3(2),S^{2n+1})$ with $f_*(\iota_{4n-3})=Q_1x_{2n-2}$, then by adjointness there is a map $g\colon S^{2n} \to map_*(P^{2n}(2),S^{2n+1})$ with $g_*(\iota_{2n})=Q_{2n-2}x_1+AQ_{2n-4}x_2+$ others. If $n>2$, then $A=0$ by commutation with Sq_*^1.

Thus $g_*(\iota_{2n})=Q_{2n-2}x_1+\Sigma(Q_Ix_1)^{2^i}+\Sigma(Q_Jx_2)^{2^j}$ where I and J are admissible with $\lambda(I)<2n-2$, $\lambda(J)<2n-4$. That $n=(2j+1)2^k$, $Sq_*^{2^{k+1}}Q_{2n-2}x_1=Q_{2n-2-2^{k+1}}x_1$ follows since the binomial coefficient $(2^{k+1},(2j)(2^{k+1})-1)=1$ if $j\geq 1$. Thus by inspection $Sq_*^{2^{k+1}}g_*(\iota_{2n})\neq 0$. But this is a contradiction.

Proof of Theorem 15.3: Let f be a self-map of $map_*(P^3(2),S^{2n+1})$ with $f_*(x_{2n-2})=x_{2n-2}$. By Theorem 15.4 it suffices to check that $f_*Q_1x_{2n-2}=Q_1x_{2n-2}$ in order to prove that f is an equivalence. A basis for the module of primitives through dimension 4n-1 is given by $\{x_{2n-2},x_{2n-1},x_{2n-2}^2,x_{2n-1}^2,Q_1x_{2n-2},Q_2x_{2n-2},\bar{Q}_1x_{2n-1}\}$.

We repeat a bit of the proof of Theorem 15.4 here. Clearly $f_*(x_{2n-2}\cdot x_{2n-1})=x_{2n-2}\cdot x_{2n-1}+\varepsilon Q_1x_{2n-2}$. Thus $f_*(x_{2n-2}^2)=x_{2n-2}^2$ by commutation with Sq_*^1. Clearly $f_*(x_{2n-1}^2)\neq 0$ by commutation with Sq_*^2 and $f_*(x_{2n-1}^2)=Ax_{2n-1}^2+BQ_2x_{2n-2}$. But B=0 by commutation with Sq_*^1 and so A=1. Evidently $f_*(Q_2x_{2n-2})=Cx_{2n-1}^2+DQ_2x_{2n-2}$. If D=1, then C=0 by commutation with Sq_*^2. Hence $f_*(Q_1x_{2n-2})=Q_1x_{2n-2}$ by commutation with Sq_*^1 and 15.3 follows. Thus we assume that D=0. This gives that C=1. We shall derive a contradiction from the statement $f_*(Q_2x_{2n-2})=x_{2n-1}^2$ if $n\neq 2,4,8$.

Let Z denote the mapping telescope of f with the canonical inclusion $i: map_*(P^3(2),S^{2n+1})\to Z$ with fibre F. Since we assume that $f_*(Q_2x_{2n-2})=x_{2n-1}^2$, it follows that $f_*(Q_1x_{2n-2})=f_*(Q_2x_{2n-2}+x_{2n-1}^2)=0$. Thus F is (4n-2)-connected with $\pi_{4n-2}F\cong\mathbb{Z}/2$ because $Sq_*^1(Q_2x_{2n-2}+x_{2n-1}^2)=Q_1x_{2n-2}$. Clearly Q_1x_{2n-2} is in the image of the Hurewicz map ϕ. This contradicts 18.1 if $n\neq 2^k$.

Next consider the composite θ, $P^{4n-1}(2)\to F\to map_*(P^3(2),S^{2n+1})$, with

$\theta_*\bar{H}_*P^{4n-2}(2)$ spanned by Q_1x_{2n-2} and $Q_2x_{2n-2}+x_{2n-1}^2$. Hence the composite $P^{4n-2}(2) \xrightarrow{\theta} \mathrm{map}_*(P^3(2),S^{2n+1}) \xrightarrow{\pi} \Omega^2S^{2n+1}$ satisfies $(\pi\theta)_*(v_{4n-2})=x_{2n-1}^2$ where v_{4n-2} generates $H_{4n-2}P^{4n-2}(2)$. Since $\Omega H\cdot\pi\cdot\theta\colon P^{4n-2}(2) \to \Omega^2S^{4n+1}$ is evidently null, there is a lift of $\pi\cdot\theta$ to give

$$\bar{\theta}\colon P^{4n-2}(2) \to \Omega S^{2n} \quad \text{with} \quad \bar{\theta}_*(v_{4n-2})=x_{2n-1}^2.$$

Thus $\bar{\theta}$ restricted to the bottom cell has Kervaire invariant one by section 11 and so this gives another proof of 15.3 in case $n\neq 2^k$. We check the remaining cases.

Since there is a fibration

$$\mathrm{map}_*(P^3(2),S^{2n+1}) \xrightarrow{\pi} \Omega^2S^{2n+1} \xrightarrow{2} \Omega^2S^{2n+1},$$

it follows that $2(\pi\cdot\theta)$ is zero in $[P^{4n-2}(2),\Omega^2S^{2n+1}]$. By the previous paragraph $\pi\cdot\theta$ restricted to the bottom cell has Kervaire invariant one. Thus there is a homotopy commutative diagram (with $n=2^k$)

$$\begin{array}{ccccc} P^{4n-2}(2) & \xrightarrow{[2]} & P^{4n-2}(2) & \xrightarrow{\pi\cdot\theta} & \Omega^2S^{2n+1} \\ \text{pinch}\downarrow & & \uparrow\text{include} & \nearrow \theta_k & \\ S^{4n-2} & \xrightarrow[\eta_{4n-3}]{} & S^{4n-3} & & \end{array}.$$

Since $2(\pi\cdot\theta)=0$, $\theta_k\cdot\eta_{4k-3}$ is divisible by 2. This is a contradiction if $2n+1\neq 5$, 9 or 17 by the following lemma and 15.3 follows.

<u>Lemma 18.2</u>. If θ_k in $\pi^S_{2^{k+1}-2}$ has Kervaire invariant one with $\theta_k\eta_{2^{k+1}-2}=2x$, then $k\leq 4$.

<u>Proof</u>: Consider the Adams spectral sequence with $E_2^S=\mathrm{Ext}_A^S(\mathbb{Z}/2,\mathbb{Z}/2)$ coverging to π_*^S. If s=3, then it is shown that E_2^S is generated by 3-fold products of

h_i's which are linearly independent except for $h_i h_{i+1} h_j = 0$, $h_i^2 h_{i+2} = h_{i+1}^3$, and $h_i h_{i+2}^2 = 0$ [A2]. The element θ_k (which we assume exists) is represented by h_k^2 and $\theta_k \eta_{2^{k+1}-2}$ is represented by $h_1 h_k^2$. Let $k \geq 5$ and assume that $\theta_k \eta_{2^{k+1}-2} = 2x$. Notice that $h_1 h_k^2$ is assumed to be a permanent cycle and since it represents an element of order 2 one of three things can happen: (1) $h_1 h_k^2 = h_o(z)$, $z \in E_2^2$, (2) $h_1 h_k^2$ is hit by a differential or (3) $h_1 h_k^2 = 2(w)$, $w \in E_2^1$. But $h_1 h_k^2 \neq h_0^2 h_{k+1}$ if $k \geq 5$. Thus (1) fails. The only choice for a differential hitting $h_1 h_k^2$ is $h_1 h_k^2 = d(\alpha)$ with α of degree 2^{k+1} but there are no elements in E_2^1 of degree 2^{k+1} and so (2) fails. Since h_k is not an infinite cycle if $k \geq 5$, the third case fails.

§19. A decomposition for $\mathrm{map}_*(P^3(2),S^5)$ and Waggoner's theorem

In this section we derive product decompositions for $\mathrm{map}_*(P^k(2),X)$ for certain values of k and X. The main point of this section is that the existence of certain low dimensional spherical homology classes implies certain global homotopy properties. The existence of these spherical classes arise from the unstable condition for the action of the Steenrod algebra.

In section 8, we gave Selick's retraction theorem which states that when spaces are localized at an odd prime p, then $\Omega^2S^3\langle 3\rangle$ is a retract of $\mathrm{map}_*(P^3(p),S^{2p+1})$. More was proven in [$S_2$]; namely $\mathrm{map}_*(P^3(p),S^{2p+1})$ is homotopy equivalent to $\Omega^2S^3\langle 3\rangle\times W_p$ where W_n is the fibre of the double suspension $E^2\colon S^{2n-1}\to\Omega^2S^{2n+1}$ localized at an odd prime p.

Understanding the difference between $\Omega^2[2]$ and 2 led to a proof of a 2-primary analogue of this theorem [C_1]. We shall give a different proof by using the classical Lie groups SU(n) and Sp(2).

Theorem 19.1 (1) There is a 2-local equivalence

$$W_2\times\Omega^2S^3\langle 3\rangle\to\mathrm{map}_*(P^3(2),S^5)\quad\text{and}$$

(2) W_2 is a retract of $\mathrm{map}_*(P^4(2),S^9)$.

Remark 19.2. More is true in 19.1. The space BW_2 exists [G_2] and $BW_2\times W_4$ is homotopy equivalent to $\mathrm{map}_*(P^3(2),S^9)$[CS].

The following theorem was proven in [W].

Theorem 19.3. W_2 is a retract of $\mathrm{map}_*(P^3(2),SU(4))$.

Theorem 19.3 gives information about the fibration $SU(4)\to SU(5)\to S^9$. Namely, the induced map $\partial\colon\pi_*(\Omega S^9;\mathbb{Z}/2)\to\pi_*(SU(4);\mathbb{Z}/2)$ is split injective on a large summand of $\pi_*(\Omega S^9;\mathbb{Z}/2)$ given by $\pi_{*-3}(W_2)$[W].

The proofs of 19.1 and 19.3 will follow directly from the existence of a 6-dimensional spherical homology class in $\Omega^2SU(4)$. Recall that $SU(4)/Sp(2)$ is S^5; let $\pi: SU(4) \to S^5$ denote the quotient map.

Lemma 19.4. There is a 6-dimensional spherical homology class, x, in $H_*\Omega^2SU(4)$ with $(\Omega^2\pi)_*(x)=x_3^2$ in $H_6\Omega^2S^5$.

The proof of this is direct, but we shall postpone it until giving the proofs of 19.1 and 19.3. Here we need maps given in [C2].

Lemma 19.5. There is a map $\sigma_n: W_n \to map_*(P^4(2),S^{4n+1})$ which induces an isomorphism on H_{4n-2}.

Finally recall that there is a map $\bar{h}: \Omega^2S^{2n+1} \to map_*(P^3(2),S^{4n+1})$ which induces an isomorphism on H_{4n-2} by Proposition 4.1.

Proof of 19.1 and 19.3: By Lemma 19.4, there is a map $\tilde{f}: S^6 \to \Omega^2SU(4)$ such that the composite $S^6 \xrightarrow{\tilde{f}} \Omega^2SU(4) \xrightarrow{\Omega^2\pi} \Omega^2S^5$ is non-zero on H_6. Since $SU(4)$ is a loop space, there is a 3-fold loop map $\Omega^3f: \Omega^3S^9 \to \Omega^2SU(4)$ which is $\tilde{f}$ on the bottom cell. Thus we get an induced map $g: map_*(P^4(2),S^9) \to map_*(P^4(2),BSU(4))$ where $g=map_*(1,f)$.

Next consider the composite α given by $W_2 \xrightarrow{\sigma_2} map_*(P^4(2),S^9) \xrightarrow{g} map_*(P^4(2),BSU(4)) \xrightarrow{\rho} map_*(P^3(2),S^5)$ where ρ is induced by $\pi: SU(4) \to S^5$ with the identification $map_*(P^4(2),BSU(4)) \simeq map_*(P^3(2),SU(4))$. Since lemma 19.4 gives $(\Omega^2\pi\cdot\Omega^3f)_*(x_6)=x_3^2$ in $H_*\Omega^2S^5$, it follows that $(\Omega^3\pi\cdot\Omega^4f)_*(x_5)=Q_1x_2$ in $H_*\Omega^3S^5$. Thus $(\rho\cdot g)_*$ is onto the module of primitives in $H_5map_*(P^3(2),S^5)$; since σ_{2*} induces on isomorphism on H_5, we have α_* induces on isomorphism on the primitives in H_5.

Next recall that there is an isomorphism $\bar{h}_*: H_2\Omega^2S^3\langle 3\rangle \to H_2map_*(P^3(2),S^5)$. Thus the composite,

$$W_2\times\Omega^2S^3\langle 3\rangle \xrightarrow{\alpha\times\bar{h}} \mathrm{map}_*(P^3(2),S^5)^2 \xrightarrow{\text{multiply}} \mathrm{map}_*(P^3(2),S^5)$$

induces an isomorphism on the module of primitives in dimensions 2 and 5. Notice that 19.1 follows from Theorem 15.4 with n=2 if we check

Lemma 19.6. $H_*(W_2\times\Omega^2S^3\langle 3\rangle)$ and $H_*\mathrm{map}_*(P^3(2),S^5)$ are isomorphic as coalgebras over the Steenrod algebra.

Notice that there is a map $p\colon \mathrm{map}_*(P^3(2),S^5) \to W_2$ such that $p\alpha$ is homotopic to the identity. Thus W_2 is a retract of both $\mathrm{map}_*(P^4(2),S^9)$ and $\mathrm{map}_*(P^4(2),BSU(4))$ and so 19.1 and 19.3 follow.

§20: Proofs of lemmas in section 19

We first give the

Proof of Lemma 19.4: Consider the diagram

$$\begin{array}{ccccc}
SU(5) & \xrightarrow{i} & SU(5)/Sp(2) & \xrightarrow{j} & K(\mathbb{Z}/2,5) \\
\downarrow{\scriptstyle \pi} & & \downarrow & & \downarrow{\scriptstyle Sq^4} \\
S^9 & \xrightarrow{=} & S^9 & \xrightarrow{k} & K(\mathbb{Z}/2,9)
\end{array}$$

where j_* induces an isomorphism on $H_5(\ ;\mathbb{Z}/2)$. Notice that this diagram homotopy commutes because $H_*SU(5)$ is isomorphic to $\Lambda[\bar{H}_*\Sigma\mathbb{C}P^4]$ as a Hopf algebra over the Steenrod algebra and $H_*Sp(2) \cong \Lambda[x_3,x_7]$, $|x_i|=i$, embeds in $H_*SU(5)$. Looping, one obtains a homotopy commutative diagram

$$\begin{array}{ccccc}
\Omega^3S^9 & \xrightarrow{=} & \Omega^3S^9 & \xrightarrow{\Omega^3k} & K(\mathbb{Z}/2,6) \\
\downarrow{\scriptstyle \Omega^2\partial} & & \downarrow{\scriptstyle \Omega^2\partial} & & \downarrow \\
\Omega^2SU(4) & \longrightarrow & \Omega^2(SU(4)/Sp(2)) & \longrightarrow & \Omega^2E \\
\downarrow & & \downarrow & & \downarrow \\
\Omega^2SU(5) & \longrightarrow & \Omega^2(SU(5)/Sp(2)) & \longrightarrow & K(\mathbb{Z}/2,3)
\end{array}$$

where the vertical columns are fibrations and E is the fibre of Sq^4: $K(\mathbb{Z}/2,5) \to K(\mathbb{Z}/2,9)$. Since Ω^3k induces an epimorphism from $\pi_6\Omega^3S^9$ to $\pi_6\Omega^2E$ and Ω^2E splits as $K(\mathbb{Z}/2,3)\times K(\mathbb{Z}/2,6)$, it follows that $\Omega^2\partial$ induces a monomorphism on H_6. Now notice that $SU(4)/Sp(2)$ is S^5 and the lemma follows.

We remark that the element in $\pi_8SU(4)$ given by Lemma 19.4 is of course the generator of the 2-primary component.

Proof of 19.5: By Lemma 4.1, there is a map $\bar{h}$: $\Omega^2S^{2n+1} \to map_*(P^3(2),S^{4n+1})$ which induces an isomorphism on H_{4n-2}. Since $map_*(P^3(2),S^{4n+1})$ is $(4n-3)$-connected, there is a map of fibrations

$$\begin{array}{ccc} W_n & \xrightarrow{\sigma_n} & \mathrm{map}_*(P^4(2),S^{4n+1}) \\ \downarrow & & \downarrow \\ S^{2n-1} & \longrightarrow & * \\ \downarrow & & \downarrow \\ \Omega^2 S^{2n+1} & \longrightarrow & \mathrm{map}_*(P^3(2),S^{4n+1}) \end{array}$$

and σ_{n*} induces an isomorphism on H_{4n-3} by a comparison of the Serre spectral sequences for these fibrations.

<u>Proof of 19.6</u>: Theorems 7.1 and Lemma 15.1 give $H_*\Omega^2S^3\langle 3\rangle$ and $H_*\mathrm{map}_*(P^3(2),S^5)$ as Hopf algebras over the Steenrod algebra. The Serre spectral for $\Omega S^3 \xrightarrow{\Omega E} \Omega^3 S^5 \to W_2$ collapses because $(\Omega E)_*$ is a monomorphism. Hence as a coalgebra over the Steenrod algebra H_*W_2 is isomorphic to $H_*\Omega^3S^5//H_*\Omega S^3$. Thus H_*W_2 is isomorphic to $\mathbb{Z}/2[Q_1^aQ_2^bx_2 \mid a+b\geq 1]$ as a coalgebra over the Steenrod algebra and 19.6 follows by inspection.

§21. Spherical homology classes and higher torsion for $\pi_* P^{2n}(2^r)$

In this section we digress to apply the previous results to give spherical homology classes in $H_*\Omega P^{2n}(2^r)$ together with some higher torsion in the homotopy groups of $P^{2n}(2^r)$. The particular elements constructed are related to the divisibility of the Whitehead product. We deliberately omit the cases for $P^{2n+1}(2^r)$ where higher torsion is also present.

Recall that $P^{n+1}(k)$ is the cofibre of the degree k map $[k]: S^n \to S^n$ and $H_*(\Omega P^{n+1}(2^r); \mathbb{Z}/2^r)$ is isomorphic to the tensor algebra $T[u_{n-1}, v_n]$ with $|u|=n-1$, $|v|=n$, $n \geq 2$.

Lemma 21.1. If $n \geq 2$, the class $[u,v]=uv-vu$ in $H_{4n-3}(\Omega P^{2n}(2^r); \mathbb{Z}/2^r)$ is spherical.

Proposition 21.2. Let α be any element in $\pi_{4n-3}\Omega P^{2n}(2^r)$ such that the mod-2^r reduction of the Hurewicz image of α is $[u,v]$. If $n \neq 2^k$, then α has order at least 2^{r+1} and if $r \geq 2$, α has order exactly 2^{r+1}.

We remark that the hypothesis $n=2^k$ in 21.2 is necessary by the following example.

Lemma 21.3. Let α be as in 21.2 with n=4 or 8. There is a choice of α such that if n=4 and $r \geq 3$, then α has order 2^r; if n=8 and $r \geq 4$, then α has order 2^r.

Notice that $[u,v]$ fails to be spherical in $H_*(\Omega P^{2n+1}(2^r); \mathbb{Z}/2^r)$ if $r \geq 2$.

Proof of 21.1: Let X denote the homotopy theoretic fibre of $f: QP^{2n}(2^r) \to K(\mathbb{Z}/2, 4n-2)$ where $f_*(u_{2n-1}^2)=\iota_{4n-2}$ and u_{2n-1} generates $H_{2n-1}QP^{2n}(2^r)$. Let $E^\infty: P^{2n}(2^r) \to QP^{2n}(2^r)$ denote the stabilization map and notice that $E^\infty \cdot f$ is null. Thus there is a lift $g: P^{2n}(2^r) \to X$ inducing an isomorphism on H_{2n-1}. Let F denote the homotopy theoretic fibre of g. Since $Sq^1_* Q_2 u_{2n-2} = Q_1 u_{2n-2}$, an inspection of the Serre spectral sequence for

Ωg gives that $(\Omega g)_*$ is onto $H_q\Omega X$ for $q\leq 4n-3$ with $[u,v]$ the element of least degree in the kernel of $(\Omega g)_*$. Since ΩF is $(4n-4)$-connected, and $[u,v]$ is in the image of $H_{4n-3}\Omega F \to H_{4n-3}\Omega P^{2n}(2^r)$, the lemma follows.

Proof of 21.2: Let ϕ denote the mod-2^r reduction of the Hurewicz homomorphism. Let α be any element such that $\phi(\alpha)=[u,v]$. Notice that $\phi(\alpha)$ has order at least 2^r because $[u,v]$ has order 2^r in $H_*(\Omega P^{2n}(2^r);\mathbb{Z}/2^r)$. If $2^r\alpha=0$, consider any choice of extension of α to give $\gamma\colon P^{4n-2}(2^r) \to \Omega P^{2n}(2^r)$. Since the mod-$2^r$ homology of $\Omega P^{2n}(2^r)$ is acyclic with respect to the 2^r-th Bockstein and $\gamma_*(u_{4n-3})=[u,v]$, it follows that $\gamma_*(v_{4n-2})=v^2$+others.

Consider the pinch map $p\colon P^{2n}(2^r) \to S^{2n}$ and observe that the composite $P^{4n-2}(2^r) \xrightarrow{\gamma} \Omega P^{2n}(2^r) \xrightarrow{\Omega(p)} \Omega S^{2n}$ is non-trivial in homology. Since there is a map $P^{4n-2}(2) \to P^{4n-2}(2^r)$ which is an isomorphism on H_{4n-2} the Pontrjagin square v_{2n-1}^2 in $H_*(\Omega S^{2n};\mathbb{Z}/2)$ is mod-2 spherical. This implies that the Whitehead product $[\iota_{2n-1},\iota_{2n-1}]$ is divisible by 2 according to 11.4. Thus $n=2^k$. Assuming $n\neq 2^k$, it follows that α has order at least 2^{r+1}. If $r\geq 2$, then the suspension order of the identity for $P^{2n}(2^r)$ is 2^r. That α has order at most 2^{r+1} follows from Proposition 13.3.

Proof of 21.3: If $r\geq 3$ and $s\geq 4$ we claim that $[2^r]\colon S^3 \to S^3$ and $[2^s]\colon S^7 \to S^7$ are H-maps. Write $S^n\{2^r\}$ for the homotopy theoretic fibre of $[2^r]\colon S^n \to S^n$. The claim implies that $S^3\{2^r\}$ and $S^7\{2^s\}$ are H-spaces if $r\geq 3$ and $s\geq 4$. Thus there are homotopy commutative diagrams

$$\begin{array}{ccc} P^3(2^r) & \xrightarrow{\text{inclusion}} & S^3\{2^r\} \\ E\big\downarrow & \nearrow_{\lambda} & \\ \Omega P^4(2^r) & & \end{array} \qquad \text{and} \qquad \begin{array}{ccc} P^7(2^s) & \xrightarrow{\text{inclusion}} & S^7\{2^s\} \\ E\big\downarrow & \nearrow_{\lambda} & \\ \Omega P^8(2^s) & & \end{array}$$

where λ is a multiplicative extension of the inclusion. Assume the above claim for the moment.

Notice that there is a homotopy commutative diagram

$$\begin{array}{ccc} S^n & \longrightarrow & * \\ {\scriptstyle [2^r]}\downarrow & & \downarrow \\ S^n & \longrightarrow & P^{n+1}(2^r) \end{array}$$

and an induced map α: $S^n\{2^r\} \to \Omega P^{n+1}(2^r)$. Thus if $r \geq 3$ and $s \geq 4$, there are maps $\lambda\cdot\alpha$: $S^3\{2^r\} \to S^3\{2^r\}$ and $\lambda\cdot\alpha$: $S^7\{2^s\} \to S^7\{2^s\}$ which are degree one. The mod-2 homology of $S^{2n+1}\{2^r\}$ is isomorphic to $\Lambda[x_{2n+1}] \otimes \mathbb{Z}/2[x_{2n}]$ as a differential coalgebra with differential given by the 2^r-th homology Bockstein. By Lemma 8.6, $\lambda\cdot\alpha$ is an equivalence.

Since λ is homotopy multiplicative, there are homotopy equivalences $\Omega P^4(2^r) \simeq S^3\{2^r\} \times F$ and $\Omega P^8(2^s) \simeq S^7\{2^s\} \times G$ where F and G are the homotopy theoretic fibres of the maps λ. Notice that $\pi_5 F \cong \mathbb{Z}/2^r$ and $\pi_{13} G \cong \mathbb{Z}/2^s$. The lemma follows after the initial claim is proven.

We omit the details of the following remark which follows immediately from [CMN1]: F is homotopy equivalent to $\Omega\Sigma(\bigvee_{k \geq 0} P^{5+2k}(2^r))$ and G is homotopy equivalent to $\Omega\Sigma(\bigvee_{k \geq 0} P^{13+6k}(2^s))$.

We must prove the claim about H-maps. Let f: $X \to Y$ be a pointed map between H-spaces and assume that the base-point is the unit (up to homotopy) and that Y has a homotopy inverse. Let μ_X denote the product $X \times X \to X$. Then $f\cdot\mu_X - \mu_Y\cdot(f \times f)$ is null-homotopic when restricted to $X \vee X$. There is an induced map $D(f)$: $X \wedge X \to Y$ which is null-homotopic if and only if f is an H-map.

Assume that n=3 or 7. Then the suspension E: $S^n \to \Omega S^{n+1}$ is of course not an H-map because ι_n has infinite height in $H_*\Omega S^{n+1}$. The H-deviation is a map D: $S^{2n} \to \Omega S^{n+1}$. Thus identify D with an element in

$\pi_6\Omega S^4 \cong \pi_6 S^3 \oplus \pi_6\Omega S^7 \cong \mathbb{Z}/4 \oplus \mathbb{Z}$ or with an element in $\pi_{14}\Omega S^8 \cong \pi_{14}S^7 \oplus \pi_{14}\Omega S^{15} \cong \mathbb{Z}/8 \oplus \mathbb{Z}$. Since n=3 or 7, the Hopf fibrations $\nu_4\colon S^7 \to S^4$ and $\sigma_8\colon S^{15} \to S^8$ induce fibrations $\Omega S^{2n+1} \to \Omega S^{n+1} \xrightarrow{\partial} S^n$ where ∂ is degree one. Since the identity is an H-map, the H-deviation lifts to ΩS^{2n+1}. Thus D is an odd multiple of ν_4 or σ_8 by inspection of the the second stage of Milnor's classifying space construction.

Next observe that there is a homotopy commutative diagram where X is a homotopy associative H-space,

$$\begin{array}{ccccc} S^n & \xrightarrow{f} & X & & \\ \Big\downarrow E & & \Big\downarrow E & \searrow 1 & \\ \Omega S^{n+1} & \xrightarrow[\Omega\Sigma f]{} & \Omega\Sigma X & \xrightarrow[r]{} & X, \end{array}$$

and the retraction r is an H-map. By the above, together with naturality, one has

Lemma 21.4. The H-deviation of f is the composite $S^{2n} \xrightarrow{D} \Omega S^{n+1} \xrightarrow{\Omega\Sigma f} \Omega\Sigma X \xrightarrow{r} X$ where D is the adjoint of ν_4 or σ_8 for n=3 or 7.

Corollary 21.5. The H-deviation of $\eta\colon S^3 \to \Omega S^3\langle 3\rangle$ the generator of $\pi_3\Omega S^3$ is non-zero. Thus η is not an H-map.

Proof: The H-deviation of η is represented by $\eta_3\cdot\nu_4$ by Lemma 21.4. By [T1], $\eta_3\cdot\nu_4 = \nu'\cdot\eta_6$, the generator of $\pi_7 S^3 \cong \mathbb{Z}/2$.

Corollary 21.6. The map $[2^r]\colon S^n \to S^n$ is an H-map if n=3 and $r \geq 3$ or n=7 and $r \geq 4$.

Proof: We give the proof for n=3 and leave the case n=7 to the reader. Let $X = S^3$ and $f = [2^r]$ in Lemma 21.4. Consider the effect of $[2]\colon S^4 \to S^4$ on $\pi_7 S^4 \cong \mathbb{Z} \oplus \mathbb{Z}/4$ with generators ν_4 and $E\nu'$. Since $E\nu'$ is a suspension,

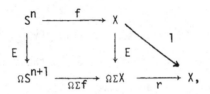

$[2]_*(E\nu')=2E\nu'$. Since $\Omega[2]=2+\Omega w_4\cdot H$ on the 2-primary component, and $H(\nu_4)=1$, it follows that $[2]_*(\nu_4)=2\nu_4+w_4$. Since $w_4=\pm(2\nu_4-E\nu')$, it follows that $[2]_*(\nu_4)=E\nu'$ or $[2]_*(\nu_4)=4\nu_4-E\nu'$. Iterating, one has $[4]_*(\nu_4)=2E\nu'$ or $[4]_*(\nu_4)=16\nu_4-2E\nu'$ and finally $[8]_*(\nu_4)=4E\nu'$ or $[8]_*(\nu_4)=4^3\nu_4$. But then $r_*[2^k]_*(\nu_4)=0$ if $k\geq 3$ and 21.6 follows.

§22. "Long" Steenrod operations and spherical homology classes

Consider the fibration $F \to E \stackrel{p}{\to} B$ together with the boundary $\partial\colon \Omega B \to F$. In this section we give some observations forcing elements in $\pi_q F$ to have non-trivial Hurewicz image in mod-2 homology. Although completely elementary, these observations give information about the homology of spaces looped beyond their connectivity and is part of joint work with F. Peterson.

Lemma 22.1. Let x be a non-trivial class in the Hurewicz image reduced mod-2 for $H_q B$. Let $y \in H^q B$ be dual to x and $p^*(y) = Sq^{q-t} w$ in H^*E with $q-t \geq 2$. If $q-2 \geq k > 2t-q$, then there is a non-zero element in the Hurewicz image $\pi_{q-k-1}\Omega^{k+1}F \to H_{q-k-1}(\Omega^{k+1}F;\mathbb{Z}/2)$.

Proof: This follows by the evident unstable conditions: There is a map of fibrations

$$\begin{array}{ccc} F & \longrightarrow & X \\ \downarrow & & \downarrow \\ E & \longrightarrow & K(\mathbb{Z}/2,t) \\ \downarrow & & \downarrow Sq^{q-t} \\ B & \longrightarrow & K(\mathbb{Z}/2,q) \end{array}$$

where X is the fibre of Sq^{q-t}. Notice that the composite $\Omega B \stackrel{\partial}{\to} F \to X$ gives an epimorphism on $\pi_{q-1}X$ if $q-t \geq 2$. Since $\Omega^k(Sq^{q-t})$ is trivial if $k>2t-q$, $\Omega^k X$ splits as a product of Eilenberg-Maclane spaces. Thus the Hurewicz map is monomorphic for $\Omega^k X$ and the Hurewicz map is non-trivial for $\Omega^k F$ since $\pi_{q-k-1}\Omega^k F \to \pi_{q-k-1}\Omega^k X$ is onto.

Let $\Omega_0^j S^n$ denote the component of the base-point.

Corollary 22.2. Let $n = 2^a(2k+1)-1$ with $k>0$. Then the Whitehead product $[\iota_n,\iota_n]$ has non-trivial Hurewicz image in $H_{n-q-1}(\Omega_0^q S^n;\mathbb{Z}/2)$ for $2n-1 > q > n+2^a$.

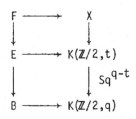

Proof: Consider the component of the base-point in $\Omega^j S^n$, say $\Omega^j_0 S^n$. There is a homotopy commutative diagram

$$\begin{array}{ccc}
\Omega^n_0 S^n & \xrightarrow{\tilde{\alpha}} & F \\
\downarrow & & \downarrow \\
\Omega^{n+1}_0 S^{n+1} & \xrightarrow{f} & K(\mathbb{Z}/2, 2^a k+2^a-1) \\
\Omega^n H \downarrow & & \downarrow Sq^{2^a k} \\
\Omega^{n+1} S^{2n+1} & \xrightarrow{\iota_{n+1}} & K(\mathbb{Z}/2, 2^{a+1}k+2^a-1)
\end{array}$$

where $f^*(\iota_{2^a k+2^a-1})=(Q_{2^a k+2^a-1}[1])^* +$ others since $Sq_*^{2^a k} Q_{2^{a+1}k+2^a-1}[1]=Q_{2^a k+2^a-1}[1]$. Apply Lemma 16.1 together with the fact that the boundary $\partial\colon \Omega^2 S^{2n+1} \to S^n$ in the EHP sequence is $[\iota_n,\iota_n]$ on the bottom cell.

Remark 22.3. Of course if $[\iota_n,\iota_n]$ is divisible by 2, the adjoint always has trivial image in mod-2 homology.

Similar calculations apply to classical Lie groups and we give a sample for SO(n). Since there is a morphism of fibrations

$$\begin{array}{ccc}
SO(n) & \xrightarrow{j} & \Omega^n_0 S^n \\
\downarrow & & \downarrow \\
SO(n+1) & \xrightarrow{j} & \Omega^{n+1}_0 S^{n+1} \\
\downarrow & & \downarrow \\
S^n & \longrightarrow & \Omega^{n+1} S^{2n+1}
\end{array},$$

the computation in 16.2 applies almost verbatim to give

Corollary 22.4. Let $n=2^a(2k+1)-1$ with $k>0$. Then $H_{n-q-1}(\Omega_0^q SO(n))$ has a non-trivial element in the image of the mod-2 reduction of the Hurewicz homomorphism for $2n-1>q>n+2^a$; this element projects to the image of the Whitehead product in $H_{n-q-1}\Omega_0^q S^n$.

Thus Lemma 22.1 has implications for $H_*\Omega_0^{n+k}S^n$.

Question 22.5. Let $k,n\geq 2$. Does $H_*\Omega_0^{n+k}S^n$ contain an infinitely generated polynomial algebra? (That is an infinite number of algebra generators) What can be said about the submodule of nilpotent elements? Is there a bound on the order of nilpotence as a function of n and k?

As examples, we state a theorem proven by F. Peterson and the author.

Proposition 22.6. $H_*\Omega_0^{n+1}S^n$ is a tensor product of exterior and polynomial algebras if $n\leq 5$.

§23: Preliminary remarks about Dickson's analogues of the Cayley numbers

The following is a very speculative beginning and is thus deliberately sketchy.

Let G be an H-space. It is a classical fact that the two natural multiplications on ΩG are homotopic. Thus the map $\beta: (\Omega^k G) \to (\Omega^k G)$ given by $(\beta f)(t)=[f(t)]^2$ is homotopic to the H-space squaring map in the H-space structure obtained from looping G k-times. There are other analogues of this process which are obtained from Dickson's description of the Cayley numbers [D]. In this section we consider a part of this information.

This situation arises in case G is replaced by S^{2^n-1} which of course is not an H-space when $n \geq 4$ [A_1]. However, by considering algebras formed by Dickson's methods [D], one finds a kind of squaring map for S^{2^n-1}.

As a special case of [D], consider a multiplication $\mu: \mathbb{R}^{2^n} \times \mathbb{R}^{2^n} \to \mathbb{R}^{2^n}$. Notice that $\mathbb{R}^{2^n}$ has a basis $e_1, \ldots, e_{2^n}$ with e_1=identity, $e_i e_j = -e_j e_i$ if $i,j>1$, $e_i^2=-1$ if $i>1$, and $e_i e_j = \lambda_{ij} e_k$ for some k and $\lambda_{ij}=\pm 1$ [S].

Proposition 23.1. Let $x \in \mathbb{R}^{2^n}$. Then $x^2=0$ if and only if $x=0$.

Proof: Write $x = \sum_{i=1}^{2^n} x_i e_i$. Then $x^2=\Sigma x_i^2 e_i^2 + 2\Sigma x_1 x_i e_i + \sum_{\substack{i \neq j \\ i,j \neq 1}} x_i x_j e_i e_j$. If $x^2=0$, then $x_1^2 - \sum_{i \geq 2} x_i^2 = 0$ and $x_1 x_i = 0$ for $i \geq 2$. Hence $x=0$ and the lemma follows.

Thus there is a map sq: $S^{2^n-1} \to S^{2^n-1}$ given by $sq(x) = \frac{x^2}{||x^2||}$. Notice that sq is the composite

$$S^{2^n-1} \xrightarrow{\Delta} S^{2^n-1} \times S^{2^n-1} - \Sigma \xrightarrow{\mu} S^{2^n-1}$$

where $\mu(a,b) = \frac{ab}{||ab||}$ and $\Sigma = \{(a,b) \in \mathbb{R}^{2^n} \times \mathbb{R}^{2^n} \mid ab=0\}$. Clearly if $ab=0$ then $(sa)(tb)=(st)(ab)=0$ for $s,t \in \mathbb{R}$ and thus Σ is contractible and is a cone with the origin as the cone point. Also notice

<u>Lemma 23.2</u>. The map sq preserves e_1 and is of degree 2.

We pose the following

<u>Question 23.3</u>. Identify the homotopy type of $(S^{2^n-1})^2-\Sigma$ and $\Omega^2[(S^{2^n-1})^2-\Sigma]$.

By section 11, one might ask whether the following diagram homotopy commutes:

$$\begin{array}{ccccc} \Omega^2 S^{2^n-1} & \xrightarrow{\Omega^2(\Delta)} & \Omega^2(S^{2^n-1}\times S^{2^n-1}-\Sigma) & \xrightarrow{\Omega^2(\mu)} & \Omega^2 S^{2^n-1} \\ & \searrow^{\Delta} & \downarrow & \nearrow_{\text{multiply}} & \\ & & (\Omega^2 S^{2^n-1})^2 & & \end{array}$$

Notice that this diagram homotopy commutes if and only if w_{2^n-1} is divisible by 2. If $n\geq 4$, then $(S^{2^n-1})^2\cap\Sigma$ is non-empty. Let F denote the homotopy theoretic fibre of the inclusion $i: (S^{2^n-1})^2-\Sigma \to (S^{2^n-1})^2$. One cannot show that the above diagram homotopy commutes by showing that the natural composite $F \to (S^{2^n-1})^2-\Sigma \overset{\mu}{\to} S^{2^n-1}$ is null if $n\geq 4$ by

<u>Lemma 23.4.</u> There is a homotopy commutative diagram

$$\begin{array}{ccccc} & & S^{2^{n+1}-3} & & \\ & \swarrow & \downarrow & \searrow^{w_{2^n-1}} & \\ F & \to & (S^{2^n-1})^2-\Sigma & \underset{\mu}{\to} & S^{2^n-1} \end{array}$$

<u>Proof</u>: Consider $(S^{2^n-1}\times\{e_1\})\cup(\{e_1\}\times S^{2^n-1})=S^{2^n-1}\vee S^{2^n-1}\subset(S^{2^n-1})^2-\Sigma$. Thus there is a commutative diagram

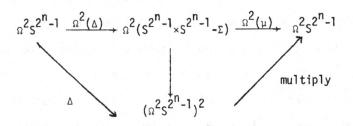

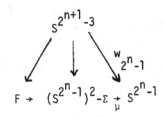

$$\begin{array}{ccc} S^{2^n-1} \vee S^{2^n-1} & \subset & (S^{2^n-1})^2 - \Sigma \\ \downarrow j & & \downarrow \\ (S^{2^n-1})^2 & = & (S^{2^n-1})^2 . \end{array}$$

The homotopy theoretic fibre of the inclusion j is homotopy equivalent to a bouquet of spheres by 2.2. Next observe that $\mu: (S^{2^n-1})-\Sigma \to S^{2^n-1}$ restricted to $S^{2^n-1} \vee S^{2^n-1}$ is the fold map. Thus Lemma 3.4 follows by definition of w_{2^n-1} [Section 1].

Instead of going further into the structure of μ, we translate some analogous constructions in terms of configuration spaces. Namely, by Proposition 11.2 w_{2^n-1} is divisible by 2 if and only if $-\Omega^2[1]$ is homotopic to 1 through the $(2^{n+1}-5)$-skeleton of $\Omega^2 S^{2^{2^n-1}}$. This last question admits a very simple description in terms of configuration spaces.

Let $c: \mathbb{R}^2 \to \mathbb{R}^2$ denote complex conjugation and let $F(M,k) = \{(m_1,\ldots,m_k)\varepsilon M^k \mid m_i \neq m_j \text{ if } i \neq j\}$. Thus c induces a homeomorphism of $F(\mathbb{R}^2,k)$. Let X be a pointed space with base-point *. Define $M_2(x) = (\mathbb{R}^2 \times X) \amalg (F(\mathbb{R}^2,2) \times_{\Sigma_2} X^2)/\approx$ where Σ_2 acts diagonally on $F(R^2,2)$ and X^2 by interchanging coordinates and $(\approx)$ is the equivalence relation $((s,t_2)(*,y)) \approx ((t,(y))$. By work of Browder [Br], and subsequently Milgram [Mi] and May [Ma], $M_2(S^{2n-1})$ is equivalent to the $(4n-1)$-skeleton of $\Omega^2 S^{2n+1}$. Evidently, complex conjugation induces a homeomorphism of $M_2(S^{2n-1})$.

<u>Lemma 17.1</u>.. The homeomorphism of $M_2(S^{2n-1})$ induced by complex conjugation is homotopic to the identity if and only if w_{2n+1} is divisible by 2.

<u>Proof</u>: The proof follows directly from properties of Boardman and Vogt's little cubes $C_n(j)$, which are explicated in [Ma]. Let $\chi: I^2 \to I^2$ be given

by $\chi(s,t)=(s,1-t)$. There is a homeomorphism $\bar{\chi}\colon C_2(k) \to C_2(k)$ given by $\bar{\chi}\langle c_1,\ldots,c_k\rangle=\langle \chi\cdot c_1\cdot \chi,\ldots,\chi\cdot c_k\cdot \chi\rangle$ [CLM, p339] where $C_2(k)$ is the space of j tuples $\langle c_1,\ldots,c_j\rangle$ for which $c_i\colon I^2 \to I^2$ is an affine, orientation preserving, axis preserving embedding with $c_i(\mathring{I}^2)\cap c_j(\mathring{I}^2)=\phi$ if $i\neq j$. It follows directly that the homeomorphism $\bar{\chi}\colon C_2X \to C_2X$ induced by $\bar{\chi}$ gives a homotopy commutative diagram

$$\begin{array}{ccc} C_2(\Sigma X) & \xrightarrow{\theta_2} & \Omega^2\Sigma^3X \\ \bar{\chi}\downarrow & & \downarrow -\Omega^2[-1] \\ C_2(\Sigma X) & \xrightarrow[\theta_2]{} & \Omega^2\Sigma^3X \end{array}$$

where $\theta_2\colon C_2(Y) \to \Omega^2\Sigma^2Y$ is given in [Ma]. Restrict to the second filtration and recall Proposition 11.2 to finish the proof.

References

[A_1] J.F. Adams, On the nonexistence of elements of Hopf invariant one, Ann. of Math., 72(1960), 20-104.

[A_2] J.F. Adams, On the structure and applications of the Steenrod algebra, Comment. Math. Helv. 32(1958), 180-214.

[AK] S. Araki and T. Kudo, Topology of H_n-spaces and H-squaring operations, Mem. Fac. Sci. Kyūsyū Univ. Ser. A., 1956, 85-120.

[B] M.G. Barratt, Spaces of finite characteristic, Quart. J. Math., 11(1960), 124-136.

[BJM] M.G. Barratt, J.D.S. Jones, and M.E. Mahowald, The Kervaire invariant problem, Contemp. Math., 19(1983), 9-22.

[BS] R. Bott and H. Samelson, On the Pontrjagin product in spaces of paths, Comment. Math. Helv., 27(1953), 320-337.

[BP] E. H. Brown and F.P. Peterson, Whitehead products and cohomology operations, Quart. J. Math., 15(1964), 116-120.

[Br] W. Browder, Homology operations and loop spaces, Ill. J. Math., 4(1960), 347-357.

[CPS] H.E.A. Campbell, F.P. Peterson, and P.S. Selick, Self-maps of loop spaces I, to appear.

[CCPS] H.E.A. Campbell, F.R. Cohen, F.P. Peterson, and P.S. Selick, The space of maps of Moore spaces to spheres, to appear.

[C_1] F.R. Cohen, Two-primary analogues of Selick's theorem and the Kahn-Priddy theorem for the 3-sphere, Topology, 23(1984), 401-421.

[C_2] F.R. Cohen, The unstable decomposition of $\Omega^2\Sigma^2X$ and its applications, Math. Zeit., 182(1983), 553-568.

[CLM] F.R. Cohen, T.J. Lada, and J.P. May, The homology of iterated loop spaces, L.N.M. v.533, Springer-Verlag, Berlin and New York, 1976.

[CM] F.R. Cohen and M.E. Mahowald, Unstable properties of Ω^nS^{n+k}, Contemp. Math., 12(1982), 81-90.

[CMN_1] F.R. Cohen, J.C. Moore, and J.A. Neisendorfer, Torsion in homotopy groups, Ann. of Math., 109(1979), 121-168.

[CMN_2] F.R. Cohen, J.C. Moore, and J.A. Neisendorfer, The double suspension and exponents of the homotopy groups of spheres, Ann. of Math, 110 (1979), 549-565.

[CMN_3] F.R. Cohen, J.C. Moore, and J.A. Neisendorfer, Exponents in homotopy theory, to appear.

[CP] F.R. Cohen and F.P. Peterson, Suspensions of Stiefel manifolds, Quart. J. Math. Oxford, 35(1984), 115-119.

[CS] F.R. Cohen and P.S. Selick, Splitting of two function spaces, to appear.

[D] L.E.J. Dickson, On quaternions and their generalizations and the history of the eight square theorem, Ann. of Math., 20(1919), 155-71.

[DL] E. Dyer and R. Lashof, Homology of iterated loop spaces, Amer. J. Math., 1962, 35-88.

[G] T. Ganea, A generalization of the homology and homotopy suspension, Commentarii Math. Helvetici, 39(1965), 295-322.

[G_1] B. Gray, A note on the Hilton-Milnor theorem, Topology, 10(1971), 199-201.

[G_2] B. Gray, On the double suspension, to appear.

[H_1] P.J. Hilton, On the homotopy groups of a union of spheres, J. Lond. Math. Soc., 30(1955), 154-172.

[H_2] P.J. Hilton, Note on the Jacobi identity for Whitehead products, Proc. Camb. Phil. Soc., 57(1961), 180-182.

[J_1] I.M. James, Reduced product spaces, Ann. of Math., 62(1955), 170-197.

[J_2] I.M. James, The suspension triad of a sphere, Ann. of Math., 63(1956), 407-429.

[J_3] I.M. James, On the suspension triad, Ann. of Math., 63(1956), 191-247.

[J_4] I.M. James, On the suspension sequence, Ann. of Math., 65(1957), 74-107.

[J_5] I.M. James, The Topology of Stiefel Manifolds, London Math. Soc. Lecture Note Series, no.24, Cambridge Univ. Press, Cambridge, 1976.

[KP] D.S. Kahn and S.B. Priddy, The transfer and stable homotopy theory, Math. Proc. Camb. Phil. Soc., 83(1978), 103-111.

[M] M.E. Mahowald, Some remarks on the Kervaire invariant from the homotopy point of view, Proc. of Symposia in Pure Math., XXII, Providence R.I., 1971, 165-169.

[MU] W.S. Massey and H. Uehara, The Jacobi identity for Whitehead products, Algebraic Geometry and Topology, A Symposium in honor of S. Lefschetz, (361-377), Princeton Mathematical series, no.12, Princeton Univ. Press, Princeton, N.J., 1957.

[Ma] J.P. May, The Geometry of Iterated Loop Spaces, L.N.M. v.268, Springer-Verlag, Berlin and New York, 1972.

[Mi] R.J. Milgram, Iterated loop spaces, Ann. of Math., 84(1966), 386-403.

[Mr] J.W. Milnor, The construction FK, Algebraic Topology, A Student's Guide by J.F. Adams, 119-135. London Math. Soc. Lecture Notes, no.4, Cambridge Univ. Press, Cambridge, 1972.

[MM] J.W. Milnor and J.C. Moore, On the structure of Hopf algebras, Ann. of Math., 81(1965), 211-264.

[N_1] J.A. Neisendorfer, Primary Homotopy Theory, Memoirs of the A.M.S., 25(1980).

[N_2] J.A. Neisendorfer, The exponent of a Moore space, to appear.

[N_3] J.A. Neisendorfer, 3-primary exponents, Proc. Camb. Phil. Soc., 90(1981), 63-83.

[NT] M. Nakaoka and H. Toda, On the Jacobi identity for Whitehead products, J. Inst. Polytech. Osaka City Univ., Ser. A (1954), 1-13.

[Ni] G. Nishida, Cohomology operations in iterated loop spaces, Proc. Japan Acad., 44(1968), 104-109.

[P] G.J. Porter, The homotopy groups of wedges of suspensions, Amer. J. Math., 88(1966), 655-663.

[Sa] H. Samelson, A connection between the Whitehead product and the Pontrjagin product, Amer. J. Math., 28(1954), 278-287.

[S] R. Schafer, On the algebras formed by the Cayley-Dickson process, Amer. J. Math., 76(1954), 435-446.

[S_1] P.S. Selick, Odd primary torsion in $\pi_k S^3$, Topology, 17(1978), 407-412.

[S_2] P.S. Selick, A decomposition of $\pi_*(S^{2p+1}; \mathbb{Z}/p)$, Topology, 20(1981), 175-177.

[S_3] P.S. Selick, A spectral sequence concerning the double suspension, Invent. Math., 64(1981), 15-24.

[S_4] P.S. Selick, A reformulation of the Arf invariant one mod p problem and applications to atomic spaces, Pac. J. Math., 108(1983), 431-450.

[S_5] P.S. Selick, 2-primary exponents for the homotopy groups of spheres, Topology, 23(1984), 97-99.

[Sn] V.P. Snaith, A stable decomposition for $\Omega^n\Sigma^n X$, J. London Math. Soc., 7(1974), 577-583.

[St] N.E. Steenrod, A convenient category of topological spaces, Mich. Math. J., 14(1967), 133-152.

[T_1] H. Toda, Composition Methods in the Homotopy Groups of Spheres, Ann. of Math. Studies, v.49, Princeton Univ. Press, Princeton, N.J., 1962.

[T_2] H. Toda, On the double suspension E^2, J. Inst. Polytech. Osaka City Univ., Ser.A, 7(1956), 103-145.

[T_3] H. Toda, Non-existence of mappings of S^{31} into S^{16} with Hopf invariant one, J. Inst. Polytech Osaka City Univ., Ser.A, 8(1957), 31-34.

[T_4] H. Toda, Complex of standard paths and n-ad homotopy groups, J. Inst. Polytech Osaka City Univ., Ser.A, 6(1955), 101-120.

[Wa] D. Waggoner, Thesis, Univ. of Kentucky, 1985.

[W_1] G.W. Whitehead, Elements of Homotopy Theory, Graduate Texts in Mathematics, Springer-Verlag, Berlin and New York, 1978.

[W_2] G.W. Whitehead, On mappings into group-like spaces, Comment. Math. Helv., 28(1954), 320-328.

[Wh] J.H.C. Whitehead, On adding relations to homotopy groups, Ann. of Math., 42(1941), 409-428.

[Wt] H. Whitney, The self-intersections of a smooth n-manifold in 2n space, Ann. of Math., 45(1944), 220-246.

[Z] H. Zassenhaus, The Theory of Groups, Chelsea, 1958.

Department of Mathematics
University of Kentucky
Lexington, Kentucky, 40506

HOMOTOPY AND HOMOLOGY OF DIAGRAMS OF SPACES

Emmanuel Dror Farjoun

0. INTRODUCTION . The present notes record a lecture series given in 1984-85 in the University of Heidelberg and in the University of Washington at Seattle. My aim here is to introduce in a somewhat informal style some of the basic ideas about the homotopy theory of diagrams of spaces developed among others by W. Dwyer, D. Kan, Elmendorf, A. Zabrodsky and myself. Dwyer and Kan introduced the *simplicial model category* of functors from a small category D to simplicial sets $S.$, or topological spaces *Top* and solved many problems in that framework. They and independently Elmendorf followed up the pioneering work of G. Bredon and G. Segal about G- equivariant homotopy equivalence and showed that diagrams of spaces with the shape of the orbit category $O_G = \{G/H\}_{H \subseteq G}$ play a crucial role in G-homotopy theory.

In this theory an intriguing step was the passage from a G-space X to the collection of the fixed point subspaces $\{X^H\}$ taken as a diagram over the category O_G of G-orbits. In an effort to understand this passage an exact generalization to arbitrary D-space for any small simplicial category was developed in [D-Z] and [D]. The crucial concept here is that of D-orbit and the category of D-orbits O_D. This concept together with other basic constructions in D-homotopy theory is given in the first section below.

The homotopy theory of W. Dwyer and D. Kan concerns *free* D-spaces and their retract: these are the cofibrant spaces in their framework. In some sense the basic work of the second and third section here is to show that any D-space can be represented by a free O-space for some other category O, *without loss of homotopical information,* thereby rendering the work of W. Dwyer and D. Kan applicable to a much wider category of diagrams of spaces. As an example of application we show in section 5 that for any two diagrams of say (topological) simplicial complexes and simplicial maps between them $X, Y : D \to$ (spaces) there exist a (simplicial) category O and two *free* O-diagrams X^O, Y^O such that there is a weak equivalence of function complexes

$$hom_D(X,Y) \approx hom_O(X^O, Y^O).$$

This weak equivalence is crucial for the construction of an equivariant homotopy spectral sequence converging to the homotopy groups of the equivariant pointed function complex between two pointed D-spaces (5.14). In the fourth section we write down a straightforward generalization of the axioms for G-equivariant (co) homology as given in [Bredon] and show how to construct all possible (classical) Bredon type (co) homology theories on D-spaces. It is interesting to note that in general these cohomology theories are *not* representable (5.9); that is an essential difference between the general case and the special case discussed in [D-K-2]. These D- cohomology theories are the correct domain for obstruction theory for extending and lifting maps between D-spaces. The spectral sequence that converges to these equivariant (co) homology theories leads to an interesting spectral sequence for the *strict direct limit* of a diagram of spaces: The E_2-term of the latter depends on the (co) homology of all the "partial inverse limits" of the given diagram (5.2).

ACKNOWLEDGEMENT The present work grew out of a common work with the late A. Zabrodsky. We would like also to thank Dan Kan for several discussions and suggestions. Of special help was H. Miller whose work has inspired my interest in equivariant homotopy and who carefully read the present manuscript and corrected several mistakes.

1. GENERALITIES

In this section we introduce some basic concepts and problems from the theory of diagrams of spaces. We start with an example:

One of the simplest diagrams often encountered in topology is that of a map $f : X \to X'$ between two spaces. To consider it as a diagram we take $J = (\cdot \to \cdot)$ to be the category with two objects and one map between them. The above map f can be thought of as a functor from J to some category of spaces. A map from the diagram f to $g : Y \to Y'$ is a commutative square thought of as a natural transformation between two functors on J. A homotopy is a commutative square

$$\begin{array}{ccc} X \times I & \xrightarrow{H} & Y \\ {\scriptstyle f \times id}\downarrow & & \downarrow{\scriptstyle g} \\ X' \times I & \xrightarrow{H'} & Y' \end{array} .$$

Now notice that even if all the spaces involved X, X', Y, Y' are, say, the unit interval, still there are many possible *homotopy types* of J-diagrams $[0,1] \to [0,1]$. For example the 3-fold map from $[0,1]$ to $[0,1]$ indicated as a projection in figure (1.1) below:

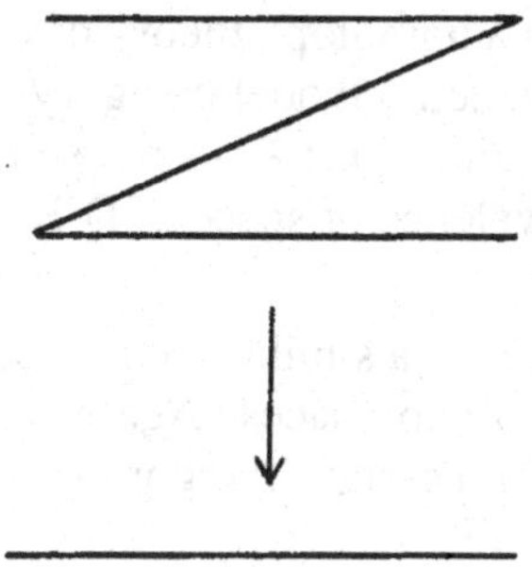

(1.1)

is *not* homotopy equivalent to the identity map $[0,1] \to [0,1]$. (See (1.4) below). The latter is, of course, homotopy equivalent to the identity map $C \to C$ on any contractible space C.

Our problematics is then to construct homotopical invariants that will help in the classification of diagrams of a given shape, as well as homotopy classes of maps between them. For example, the following is an interesting problem [5.12]: Given a map $f : K \to L$, say a simplicial map between two simplicial complexes, and another map $g : X \to Y$ - find a method to understand the function complex map (f,g); namely, the space of commutative squares:

$$\begin{array}{ccc} K & \dashrightarrow & X \\ \downarrow & & \downarrow \\ L & \dashrightarrow & Y \end{array}$$

Another typical problem is to classify up to fibre homotopy equivalence (nice) maps $E \to B$ such that the $f^{-1}(b)$ has the homotopy type of a fixed space F for all $b \in B$. It turns out that these fibre homotopy types are given as a "principle-pseudo-fibration" namely by an action of the monoid $End(F)$ on some space C.

A common example of a diagram of spaces occurs whenever a group acts on a space X. Suppose for a moment that the group G is discrete (or more generally simplicial). Then we think about X as a functor from the category $D = G$ with one object whose self morphisms are the elements of G, to the category of spaces. It was discovered by Bredon [Br] and others that in order to understand the equivariant homotopy type of such G-spaces, with the obvious associated notions of equivariant homotopy and equivariant homotopy equivalence one must pass from the category G, to a much larger category: namely O_G. This O_G is the category whose objects are the G-sets G/H for all $H \subseteq G$ subgroup, and whose morphisms are G- maps of these orbits. With every G-space X one associates an O_G^{op}-space X^O whose value on G/H is given by X^H - the space of H-fixed points of X - thought of, for the sake of constructing maps, as the space of G-maps $\{G/H \to X\}$. As usual C^{op} denotes the opposite category to C. Using O_G

several people have constructed a homotopy theory of G- spaces. For example, W. Dwyer and D. Kan have constructed a model category structure on G-spaces with weak equivalence being a map $f: X \to Y$ such that for all $H \subseteq G$, $f^H: X^H \to Y^H$ is a weak equivalence of spaces. [D-K-1]. See also [El., May].

One of our aims is to construct a similar theory for general diagrams, i.e. for functors from a fixed category D to spaces. Again the main step is the passage from D-spaces to O_D^{op}-spaces for a certain category O_D associated to D.

1.2 D-SPACES , D-MAPS We are interested both in diagrams of topological spaces and of simplicial sets. Let *Top* and *S.* be the categories of topological spaces and simplicial sets. Let D be a small (simplicial) category. A (topological-) D-space is a functor $D \to S.$ (or $D \to Top$). A D-map $X \to Y$ between two D-spaces in an assignment of a map $X(d) \to Y(d)$ to each $d \in obj\, D$ such that for any two objects $d,e \in obj\, D$ the composition $D(e,d) \times X(e) \to X(d) \xrightarrow{f(d)} Y(d)$ is equal to the composition $D(e,d) \times X(e) \xrightarrow{1 \times f(e)} D(e,d) \times Y(e) \to Y(d)$. This definition holds in either S^D or Top^D whether D is a simplicial or a topological category. For any two D-spaces X,Y let $hom_D(X,Y)$ denote the space of all D-maps $f: X \to Y$ (namely all natural transformation between these functors). Simplicially in dimension n of $hom_D(X,Y)$ we have all the maps $X \times \Delta[n] \to Y$ where $\Delta[n]$ is the constant diagram with the standard n-simplex in each place and $-\times-$ is the objectwise *product* of two D-space; where maps are taken on each factor separately. Topologically, the space $hom_D(X,Y)$ is any construction that is adjoint to the product or satisfies the exponential law:

$$hom_D(X \times C, Z) = hom_D(X, hom(C,Z))$$

$$hom_D(C \times X, Z) = hom(C, hom_D(X,Z))$$

for X,Z any D-spaces and C any topological space. For example if *Top* is the Steenrod category of compactly generated Hausdorff spaces, the corresponding $hom_D(X,Y)$ will do. [MacLane]. Therefore in $S.^D$ and Top^D one has products, limits inverse and direct and hom spaces. Notice that one can naturally define an *internal* hom. Namely for any two D-spaces X,Y a D-space $hom(X,Y)$ (without subscript D), whose inverse limit is $hom_D(X,Y)$: We shall not use this construction in any important way - see (2.17) below for its definition. One also has the *D-point:* Namely the D-space denoted by (pt.) whose value on each $d \in obj\, D$ is the one-point-space. It will turn out that there is a canonical homeomorphism (isomorphism): $hom(pt.,X) \approx X$. The D-point *pt.* also serves to define the inverse limit of a D- space: This is the space of "all the D-points in X" $\lim_D X = \lim_{\leftarrow D} X = hom_D(pt.,X)$. For example, for $D = G$ a discrete group

$\lim_{\leftarrow G} X$ is the space of all the G-fixed points in X namely $hom_G(G/G, X) = X^G$. Dually, one considers the direct limit of X over D: $dirlim_D X = colim_D X = \lim_{\rightarrow D} X$ to be the equivalence classes in $\amalg_d X(d)$ under the equivalence relation: $x \sim f(x)$ for all $x \in X(d)$, $f: d \to d'$ in $mor\ D$. One has canonical maps $\lim_{\leftarrow D} X \to X$ and $X \to \lim_{\rightarrow D} X$.

1.3 D-homotopy A *homotopy* between two D-spaces is a map $H: X \times I \to Y$ where I denotes the constant D-space $I(d) = [0,1]$ = the unit interval. A homotopy equivalence $f: X \to Y$ is a map with a (two-sided) D- homotopy inverse. It is not hard to check by naturality that a homotopy $H: X \times I \to Y$ gives rise to homotopies of spaces:

$$(\lim_{\leftarrow D} X) \times I \to \lim_{\leftarrow} Y$$

$$(\lim_{\rightarrow D} X) \times I \to \lim_{\rightarrow} Y$$

$$hom_D(Z, X) \times I \to hom_D(Z, Y)$$

for all D-spaces Z.

Therefore a D-homotopy equivalence induces a usual homotopy equivalence on limits, colimits and function spaces.

1.4 Let us show that the 3-fold map given in (1.1) above is not J-equivalent to the J-point $* \to *$. To do that let P_2 be the J-space $\{0,1\} \to \{0\}$. By a direct inspection one gets $hom_J(P_2, F_3) \simeq S^1 \vee S^1$ where F_3 denote the 3-fold map in (1.1) $F_3: J \to Top$, and S^1 denotes the circle. Since it is clear that $hom_J(P_2, pt) = *$ we get the non-equivalence from the above homotopy invariance of D-function complexes.

1.5 D-orbits. We now introduce the central concept for D-homotopy theory - that of a D-orbit: A D-orbit is a D-space $T: D \to$ (spaces) whose direct limit over D is the one point space: $\lim_{\rightarrow D} T = \{*\}$. The collection of D-orbits play a strictly analogous role in D-homotopy theory to the collection of G- spaces $\{G/H\}$ in G-equivariant homotopy theory, for $H \subseteq G$ a subgroup of G.

For example: Just as any G-set (a set with some G-action) is a disjoint union of G-orbits so any D-set - a functor $D \to$ (sets)- breaks up naturally as a disjoint union of its D-orbits: We have in a D-set $S: D \to$ (sets) exactly one D-orbit over each point in $\lim_{\to D} S$: One can extract this D-orbit as a pull back S_x, in the category of D-sets, over $x \in \lim_{\to D} X$:

$$\begin{array}{ccc} S_x & \longrightarrow & S \\ \downarrow & pull & \downarrow \\ * & \xrightarrow{x} & \lim_{\to} S \end{array}$$

If x_d is any element of $S(d)$ for some $d \in Obj\ D$, there is a corresponding D-orbit - the orbit of x_d - it is the D- orbit over the image of x_d in $\lim_{\to} S$. Roughly speaking the orbit of some point x_d is the subdiagram of S that contain all the points in $S(d')$, for each $d' \in obj\ D$, that can be connected by maps in S to the given point x_d.

1.6 Definition . The *orbit category* O_D *of* D *is the full simplicial subcategory of all* D*-orbits:* $O_D \subseteq S.^D(Top^D)$.

1.7 Remark. This definition is inconsistent with the usual notation of O_G- that stands for the *set* of G-orbits $\{G/H\}$ one for each $H \subseteq G$, and not for all G- sets S with $S/G \equiv colim_G S = \{*\}$. Our O_G contains a large class of orbits $\{T_H\}$, all of which are isomorphic to G/H, for each $H \subseteq G$. This abuse of notation makes no difference in the following presentation. The problem is that for a general category D the category O_D is *essentially* large, it contains a large subcategory of mutually non-isomorphic members: Recall the category $J = (\cdot \to \cdot)$ from (1.1) above. A J- orbit is just a map $X \to \{*\}$ of any space to the one point space: (since $\lim_{\to J} (X \to X') = X'$). Therefore there are as many non-equivalent J-orbits as there are non- equivalent spaces. For each space in $S.$ (or Top) one gets one J-orbit in $S.^D$ (or Top^D). To get a good hold on the D- homotopy types we consider in (1.6) the full *simplicial* subcategory O_D of $S.^D$ (or Top^D). In O_D the set of morphism $hom(T,S)$ is a simplicial set and composition map is a simplicial map $hom(T,S) \times hom(S,V) \to hom(T,V)$. Notice that O_D is not, strictly speaking, a category but rather a category enriched over S. [MacLane]: One can compose maps only in the same dimension. Thus a typical map in O_D is a simplicial map $T \times \Delta[n] \to S \times \Delta[n]$ over $\Delta[n]$. The identities are given by simplicial maps of the one point space into $hom(S,S)$ for each object S in the simplicial category. For example, for $D = J$ as above O_J is isomorphic to $S.$ (or Top) the simplicial category of simplicial sets (or of topological spaces).

1.8 EXAMPLES OF DIAGRAMS. Here we list some examples of D-spaces. Some of them simplicial others topological.

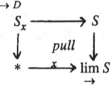

1.9 Action of simplicial categories and simplicial groups. Let D be a small simplicial category, then for each $d, d' \in obj\ D$ the set $hom(d,d')$ has a simplicial structure. A diagram over D assigns a simplicial set $X(d)$ to each $d \in obj\ D$ and a simplicial map $hom(d,d') \times X(d) \to X(d')$ for each $d, d' \in obj\ D$ that respects composition in D in the obvious sense. In particular a simplicial group action on a space X is given by a simplicial map $G \times X \to X$ where G is some simplicial group. A discrete group action is a special case.

1.10. Let G be a discrete group and let O be a set of G- orbits. Then O can be thought of as a small category with G-maps as morphisms. With each G-space X one associates an O^{op}-space X^O given as above by $X^O(T) = hom_G(T,X)$ where T is a typical element of O.

1.11. Let K be a (geometric) simplicial complex. Denote by the same letter K the category whose objects are the simplices of K and whose morphisms are face inclusions of these simplicies. The nerve of K is just the subdivision of the complex K. Given a continuous map of some topological space X to K namely $f: X \to K$ we associate to f a diagram $\underset{\sim}{X}$ over the category K:

$\underset{\sim}{X}(\sigma) = f^{-1}(\sigma) \subseteq X$ for $\sigma \in K$. For each inclusion $\sigma < \tau$ one gets a map $\underset{\sim}{X}(\sigma) \subseteq \underset{\sim}{X}(\tau)$. The study of spaces over K is very close to the study of K-spaces - namely of diagrams with the shape of the category K. Notice that the colimit of $\underset{\sim}{X}$ over the category K namely $\lim_{\to K} \underset{\sim}{X}$ is just X.

1.12. Let $\ldots\ldots X_{s+1} \to X_s \to \ldots . X_0$ be an infinite tower of spaces. It is a diagram over the obvious small category. An interesting orbit over this category is given by $X_s = \{*\}$ for all s. This is the *point* for the diagram and $hom(pt, \{X_s\}_s) = \lim_{\leftarrow} X_s$ in the usual sense.

1.13. Let F be the category of the finite pointed sets $\{*, 1, 2, 3, \cdots n\}_{n \geq 0}$ with set maps between them, an F-space is a collection of pointed spaces X_n $(n \geq 1)$ and maps $X_n \overset{\Leftarrow}{\Rightarrow} X_m$ between them. This structure plays a central role in Segal's theory of infinitely commutative H-spaces.

1.14. Let $X = \bigcup_A X_\alpha$ be any decomposition of a space X into subspaces. There is a partial ordering of A by the inclusions $X_\alpha \subseteq X_\beta$ for $\alpha, \beta \in A$. This turns A into a small category. Then the assignment $\alpha \mapsto X_\alpha$ gives an A-space $\{X_\alpha\}$. We have $X = \lim_{\to A} \{X_\alpha\}$.

Problem: When is the A-space $\{X_\alpha\}$ A-homotopy equivalent to the A-space $\{Y_\alpha\}$ with $Y_\alpha = \{*\}$ for all $\alpha \in A$. In other words when is there a contraction h of X that restricts to a contraction h_α for each $X_\alpha \subseteq X$? The theory will imply that such an h exists if and only if for each subset $B \subseteq A$ the space $\bigcap_B X_\alpha$ is contractible - with *no* relations assumed amongst the various contractions.

1.15. Let G be a group. Consider a lattice $L = \{H_\alpha\}$ of subgroups $H_\alpha \subseteq G$. This lattice defines a small category with morphism being the inclusion maps $H_\alpha \subseteq H_\beta$. A typical L-space is the diagram of classifying spaces $\{B\, H_\alpha\}_\alpha$.

1.16. D-CW-complexes In order to be able to invert D-maps up to D-homotopy we shall need to restrict attention to certain D–$C.W.$ complexes. They are built by glueing certain D-cells in much the same way topological $C.W.$ complex are constructed. Compare [D-Z].

A D-cell is a D-space of the form $T \times e^n$ where T is a D-orbit (1.5) and e^n is the standard n-cell. An attaching map of this D-cell to some D-space X is a map $T \times \partial e^n \to X$. (Simplicially we take $T \times \Delta[n]$ where T is a simplicial D-space with $\lim_{\to D} T = \{*\}$.)

A D–$C.W.$-complex is a D-space X with the weak topology with respect to a filtration $\cdots X_n \subseteq X_{n+1} \subseteq \cdots X$, such that X_{n+1} is gotten from X_n by attaching a set of n-dimensional D-cells. Namely one has a push out diagram of D-spaces:

$$\begin{array}{ccc} \coprod_T (T \times \partial e^n) & \to & \coprod_T (T \times e^n) \\ \downarrow & & \downarrow \\ X_{n-1} & \to & X_n \end{array} \qquad \text{where } X_{-1} = \emptyset \text{ and } n \geq -1.$$

Relative D-C.W.-complex are defined in the same way: The only change from the absolute case is that we do not assume $X_{-1} = \emptyset$. In that case the pair (X, X_{-1}) is a relative D–$C.W.$-complex.

As might be expected there are plenty of D–$C.W.$-complexes around: The realization of any diagram of simplicial sets is a D–$C.W.$-complex and in particular any diagram of simplicial complexes and simplicial maps between them gives a D–$C.W.$-complex in a natural way after applying the barycentric subdivision functor to all the complexes in the diagram:

1.17 Theorem: *Let X be a D-space: $X : D \to S_.$. The geometric realization of X, namely $|X|$, gotten by applying $|\ |$ to each place in the diagram X, has the structure of a D–$C.W.$-complex.*

Proof (See [D-Z]). We will not give here the proof. The basic idea is to notice that the geometric realization of any simplicial map $\Delta[m] \to \Delta[n]$ is a J–$C.W.$ complex in a natural way where J is the small category $(\cdot \to \cdot)$ from (1.1) above.

1.18. In the next section we shall discuss free D-spaces. They all turn out to be D–$C.W.$-spaces (2.5).

1.19. An important open problem is to determine whether any $X \in S.^D$ is a retract of a diagram $W \in S.^D$ with a D–$C.W.$-structure.

1.20. Weak equivalences: We will use two different concepts of weak equivalence: A map $f : X \to Y$ between D-spaces is a *local weak equivalence* if for all $d \in D$, $f(d) : X(d) \to Y(d)$ is a weak equivalence of spaces. The map f will be called a weak equivalence if for every D-C.W.-complex K, $hom(K, f)$ is a weak equivalence of spaces. The interplay between these two concepts is one of the main themes of the present exposition. See e.g. 2.9, 2.14, 3.4.

2. FREE DIAGRAMS AND HOMOTOPY LIMITS

The concept of a free D-set is the natural generalization to small categories of that of a free G-set - a set on which a group G acts freely (i.e. $\forall x \forall g$ $(x_g = x \Rightarrow g = e)$). While there is only *one* (up to an isomorphism) free G-set namely the free orbit $G/\{e\}$, for a more general category D there may be as many free orbits as objects in a small category D. Using these free D-sets one considers the natural generalization ED of the universal space EG. This has the desired uniqueness properties and gives rise to the concept of *homotopy limit* [B-K]. Function complexes $hom_D(F, X)$ with F free have nice properties, therefore we will attempt to "model" any D-space X by a free O-space for a certain O without losing its D-homotopy type.

2.1 Free orbits, free D-spaces. Given a small category D consider [with MacLane p.61] the following D-set $F^d \in Top^D$ associated to any object $d \in obj\, D$, $F^d : D \to Top$ is given by $F^d(d') = hom(d, d')$, and $F^d(d' \to d'')$ is given by the composition. This F^d is the *free D-set generated at d* or the *free D-orbit generated* at D.

If D is a simplicial category then $F^d(d')$ is the *space* $hom(d, d)$ and the action of D on F^d is the composition map $D(e, f) \times F^d(e) \to F^d(f)$ in D. In that case F^d is the free D-space generated at d.

Clearly F^d is a D-orbit (1.3) since there is a map $d \to d'$ connecting $id : d \to d$ to any other element in $F^d(d')$. So $\lim_{\to D} F^d = \{*\}$.

2.2 Lemma. (Fundamental property of free D-orbits) *For any space there is a natural isomorphism* $X(d) \tilde{=} hom_D(F^d, X)$.

Proof: (See [Mac] p. 61). One simply sends $F^d \to X$ to its value on $id \in F^d(d)$, and similarly in higher dimensions for $X \in S.^D$.

A *free* D-set is any D-set that is isomorphic to a disjoint union of free D-orbits. Given any functor $S : D^\delta \to X$, where D^δ is the discrete category associated to D, one can assign to it a natural free D-set $F(S) = \coprod_{d \in S} F^d$.

2.3 Examples. If $D = G$ one recovers the usual concept of a free G-set. In general however, the *product* of two free D-sets is not necessarily free. Consider the small category given by the diagram $(d \overset{\to}{\underset{\to}{\leftarrow}} e)$ and no other non- identity arrow. $F^d \times F^d$ is not free. It is clear from the example of a discrete group G that given a free D- set there is no canonical choice of generators. In the general case even the location of the generators in $Obj\ D$ is not determined. In other words it is easy to construct examples with $F^d \tilde{=} F^e$ but $e \neq d$.

Let J be the category $(\cdot \to \cdot)$ with two objects and one map. A free J-set is simply an inclusion of sets $A \to X$. Let $\Lambda = (\cdot \to \cdot \leftarrow \cdot)$ be the obvious category. A free Λ set is given by set inclusions $A \overset{f}{\to} X \overset{g}{\leftarrow} B$ with non-intersecting images $\mathrm{Im}(f) \cap \mathrm{Im}(g) = \varnothing$. A telescope of inclusions of sets $A_0 \to A_1 \to A_2 \cdots \to .... A_n \to A_{n+1} \to \cdots$ is clearly a " free diagram", (i.e. a free D-set for the obvious category D.) An inverse tower of sets $X_{n+1} \to X_n ..... \to X_0$ is free if and only if all maps are inclusions and its *inverse limit* is empty. The same holds for towers in $S.^D$.

2.4 FREE (topological and simplicial) D-SPACES

A free topological D-space X is one for which the underlying D-set, also denoted X, is free. To define a free D-space $X \in S.^d$ one must be slightly more careful, since we want the skeleta to be free in a consistent fashion:

A free D-diagram of simplicial sets $X \in S.^D$ is a D-space together with explicit sets of generators $Q = \{q_\alpha(n,d)\}_\alpha$ with $q_\alpha(n,d) \in X_n(d)$ such that Q is closed under degenerecies in $X(d)$ and for each n, $\{q_\alpha(n,d)\}_{d,\alpha}$ freely generate

X_n as an D_n-set where D_n is the simplicial category D in dimension n. In other words for each $y \in X_n(d)$ there exist a *unique* $q_\alpha(n,e)$ and a *unique* arrow $\phi \in hom(e,d)_n$ with $\phi(q_\alpha(n,e)) = y$. Compare [D-K-5].

In an appropriate model category structure in the sense of Quillen [Q], the free D-space and their retracts are precisely the *cofibrant spaces.* Compare [D-K-1].

Example: Consider a continuous map $W \to |K|$ to the geometric realization of a simplicial complex K. Let $\underline{W}$ be the diagram subspaces of W over simplices of K given in (1.11). Then $\underline{W}$ is always K-free, as a topological diagram. This follows easily from the usual properties of the simplicial structure of K.

2.5 Proposition *Let $X \in S.^D$ be a free D-space for the simplicial category D. Then X has a D–C.W.-structure.*

Proof: One shows that the D–$C.W.$ filtration is determined by that of $\lim_{\to D} X$ via the pullback

$$\begin{array}{ccc} sk_n X & \longrightarrow & X \\ \downarrow & & \downarrow \\ sk_n \lim_{\to D} X & \longrightarrow & \lim_{\to D} X \end{array}$$

It is therefore sufficient to consider the case $\lim_{\to D} X = \Delta[n]$. One must find $d \in obj\ D$ and a map $\phi : \Delta[n] \times F^d \to X$ so that X is the pushout in the square

$$\begin{array}{ccc} \dot{\Delta}[n] \times F^d & \longrightarrow & \Delta[n] \times F^d \\ \downarrow & & \downarrow \phi \\ sk_{n-1} X & \longrightarrow & X \end{array}.$$

Let T_n be the orbit in X of the canonical $\iota_n \in \Delta[n]$. By assumption this orbit is free so let $p_n \in X_n(d)$ be its generator. To define ϕ let $\sigma\delta\iota_n = y_m$ the canonical representation of an element $y_m \in \Delta[n]_m$ and $f_m = f_m(d,e) \in F^d_m(e) = D_m(d,e)$ be any two elements. Define $\phi(y_m, f_m) := X(f_m)(\sigma\delta p_n) \in X_m(e)$, where $X(f_m)$ is the function $X_m(d) \to X_m(e)$ that corresponds in X to f_m. One must show that ϕ is one-to-one onto away from the $(n-1)$ skeleton which means $\delta = id$ i.e y_m is some degeneracy of $\iota_n \in \Delta[n]$. This follows easily from the closure under degeneracies of the generating sets for X: which implies that the degeneracies σp_n generate the corresponding orbits.

2.6 Resolution by a free D-space: Given any D-space $X : D \to Top$ one can"resolve" it by a free space via a "weak equivalence" in the sense of 1.20: Namely one can find a free D-space $\bar{X}$ and a map $\bar{X} \to X$ such that for all $d \in obj\ D$ the map $\bar{X}(d) \to X(d)$ is a weak homotopy equivalence of spaces in the usual sense. Clearly this map will not, in general, be a D-homotopy equivalence. In the case $D = G$ a group the projection $EG \times X \to X$ is a possible resolution. As a result of 2.3 above one may not in general take $ED \times X \to X$, since $ED \times X$ may fail to be D-free. A general construction is given in (3.12) below.

2.7 The free diagram associated to a group action on a space. The most important diagram in the theory of G-spaces is that of the fixed point subspaces: Let S be a G-set. Recall the O_G^{op}-set S^O of fixed points (1.10 above).

Claim *For any G-set S the diagram S^O is O_G^{op}-free.*

Proof: We must show that the orbit of any point over $\lim_{\to O_G^{op}} X^O$ is O_G^{op}-free. In other words we must find a generator for such a O_G^{op}-orbit. (See 2.1 above). First notice that there is a natural isomorphism of sets

$$\lim_{\to O_G^{op}} S^O \cong \lim_{\to G} S$$

This follows from the fact that for any $G/H \in O_G^{op}$ there is a map $S^O(G/H) \equiv S^H \to S$, the canonical inclusion. Notice also that the self maps of $S^O(G/e) = S^{\{e\}} = S$ in S^O are given exactly by the action of G and S. Now given $\bar{x} \in \lim_{\to O_G^{op}} S^O$ consider it as an element of $\lim_{\to G} S = S/G$. Let $\bar{S} \subseteq S$ be the G-orbit lying over $\bar{x} \in S/G$. Now choose a point $[g] \in \bar{S}$, a subgroup $H \subseteq G$ and a G-isomorphism $G/H \cong \bar{S} \to S$. This gives us a point $x \in S^H = S^O(G/H)$. It is easy to see that x freely generates the orbit $T_{\bar{x}}$ over $\bar{x}$. Namely $T_{\bar{x}}(S/K) \cong hom_G(G/K \to G/H)$. This is so because every map $f : G/K \to S$ with $\lim f = * \to S/G$ being $\bar{x}$, factors *uniquely* through our map $\bar{S} \to S$, so also through $G/H \to S$. This completes the proof of the claim.

If a group G acts on a topological space X, the topological O_G^{op}-space X^O will be free too since this property depends only on the underlying set. Similarly it is not hard to check that if X is a simplicial set in $S.$ on which a simplicial G acts, then the O_G^{op}-space X^O is free in the sense of (2.4) above.

2.8 Homotopy equivalences between free D-spaces: The basic property of free diagrams of spaces is that they behave nicely when mapped to other diagrams. The following are immediate consequences of results that are proven later on (see 3.4 below). Recall (1.20) that a *local weak equivalence* is a map f with $f(d)$ weak equivalence for all d.

2.9 Proposition: $f : X \to Y$ *be a map of topological* D*-spaces. Let* F *be a free* D*–C.W.-complex (1.16). Assume* f *is a local weak equivalence. Then the induced map on function spaces* $hom(F,f) : hom(F,X) \to hom(F,Y)$ *is a weak homotopy equivalence of spaces. Therefore a local weak equivalence* f *between two free D-C.W. complexes has a* D*-homotopy inverse* $g : Y \to X$.

Proof: This follows from 3.4 below. Take O, the orbit category, to be the small category of all the free D-orbits. This category is isomorphic to D^{op}. Now since $hom(F^d, X) = X(d)$ for any D-space X, our assumptions imply that for any free D-orbit $F^d \in O$ the map f induces a weak equivalence $hom_D(F^d, X) \to hom_D(F^d, Y)$. Now since by assumption on F the orbits in F belong to O, the conclusion of 3.4 applies.

2.10 Remark: To formulate 2.9 for diagrams of simplicial sets one must assume that X, Y are *fibrant* i.e. $X(d), Y(d)$ are Kan complexes for each $d \in obj\ D$. On the other hand by 2.5 any free F has a D–$C.W.$ structure. The virtue of 2.9 is that we do not assume f to be a D-homotopy equivalence. Thus as we saw in (1.1) and (1.4) for $D = J = \{\cdot \to \cdot\}$ the fact that for all $d \in obj\ D$ the map $f(d)$ is a homotopy equivalence, does not imply that $hom(W, f)$, for a general D–$C.W.$ complex W, will be a weak equivalence of spaces. Only mapping in a *free* D-space converts a *local* D-equivalence into a (weak) homotopy equivalence.

2.11 Locally contractible free D-spaces: Of special importance are free D-spaces E with contractible $E(d) \simeq *$ for each $d \in Obj\ D$. These spaces of course are the exact analogs in the diagram case to contractible spaces on which a group G acts freely. Just as in the group case all D–$C.W.$ complexes W that are free and locally contractible (each $W(d) \simeq *$) are D-homotopy equivalence to each other (Assuming they are made of Kan complexes in $S.^D$) The classifying space of D denoted by BD is nothing but the direct limit $BD \cong \lim_{\to D} W$. One can define a canonical free locally contractible D-space (or D–$C.W.$ complex) using (2.6) above. Namely, let $pt : D \to S.$ (or $D \to Top$) be the D-point with $pt(d) = \{*\}$. Then ED is the canonical free resolution (2.14) above $E = \overline{pt} \to pt$. In this case however [B-K, XI] gives us an explicit example for the space $ED(d) = D/d$ where D/d is the "over category" of objects in D over $d \in obj\ D$. It is not hard to check the resulting ED, denoted by $D/-$ is a free locally contractible D–$C.W.$ complex. It is not hard to construct for any two locally contractible D-spaces E', E'' a relative D-C.W.-complex $(C, E' \amalg E'')$ with C also locally contractible. Therefore any two such E', E'' are locally weakly equivalent in the sense of (1.20) via maps $E' \to C \leftarrow E''$.

2.12 Homotopy limit and co-limit of a diagram X is by definition the function space $\underset{\rightarrow D}{holim}\ X = hom_D(ED, X)$. If E is any free locally contractible D-space, and $X(d)$ is a Kan complex for all $d \in obj\ D$, then there is a homotopy equivalence $hom_D(ED, X) \simeq hom_D(E, X)$.

2.13 Example : Let C be any contractible space with at least two *distinct* points: $x_0 \to C \leftarrow x_1$. Then the homotopy pull back of a $X_0 \to X \leftarrow X_1$ is equivalent to $hom_\Lambda(x_0 \to C \leftarrow x_1, X_0 \to X \leftarrow X_1)$ since the domain is a free Λ-space (2.3).

The general proposition 2.9 has the following useful consequences:

2.14 Proposition: *Let* $f : X \to Y$ *be a map of* D*-spaces (in* $S.^D$ *assume* $X(\mathrm{d}), Y(\mathrm{d})$ *are Kan). If* f *is a local (weak) equivalence i. e.* $f(\mathrm{d})$ *is a (weak) equivalence, then the map* $\underset{\leftarrow}{holim}_D\, f : \underset{\leftarrow}{holim}_D X \to \underset{\leftarrow}{holim}_D Y$ *is a (weak) equivalence.*

Thus $\underset{\leftarrow}{holim}$ "converts a local equivalence into a homotopy equivalence".

2.15 Example: If X is a G-space for a group G then $holim_G X = hom_G(EG, X)$ is usually refered to as the homotopy fixed points of X. An interesting conjecture of D. Sullivan (1971), relates the homology of $hom_G(EG, X)$ to that of the space of actual fixed points X^G, for a *finite* complex X, and a finite group G. This conjecture has been proven by H. Miller [M] in many interesting cases. In many other situations the relation between the homotopy limit and the limit, namely properties of the map: $\underset{\leftarrow D}{\lim}\ X = hom_D(pt, X) \to hom_D(ED, X)$ is of great interest.

2.16 Homotopy direct limit: We give a somewhat novel definition for this concept originated in [B-K]: We use a strictly adjoint construction to the definition of the homotopy inverse limit via $hom_D(-,-)$. Namely let ED^{op} be the free contractible D^{op}-space. Then $\underset{\rightarrow}{holim}\, X$ is defined to be

$$\underset{\rightarrow}{holim}\, X = X \times_D ED^{op}.$$

Here $(- \times_D -)$ is the product of a D-space with a D^{op}- space (compare [Watts]). It is a dual to hom: For any space $A \in Top$ (or $S.$) one has the homeomorphism (isomorphism).

$$hom(X \times_D Y, A) = hom_D(X, hom(Y, A)).$$

The product $(X \times_D Y)$ can be defined as a push-out or equalizer of the two obvious maps (for a simplificial D this is a dimensionwise formula)

$$\coprod_{(d \to e) \in mor\, D} X(d) \times Y(e) \rightrightarrows \coprod_{d \in obj\, D} X(d) \times Y(d).$$

For example, the product of two pointed spaces taken as diagrams over $J = (\cdot \to \cdot)$ is the wedge product

$$(* \to X) \times_J (Y \leftarrow *) = X \vee Y.$$

The product of $(X \to *)$ with $(* \leftarrow Y)$ is the one point space. It is not hard to see that $\underset{\to}{holim}\, X$ is canonically equivalent to $\lim_{\to D} \bar{X}$ from 2.6.

2.17 Internal function complex: For the sake of completeness we show how to define a dual to the product of two D-spaces $X \times Y \in Top^D$ (or $S.^D$). This is the internal function complex that gives for any two D-spaces X, Y a D-space $hom(X,Y)$. In the case $D = G$ a group like this gives the usual G-action on the space of *all* maps from X to Y. The definition uses the Yoneda lemma (2.2) above. Namely we want to identify $hom(X,Y)(d)$ for $d \in obj\, D$ but by (2.2) this is canonically homeomorphic (isomorphic) to $hom_D(F^d, hom(X,Y))$. By adjuction we get the definition: $hom(X,Y)(d) = hom_D(F^d \times X, Y)$. The value of $hom(X,Y)$ on maps $d \to e$ is given by composing with $F^e \to F^d$. For example it is immediate from the definition that $hom(pt, X) = X$. It is also clear that $\lim_{\leftarrow D} hom(X,Y) = hom_D(X,Y)$.

3. SPACES OF 'FIXED POINTS' FOR DIAGRAMS OF SPACES

In the previous section we emphasized the good properties of the free D-spaces. We mentioned that any D-space X can be resolved by a free D- space via $r : \bar{X} \to X$. However, this map is not a D-equivalence unless X itself is D-free. Thus by this passage from X to $\bar{X}$ the D-homotopy type of X is lost and cannot be recovered: Notice that if $f : X \to Y$ is a local D-equivalence $(f_d : X(d) \to Y(d))$ then $\bar{X} \xrightarrow{\sim} \bar{Y}$ is clearly a D-homotopy equivalence, by 2.9, so there is no going back from $\bar{X}$ to X.

On the other hand we saw in (2.7) that one can associate with every G- space X a free O_G^{op}-space X^O from which the G- homotopy type of X can be recovered. It turns out that *this construction generalizes to an arbitrary small category D.* So the original D-homotopy type of X can always be recovered by a functorial construction from that of its free version X^O. In this section we discuss the construction of X^O out of X and vice versa. In order to get some insight the situation we begin with a generalization of an important result of Bredon's to arbitrary small categories [D-Z]. Recall that Bredon proved in [Br] the following fundamental result.

3.1 Theorem: *Let G be a discrete group and $f: X \to Y$ a map between G–C.W.- complexes. Assume that for any $H \subseteq G$, the induced map on the fixed points spaces: $f^H : X^H \to Y^H$ is a homotopy equivalence. Then f has a G-homotopy inverse $g: Y \to X$, i.e. f is a G- homotopy equivalence.*

This theorem illustrates the importance of the diagram of fixed points sets in G-equivariant homotopy theory. To generalize 3.1 to D–C.W. complex for an arbitrary small category D we must construct an analog of the " fixed point space". This is the "space of orbit points".

3.2 Definition : *Let X be a D-space and T be a D-orbit, the space of T-orbit points of X is the function complex $X^T = hom_D(T, X)$.*

3.3 Remark: Definition (3.2) can be motivated by recalling that for a G-space Y, there is a canonical isomorphism (homeomorphism) $hom(G/H, Y) \xrightarrow{\cong} Y^H$. Furthermore, a D–C.W.-complex is built in (1.16) out of D-orbits in complete analogy to the assembly of a G- complex out of the usual G-orbits. Recall that a D-space X is called of type O for an D-orbit category O, if every orbit of X is represented (up to D-homotopy equivalence) in O.

We are now ready to state and indicate the proof of:

3.4 Theorem: *Let $f: X \to Y$ be a map of D–C.W. spaces of type O for some category of D-orbits O. Assume that $X(d)$, $Y(d)$ are Kan spaces and that for all $T \in O$ $f^T : hom_D(T, X) \to hom_D(T, Y)$ is a homotopy equivalence then f has a D-homotopy inverse $g: Y \to X$.*

Remark: The same theorem holds both for diagrams of simplicial sets and for diagrams of topological spaces. In the latter case one must be careful to use a topological category and $hom_D(-,-)$ construction for which the exponential law holds. In both cases the argument is by rather formal duality of homotopy push-outs and homotopy pull-backs: A more substantial difficulty not resolved here (compare 1.19) is what kind of D-spaces are D–C.W.-complexes or their retracts in $S.^D$.

Proof of 3.4. One proceeds by induction on the skelata filtration of W:

$$\begin{array}{ccc} i : T \times \dot{\Delta}[n+1] & \rightarrow & T \times \Delta[n+1] \\ \phi_{n+1} \downarrow & & \downarrow \\ W_n & \longrightarrow & W_{n+1}. \end{array}$$

When mapping this push-out square into X and Y, one uses the exponential law to get a fibre map out of i, namely

$$\begin{array}{c} hom(\Delta[n+1], \quad hom_D(T,X)) \\ \downarrow \text{restriction} \\ hom(\dot{\Delta}[n+1], \quad hom(T,X)) \end{array}$$

Since by induction $hom_D(W_n,X) \rightarrow hom_D(W_n,Y)$ is a weak homotopy equivalence, and the desired map f^T is the strict pull back over the fibre map above into X and Y one uses the exponential law to observe that the restriction map r in the diagram below is a fibre map.

$$\begin{array}{ccccccc} hom_D(W_{n+1},X) & & & \longrightarrow & & & hom_D(T\times\Delta[n+1],X) \\ \downarrow & \searrow & & & & \swarrow & \downarrow \\ & & hom_D(W_{n+1},Y) & \rightarrow & hom_D(T\times\Delta[n+1],Y) & & \\ & & \downarrow & & \downarrow r & & \\ & & hom_D(W_n,Y) & \rightarrow & hom_D(T\times\dot{\Delta}[n+1],Y) & & \\ \downarrow & \nearrow & & & & \nwarrow & \downarrow \\ hom_D(W_n,X) & & & \longrightarrow & & & hom_D(T\times\dot{\Delta}[n+1],X) \end{array}$$

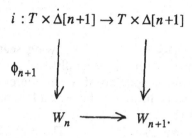

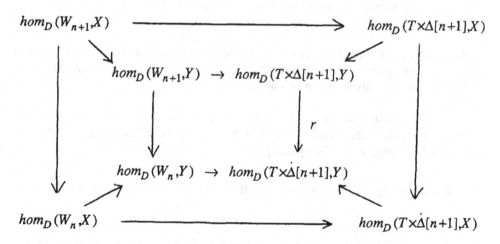

Thus each square is both a pull-back and a homotopy pull back. Th proof can now be completed by observing that by induction the lower left corner map is a weak equivalence while by assumption the maps on the right taken by exponential law to be, say, $hom_D(T,hom(\Delta[n+1],X)) \tilde{\rightarrow} hom_D(T,hom(\Delta[n+1],Y)$ are weak homotopy equivalences. This is so since T has the same D-homotopy type as some orbit in O and for all these orbits we assume that the said map is a (weak)

equivalence).

3.5 Free O-resolution of a D-space In the following sections we discuss the assembly of "fixed points data" or more generally "orbit point data" into a D-space. This is a straightforward adaptation of a procedure described by [D-K-3 ,Elmendorf]. In both cases the realization is built as a homotopy direct limit over a diagram containing $T \times F(T)$ and $T \times F(T')$ for $T \in O_G$ and maps coming from maps in the simplicial category $O \subseteq O_G$. There results a G-space whose only G-orbits are those of O_G and whose fixed point diagram is the given one F up to a local weak equivalence. Similarly for a small category D we are given a simplicial category of D-orbits, O, namely a full subcategory of O_D and orbit-points data in the form of a functor $F : O^{op} \to$ (spaces). We want to "realize" F as the actual diagram of orbit points of some D-space $|F|_D = |F|$. Namely, we want a functor $|\ |$ and a natural transformation $F \to |F|^O$ so that in favorable cases $F(T) \to hom_D(T, |F|)$ is a weak equivalence of spaces. To do that we first establish in (3.7) that *free O^{op}-spaces* arise as the orbit-points diagram X^O of a D-space X provided the small orbit category O is large enough.

3.6 Free O-resolution: To every D-space X and every small full simplicial subcategory O of the simplicial category O_D (1.6), we associate a canonical O^{op}-space $X^O : O^{op} \to$ (spaces) using (3.2) above: $X^O(T) = X^T$ as a simplicial set. The composition map in each simplicial dimension of O namely

$$hom(S,T)_n \times X_n^T \to X_n^S$$

is defined by the obvious composition. In the same manner we deal with diagrams of topological spaces. We now note that if all the orbits of X appear up to isomorphism (or homeomorphism) in O then X^O is O-free:

3.7 Lemma *Assume $X : D \to Top$ is of type O for some orbit category O over D, then the associated O^{op}-space $X^O(T) = hom_D(T,X)$ is O^{op}-free. The same holds for diagrams of simplicial sets.*

Before giving the proof let us illustrate the lemma by a simple example.

3.8 Example : Consider the Λ-space of inclusions $A \xrightarrow{f} X \xleftarrow{g} B$. If $Imf \cap Img \neq \varnothing$ then this is not a free Λ-space. To make it free we notice that in $(X;A,B)$ there are in general four types of Λ-orbits (for inclusion maps) $L_{00} = \varnothing \to * \leftarrow \varnothing, L_{10} = * \to * \leftarrow \varnothing, L_{01} = \varnothing \to * \leftarrow *, L_{11} = * \to * \leftarrow *$.

The first three are the three free Λ-orbits and the forth represent an intersection point in $Imf \cap Img = A \cap B = hom(L_{11}, A \to X \leftarrow B)$. Notice that L_{11} is *not* free. Now all such diagrams of inclusions are Λ-spaces of type O^{op} where O

is the category

$$\begin{array}{ccc} L_{00} & \rightarrow & L_{01} \\ \downarrow & & \downarrow \\ L_{10} & \rightarrow & L_{11} \end{array}$$

The associated O^{op}-space to $(X:A,B)$ is the diagram

$$\begin{array}{ccc} A\cap B & \rightarrow & B \\ \downarrow & & \downarrow \\ A & \longrightarrow & X \end{array}$$

This last diagram is clearly free over O^{op}.

This is a special case of the general phenomenon: whenever one has any diagram of *inclusions*, it is *freed* by adding to it all the possible intersections - thought of as "partial inverse limits" - to get a free diagram. From its O^{op}-homotopy type the D-type of the original diagram can be recovered. Further examples are given after the proof.

Proof: We first notice that if X itself is a D-orbit in O^{op} then X^O is free: In that case for any orbit $T \in obj\ O^{op}$, $X^O(T) = hom(T,X) = O^{op}(X,T) = F^X(T)$ (see 2.1) so that X^O is precisely the free O^{op}-orbit at X. (The same argument holds if D is a simplicial category and X a D-orbit in $S.^D$). For a general D-space X of type O we must show that the natural decomposition of X^O into O^{op}-orbits over $\lim_{\to O^{op}} X^O$ is a decomposition into free orbits. Since there is clearly a canonical isomorphism $\lim_{\to D} X = \lim_{\to O^{op}} X^O$ the following claim whose verification is straigthforward makes sense. Let s denote an element in either one of these direct limits, let Y_s denote the orbit over s, then for a D-space of type O one has the equality $(X_s)^O = (X^O)_s$ (again X may be in $S.^d$ or Top^D). Since this equality holds for every s and the left hand side is free by the above, while the right hand side is an arbitrary orbit in X^O, the equality proves that all orbits are free (on canonical generators X_s).

Remark: From the proof it is clear why we need O to be full subcategory: otherwise X_s^O may be not *free*.

3.9 Example : As another example, now with non-inclusion maps, consider the J-diagram $f: X \to B$ where $\#f^{-1}(b) \leq 2$ i.e. f is at most 2-to-1. Then $X \to B$ has at most three orbits types: $P_0 : \varnothing \to *, P_1 : * \to *$, $P_2 : \{*,*\} \to \{*\}$. Now if f is not 1–1 it is not free since P_2 is not a free orbit. Notice that $hom_J(P_2,f) = X \times_B X$ and therefore the free O^{op}-diagram associated to f over the category $\{P_0,P_1,P_2\} = 0$ has the form:

$$X \times_B X \overset{\rightarrow}{\underset{\rightarrow}{\leftarrow}} X \rightarrow B.$$

Again this example generalizes: If $g : X \rightarrow B$ is any finite-to-one map then we take the orbit category O to be all maps of a set of n- element $\{1,2,\ldots,n\} \rightarrow \{O\}$. The associated diagram over the category O^{op} will be powers of X over B, indexed by the *small category of all finite sets* since $hom_J(P_n,P_k)$ = set of all maps from $\{1,\ldots,n\}$ to $\{1,\ldots,k\}$. Namely:

$$\cdots \overset{\rightarrow}{\underset{\rightarrow}{\leftarrow}} X_B^k \ldots \equiv X_B^3 \overset{\rightarrow}{\underset{\rightarrow}{\leftarrow}} X_B^2 \overset{\rightarrow}{\underset{\rightarrow}{\leftarrow}} X \rightarrow B.$$

This last diagram is free over the category $O^{op} = \{\{1,2,\ldots,n\}\}_{n \geq 0}$ of all finite sets. We see that the category of finite sets plays in the theory of finite-to-one maps the same role played by the category G-orbits $\{G/H\}$ for the homotopy theory of G-space. We now turn to the opposite process: Starting with a diagram over the category of finite sets the problem is to find a map whose powers (or spaces of orbit-points) are given by this diagram: Namely we want a functor that *realizes* a given free diagram over the opposite category to finite sets as a single, finite-to-one map whose powers are equivalent as a diagram to the given one.

3.10 Realization: We briefly indicate how to "realize" orbit-point data by assembling it into a D-space. This D-space will have the prescribed orbit-point data - namely its associated O^{op}-space will be the given one provided we start with a free O^{op}–$C.W.$-space. The relation between the functor $X \rightarrow X^O$ of orbits-points spaces for $X : D \rightarrow$ spaces and $W \rightarrow |W|$ of realization for $W : O^{op} \rightarrow$ (spaces) is rather like that between the "singular complex" functor on spaces $Sing : Top \rightarrow S.$ and the realization functor $| \ | : S \rightarrow Top$. Namely the latter is a *left adjoint* to the former. For a further explanation on this point see [D-K-3, DF].

Now let O be the full simplicial subcategory of O_D spanned by a set $\{T\}$ of D-orbits.

3.11 Theorem: *There exists a functor* $| \ | = | \ |_D : (O^{op}$*- spaces)* $\rightarrow$ *(D spaces) left adjoint to the spaces of O-orbits functor* $X \rightarrow X^O$ *with natural transformations* $\mu : |(-)^O|_D \rightarrow id$ *and* $\varepsilon : id \rightarrow (|-|_D)^O$ *such that* μ *gives a local weak equivalence of D-spaces (In* $S.^D$ *if one restrict to diagrams of Kan complexes) and* ε *given a local weak equivalence when restricted to free* O*–C.W. complex and their retracts.*

Remark: Roughly speaking the theorem means that if we restrict our attention to D–$C.W.$ complexes of type O then the "strict homotopy theory" with weak equivalence being just D-homotopy equivalence is the same as the "local homotopy theory" for the associated O-spaces where weak equivalence is local weak equivalence, (1.20) in the spirit of [D-K-3,4]. Since the latter theory is much simpler the theorem can be read as reducing "hard" homotopy problems for D-space of type O to "simpler" ones for O-spaces. The price one pays is of course the complexity of the simplicial category O.

Proof: We proceed by noticing that we have here a special case of a general procedure described in category theory as co-end construction [MacLane p. 222]. The realization $|\underline{X}|$ for $\underline{X} : O \to$ (spaces) can be thought of as a tensor product of the cofunctor $\underline{X}^{\#} : O^{op} \times \Delta \to (D\text{-spaces})$ and the inclusion functor $t : O \to (D\text{-spaces})$ denoted by $\underline{X}^{\#} \otimes_O t$. Here Δ denotes the simplicial category of standard maps $\Delta[n] \to \Delta[m]$, $n,m \geq 0$ [MacLane] and $\underline{X}^{\#}$ denotes the co-functor on O that assigns to each $T \in O$ and each $n \geq 0$ the constant D-space $\underline{X}(T) \times \Delta[n]$. In more concrete terms $|\underline{X}|$ is the direct limit of a diagram of D-space in $S.^D$ (or Top^D) that contains for each $T \in O$ and $n \geq 0$ the D-space $T \times \Delta[n] \times \underline{X}(T)$ and for each map $g \in O_n(T,T')$ (where $g : T \times \Delta[n] \to T'$ a D-map) a pair of maps $T \times \underline{X}(T) \xrightarrow{1 \times \underline{X}(g)} T \times \Delta[n] \times X(T') \xrightarrow{g \times 1} T' \times X(T')$. [D-K-3]. Using the standard techniques from [MacLane] [D-K-3] it is not hard to check that $| \;\, |$ has the desired properties.

3.12 Free D-C.W.-resolution of a D-space In order to utilize the realization functor, and for other purposes we would like to resolve an arbitrary D-space X by a free $C.W.$-complex $\bar{X}$ via a local weak equivalence $\bar{X} \to X$: Namely, for all $d \in obj\ D$ the map $\bar{X}(d) \to X(d)$ is a weak equivalence. The construction is the exact extension to D-spaces of the usual $C.W.$- approximation of an arbitrary topological space A by a weak homotopy equivalence $\bar{A} \to A$ with $\bar{A}$ a $C.W.$-complex. This is the usual process of killing the relative homotopy groups of the initial approximation by the empty set $\varnothing \to A$. Here we construct inductively the D–$C.W.$-space $\bar{X}$ out of X using only free D-orbit. Again the price of making $\bar{X}$ a functor in X is that $\bar{X}$ is very large and "wasteful". We start by taking $\bar{X}_0$ to be the disjoint union $\coprod_{F^d \to X} F^d$ over all maps of all the free D-orbits F^d, $d \in obj\ D$. There is a canonical map $p_0 : \bar{X}_0 \to X$. Given $\bar{X}_n$ we glue on $\coprod F^d \times e^{n+1}$ where the union is over all commuting squares

$$\begin{array}{ccc} F^d \times \partial e^{n+1} & \to & e^{n+1} \\ \phi\downarrow & & \psi\downarrow \\ \bar{X}_n & \hookrightarrow & X \end{array} , \quad \bar{X}_{n+1} = \bar{X}_n \cup \frac{\coprod}{(\phi,\psi)} F^d \times \partial e^n .$$

In the limit we get a free D-$C.W.$-space $\bar{X} = \cup \bar{X}_n$. Using the exponential law the desired weak equivalence $hom_D(F^d, \bar{X}) \to hom_D(F^d, X)$ for all $d \in obj\, D$, is immediate.

We now proceed to give a simple application of the realization functor for arbitrary orbit categories:

3.13 *Up to a weak equivalence any space is the (strict) inverse limit of any D-space.* Suppose we are given a tower of maps $Y = \{Y_i\}_i = \cdots Y_{n+1} \to Y_n \to \cdots Y_0$ and a space Y_∞ together with a map $Y_\infty \to \{Y_i\}_i$ into the tower. Suppose one would like to modify Y by a local weak equivalence into another tower $\{X_i\}_i$ such that the inverse limit of $\{X_i\}_i$ is weakly equivalent to the space Y_∞ - that was arbitrarily assigned to $\{Y_i\}_i$. (In a different context this is the problem of realizing a map $A \to X^G \subseteq X$ as the fixed point $A = Y^G \subseteq Y$ of a G-space Y weakly equivalent X.) One can modify $\{Y_i\}$ as follows: Let J_∞ be the indexing category of the tower $\{Y_i\}_i$ and J_∞^+ that of the extended tower $Y_\infty \to \{Y_i\}_i$. One constructs two local weak equivalences of J_∞^+-spaces:

$$\begin{array}{ccccc} \lim\limits_{\leftarrow}\{X_i\}_i & \overset{\varepsilon_\infty}{\leftarrow} & \bar{Y}_\infty & \overset{f_\infty}{\to} & Y_\infty \\ \downarrow & & \downarrow & & \downarrow \\ \{X_i\} & \underset{\varepsilon_i}{\leftarrow} & \{\bar{Y}_i\} & \underset{f_i}{\to} & \{Y_i\} \end{array}$$

in which the tower $\{X_i\}$ is the desired tower in which the preassigned Y_∞ has become the actual inverse limit via weak equivalences. The maps (f_∞, f_i) is the free J_∞^+-resolution of the J_∞^+-space $Y_\infty \to \{Y_i\}_i$ given in 3.12. To get $(\varepsilon_\infty, \varepsilon_i)$ one notices that J_∞^+ is an orbit category for J_∞ containing all the free orbits plus the J_∞-point. Therefore one can realize using 3.11 the free J_∞^+-space in the middle as a J_∞-space $\{X_i\}_i$ and the map $(\varepsilon_\infty, \varepsilon_i)$ is the adjunction given by 3.11, that for free J_∞^+-space is known to be a local weak equivalence of J_∞^+-spaces as needed.

It is plain that this example can be generalized: Any "candidate" Y_∞ for an inverse limit of a D-space $\{Y_d\}$ can be realized as the actual inverse limit of the weakly equivalent diagram $\{Y_d\}$ gotten by taking the free $C.W.$- version of the combined diagram $\{Y_\infty, Y_d\}$ and then restricting to its D-part. By a "candidate" we mean of course a $D \cup \{pt\}$ -space $\{Y_\infty, Y_d\}$.

It may be of some interest to note that no such easy control of the possible *direct limit* of a given diagram is possible. I have no idea what is the condition for such a realization.

4. BREDON HOMOLOGY AND COHOMOLOGY

Two typical questions that arise in D-homotopy theory are that of obstruction theory for extension problems and that of a spectral sequence converging to the homology of the colimit of X over the category D. It turns out Bredon's approach [Br] when combined with [D-K] and the material in §3 above gives rather complete answers to such questions. There is an interesting phenomenon that distinguishes cohomology theories for general D-spaces: In general these theories are not representable by any D-space. (See 5.8 below).

A special case of obstruction theory was considered in [D-K-2]. In some sense it is shown here and in (5.10) below how to reduce the general case to their special one. The following diagram gives the typical extension problem

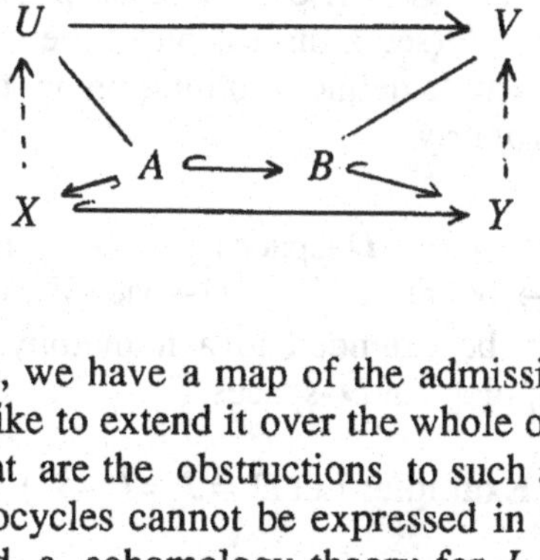

Given the solid arrows diagram of maps, we have a map of the admissible subpair $(A \to B)$ into $(U \to V)$ and would like to extend it over the whole of $X \to Y$ so that the diagram will commute: What are the obstructions to such an extension? It turns out that the obstruction cocycles cannot be expressed in terms of classical homology of pairs. We need a cohomology theory for J-diagrams that is defined for the J- space: $X \cup CA \to Y \cup CB$ with coefficient in the map $\pi_k U \to \pi_k V$. It turns out that the cohomology of any pair $X \to Y$ with coefficient in any map $M \to N$ of modules is simply the cohomology of the co-chain complex

4.2 $$C^n_J(X \to Y, M \to N) = hom_J \left(\begin{bmatrix} C_n(X) \\ \downarrow \\ C_n(Y) \end{bmatrix} , \begin{bmatrix} M \\ \downarrow \\ N \end{bmatrix} \right)$$

where C_n is any reasonable (singular or not) chain functor for spaces. Dually one constructs homology groups as the homology of the chain complex obtained by a "tensor product" over J.

4.3 $$(C_n(X) \to C_n(Y)) \otimes_J (N \leftarrow M) \equiv$$
$$coker\ (C_n(X) \otimes M \xrightarrow{\Delta} C_n(x) \otimes N \oplus C_n(Y) \otimes M)$$

where Δ is the obvious induced map. These (co)-homology theories are J-homotopy invariant and have long exact sequences corresponding to union (Mayer-Vietoris sequence) and to short exact sequences of the coefficients. Moreover the coefficients groups are precisely the values of the (co) homology on the J-orbits $P_0 = \varnothing \to *$, $P_1 = * \to *$. If $X \to Y$ is not a cofibration, then more involved J-cohomology groups must be employed reflecting the "J-orbit structure" of $X \to Y$.

4.4 D-cohomology theories and cofibrations. The axioms given in [Bredon] for "classical" equivariant cohomology easily generalize to axioms for homology and cohomology theories for the category of D-spaces. Since one would like to have homology groups for a pair of diagrams it is necessary to consider admissible pairs. For this one would like to have a notion of admissible pairs. This is related to the notion of a cofibration map $A \to X$ where the resulting pair (X,A) is to be admissible. The various possibilities of defining a cofibration come from and give rise to various possibilities of defining a *model category* structure on S^D (see a discussion of these structures in [DF]). There is however one natural way to define a cofibration in both $S.^D$ and Top^D using the *homotopy extension property*.

A map of D-spaces $A \to Y$ is a cofibration if any homotopy $H : A \times I \to W$ into a (Kan) D-space W, with a given extension of $H_0 = H(-,0)$ over Y, can be extended to a homotopy $\tilde{H} : Y \times I \to W$. The same definition holds for topological D-spaces.

4.5 Example: Let $D = J = (\cdot \to \cdot)$ then the map $P_0 \to P_1$ (i.e. the square

$\varnothing \to *$

) is not a cofibration in Top^D: take the J- space W to be any inclusion

$* \to *$

$\{0\} \to [0,1]$ and H any non- constant homotopy. For example, take H to be the only map $P_0 \times I \to W$ taking the range of $P_0 \times I$ to the range of W by the identity map: then $H(-,0)$ extends to P_1 but clearly H cannot be extended over $P_1 \times I = (I \xrightarrow{id} I)$.

It is easily seen that the present definition of cofibration is much less restrictive than [D-K-5, 2.4]. For example the natural inclusion $P_1 \to P_1 \amalg P_2$ has the homotopy extension property but it is not a cofibration in the sense of [D-K-5]. Note also that in general a map $A \to X$ with $A(d) \to X(d)$ a cofibration is not a cofibration. It is not hard however to show:

4.6 Lemma: *If (X,L) is a relative D–C.W. pair then the inclusion $L \to X$ has the homotopy extension property.*

Proof: This is a straightforward argument by induction on D-cells of the form $T \times e^n$ for a D-orbit T: One uses the exponential law to reduce the problem to that of constructing a map $e^n \to hom_D(T,W)$, which is solved using the cofibration in spaces $\partial e^n \to e^n$.

It seems to reasonable to conjecture that in the simplicial category a cofibration $A \to X$ is a retract of an "isovariant inclusion": of a D-map $i : A \to X$ such that for any simplex τ_n in A, the orbit of τ_n in A is mapped by isomorphism via i to the orbit of $i(\tau_n)$ in X - so that i preserves orbits. That exactly what went wrong in example 4.5 above. A map f between two orbit is a cofibration in $S.^D$ if and only if it is an isomorphism. The same goes for a map between two D–C.W. orbits. An isovariant inclusion $A \subseteq X$ gives rise to the notion of

4.7 Admissible pair. An inclusion $L \subseteq K$ of D-spaces in $S.^D$ is an *admissible pair* if, roughly speaking, for every simplex $\sigma \in L_n$ the full D-orbit of σ is also in L_n. Formally we define (K,L) to be admissible if the following diagram is a pullback one:

$$\begin{array}{ccc} L & \subset & K \\ \downarrow & & \downarrow \\ \varinjlim L & \hookrightarrow & \varinjlim K \end{array}$$

This comes very close to say that (K,L) is a retract of a relative D–C.W. pair (1.16). Certainly any relative C.W. complex $A \subseteq X$ defines an admissible pair (X,A). Also $(X,\varnothing)$ is admissible for any $X \in S.^D$.

4.8 AXIOMS FOR HOMOLOGY. The basic properties of such homology can be taken directly from the case of $D = G$ a discrete group. For that case a set of axioms is given in [Bredon] for relative homology of a pair of G-spaces $A \subseteq X$. As we saw above for D-space one must be careful to admit as pairs of diagrams $A \subseteq X$ only "admissible" inclusions that have the homotopy extension property for D-spaces. It is also possible, just as in classical homology to write down axioms for the absolute homology groups. The axioms falls into three groups: The homotopy axiom, exact sequence axioms (for relative homology of an admissible pair, excision for union along a cofibration), and the dimension axiom.

A homology theory for D- spaces is a Z graded, R-modules valued functor h_*^D on admissible pairs of D-spaces (X,A) together with a connecting morphism $\partial : h_*^D(X,A) \to h_*^D(A,\emptyset)$ that satisfies the following four axioms:

4.9 Homotopy axiom. If $f \sim g$ are two homotopic D-maps then $h_*^D(f) = h_*^D(g)$.

4.10 Excision axiom. The natural inclusion $(L, L \cap K) \to (L \cup K, K)$ of admissible D-pairs induces an isomorphism on h_D^* that is compatible with ∂ in the obvious sense.

4.11 Exact sequence axiom. The map ∂ fits into long exact sequence for an admissible pair (K,L) given by $\dots \to h_*^D(L) \to h_*^D(K) \to h_*^D(K,L) \xrightarrow{\partial} h_{*-1}^D(L) \to \cdots$, where the maps are the obvious ones.

We say that h_*^D is a classical homology theory if it also satisfies the dimension axiom.

4.12 Dimension axiom. For any D-orbit T one has $h_i^D(T) = 0$ for all $i \neq 0$.

4.13 The only noteworthy aspects of these axioms is the concept of admissible pair and the form of the dimension axiom. These are strongly related: In both the concept of D-orbit plays the major role. If we change our concept of admissible pair, or cofibration we get a different concept of homology theory. Another possible definition of cofibration $A \subseteq X$ is gotten by requiring that the orbits of all points (simplicies) in $X - A$ are free. See [D-K-5] the corresponding cohomology theory is given in [D-K-2].

It turns out that since all D–$C.W.$-complexes can be built as homotopy push-outs along orbits, the value of any classical homology theory on D- orbits and maps between them gives, up to an isomorphism, its value on all D–$C.W.$-complexes. The values $h_0^D(T)$ on all orbits determine then the classical homology. These values form together the *coefficient system* of h_*^D. We shall see shortly that any homotopy functor $M : \{D\text{-orbits}\} \to \{R\text{-modules}\}$ can be extended in the spirit of [Andre] and [Bredon] to a classical homology theory on D-spaces (topological or simplicial) with a canonical isomorphism $H_0^D(-) \approx M(-)$ of functors on O_D.

Notice however that in general the category O_D of *all* D-orbits is not a small category (1.7). So that one must assign values to a large category of D-spaces in order to determine a general homology theory. Of course M might be determined by a *small* subcategory of O_D as a left Kan extension: (see (4.25) below). Any homotopy functor F on a subcategory of O_D, can be extended to a coefficient system $LF : O_D \to R$-module.

In spite of h^D being determined in general by its value on a large category O_D, its value on any particular space X depends only on its values on a small subcategory of O_D - the subcategory of orbits that actually appear in X and $X^{\Delta[n]}$ for all $n \geq 0$.

4.14 Cohomology theory. A general D-cohomology theory is defined by direct dualization of 4.8. In this case a classical cohomology is determined by its value on the opposite category to O_D, namely by the functors $h_D^0 : O_D^{op} \to R$-modules.

4.15 Relations to other homology theories. In [D-K-2] the authors construct cohomology groups for D-spaces, that depends on a coefficients functor $M : D \to (K$-modules) by using the associated diagram of Eilenberg-MacLane spaces. No homology groups, chains or co-chain are given there. That cohomology theory is a special case of classical (co)-homology theory, (4.16). It is the most general cohomology theory that corresponds to the restrictive definition of an admissible pair given in 4.7 above. The present notion of cohomology reduces to that restrictive notion if the coefficient system $M : O_D^{op} \to (R$-module) is a left Kan extension of a functor $\{F^d\}^{op} \to (R$-module) where $\{F^d\}^{op}$ is the opposite category to that of *free* orbits: Notice that it is isomorphic to D^{op}. It will follow from the discussion below that a classical D-cohomology is representable if and only if its coefficient system h_D^O is a Kan extension from a small subcategory of O_D.

4.16 CONSTRUCTION OF CLASSICAL D-HOMOLOGY THEORY: We now proceed to associate a D-homology theory to any homotopy functor $M : O_D \to (R$-modules), where O_D is the category of all D-orbits and M factors through the D-homotopy category hO_D of O_D, we have $obj\ hO_D = obj\ O_D$ while $hO_D(d,e) = \pi_0 O_D(d,e)$. A D- homology theory associated to such a functor on hO_D is denoted by $H_*^D(\ ,M)$. We will make the construction in the simplicial case for $X \in S.^D$ and then one can use the singular functor $sing : Top^D \to S.^D$ to get homology groups for topological diagrams. If the topological diagram K has a nice cellular structure such as simplicial complex then one can easily obtain a combinatorial definition using that structure.

4.17 Local O-chains and O-chains with coefficients for $O \subseteq O_D$. We first associate to any homotopy (co-)functor $M : hO \to R$-modules (or $M : hO^{op} \to R$-modules) where $O \subseteq O_D$ is a full small subcategory of D-orbit, a chain (co-chain) complex functor on D-spaces. Given a D-space $X \in S.^D$ one constructs the associated O^{op}-diagram of O-orbit points of X denoted by X^o (compare 3.2 above) $X^o \in S.^{O^{op}}$ with $X^o(T) = hom_D(T,X) = X^T$ for any $T \in O^{op}$. Next let $R\,X_n^o : O^{op} \to (R\text{-module})$ be the free R- module generated by the n-simplices of X^o : $(R\,X_n^o)(T) = R(X_n^o(T))$ for $T \in O^{op}$. We call $R\,X^o$ the *local chain complex* of X associated to O and denote it also by $C_*^o(X)$. The boundary map comes from the simplicial structure as usual. So $C_n^o(X)$ is a chain complex in the category of O^{op}-diagrams of R-modules. For an admissible pair (X,A) we define the local chains $C_*^o(X,A)$ to be the quotient group of $C_*^o(X)$ by the subgroup $C_*^o(A)$.

We will notice 4.33 that admissibility guarantees that $C_*^o(A)$ is a direct summand of $C_*^o(X)$. To obtain the chain complex with coefficient in M we take the tensor product of a functor with a cofunctor, dually we take $hom_D(-,-)$. [compare MacLane X.4.]

4.18 $\qquad C_n^o(X,M) = C_n^o(X) \otimes_{O^{op}} M$

4.19 $\qquad C_o^n(X,M) = hom_{O^{op}}(C_n^o(X),M).$

The tensor product is the usual dual to the $hom_O(\ ,\ )$ functor. The former is a colimit over a diagram involving the products $C_n^o(X)(T_1) \otimes M(T_2)$ for all maps $T_1 \to T_2$, just as the latter is a limit over a similar diagram involving $hom(C_n^o(X)(T_1),\ M(T_2))$. In the case $D = J = (\cdot \to \cdot)$ discussed in (4.1) if one takes the orbit category $\{P_0,P_1\}$ we recover definitions (4.2) and (4.3) as special cases. We now take the homology and cohomology of X with coefficients in M to be the (co)-homology of the (co)chain complexes (4.18), (4.19). Denote these groups by $H_*^O(X,M)$ and $H_O^*(X,M)$.

4.20 Let P consist of all the orbits in the D-set pt with $pt(d) = \{*\}$ for all $d \in obj\ D$. If the classifying space of D is connected then pt and therefore P consist of a single orbit. It is then clear that

$$H_*^P(X,M) = H_*(\underleftarrow{\lim}\, X,M)$$

where $M : P \to (R\text{-modules})$ is now just an R-module since P has only an identity map.

4.21 Let O_d consists of the single principle orbit $\{F^d\}$ defined in (2.1) above.

Then X^O consists of a single space $hom(F^d,X) = X(d)$ and it is not hard to see that if we choose a coefficient system with $M(F^d) = \bigoplus_{F^d \to F^d} M$ for some fixed R-module M and $M(f : F^d \to F^d)$ is given on a generator $m_g \in M$, $g : F^d \to F^d$ by its shift $m_{f \circ g}$ then: $H_*^{O_d}(X,M) \tilde{=} H_*(X(d),M)$.

4.22 Bredon (co)homology. Let $D = G$ a discrete group and $O = O_g = \{G/H\}$ the category of all G-orbits. Then we recover the definition of chains and cochains given by [Bredon] and [Illman].

In that case every G-orbit is G-isomorphic to a quotient G-set of the form G/H for some $H \subseteq G$. Thus O_G can be taken to be the category of these quotients with G-maps $G/H \to G/K$ as morphisms. For a given G-space X the associated $O = O_G^{op}$-space X^O is just the diagram of all the fixed points subspaces (1.10) the (classical) Bredon cohomology of X with coefficients in a functor $O_G^{op} \to (R$-modules), denoted by $H_G^*(X,M)$ is given as the cohomology of the cochain complex $hom_{op}(C_*^o(X),M)$ where $C_*^o(X)$ is the diagram of the free chain complexes $\{C_*(X^H)\}_{H \subseteq G}$.

4.23 Free diagrams of R-modules. The local chains $C_n^o(X)$ associated to some category of orbits $O \subseteq O_D$ is, in general not *free* in the category of diagrams of R-modules. In order to prove various exactness properties we need $C_*^o(X)$ to be free. This can be guaranteed using (3.7) by taking O to be big enough i.e. O must contain $O(X)$ that is all the D-orbits in X and $X^{\Delta[n]}$. A free D-diagram of R-modules $N : D \to (R$-modules) is a diagram together with a set $R \subseteq \underline{\underline{N}}$ of generator $Q = \{q_\alpha(d)\}$ with $q_\alpha(d) \in N(d)$, and where the map $\bigoplus_{\alpha,d} I(q_\alpha(d)) \to N$ is an $I(q_\alpha(d) : D \to R$-modules is given at e by $I(Q_\alpha(d))(e) = \bigoplus_{d \to e} R$. In other words $I(q_\alpha(d)(e)$ is the free R module generated by $F^d(e)$ (See 2.1). It is immediate from this definition that if X^O is free O-space for some $O \subseteq O_D$ and $X \in S.^D$ then $C_*^o(X)$ is a free R-module. Similarly, if (A,X) is admissible then $C_*^o(A)$ is a direct summand of $C_*^o(X)$ and the quotient $C_*^o(X,A)$ is a free diagram of R-modules: This is because every map of an orbit $T \to X$ whose direct limit $* \to \lim_{\to} X$ lies in A is generated in X^O by an orbit $T_1 \to A$ where T_1 is the full orbit of T in X (3.8). By the admissibility condition $T_1 \to X$ factors through $T_1 \to A$.

In order to get chain complex with coefficients that does not depend on the choice of $O \subseteq O_D$ we must construct the chain complex $C_*^D(X,M)$ for the full orbit category O_D with respect to a homotopy functor M. This is done by restricting M to some O that contains $O(X)$ for each X. To make sense the result must be shown to be invariant under the choice of O as long as it contains $O(X)$.

4.24 Changing orbit categories. With every D-space $X \in S.^D$ we associated a small category $O(X)$ namely the full subcategory of O_D of all orbits in $X^{\Delta[n]}$ for all $n \geq 0$. We say that T is an orbit in $Y \in S.^D$ if T lies over some vertex of $\lim\limits_{\rightarrow} Y$. Let $O_1 \supseteq O(X)$ be an orbit category containing $O(X)$. Then the diagram of O_1-orbit points can be obtained from that of $O(X)$-orbit points as a *left Kan extension.*

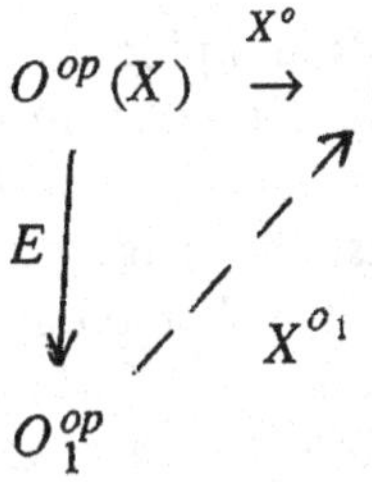

Namely $X^{O_1} = L\, X^O$ where L denotes the Kan extension along the inclusion E: Let us recall the Kan extension [MacLane]:

4.25 The Kan extension. Let $F : O \rightarrow Ab$ or $F : O \rightarrow$ (sets) be a functor from a small category O (say of D-orbits) to some abelian category (with arbitrary direct limits) or to the category of sets. Then if $E : O \subseteq O_1$ is any inclusion of O in a larger category there is a canonical extension of F to all of O_1 (which may be large). The value of the extension $L_E F = LF$ on an element $T_1 \in O_1$ is given as a colimit $LF(T_1) = \lim\limits_{\rightarrow} (O/T_1)F$ of the diagram that assigns to any map in $O_1 : T \rightarrow T_1$ with $T \in O$, the value $F(T)$ and any map over T_1 i.e $T' \rightarrow T \rightarrow T_1$ the map $F(T') \rightarrow F(T)$. So that the colimit is taken over the category O/T_1. Huristically, LF puts together all the possible ways of "approximating" an element T_1 of O_1 by elements of O. Clearly for $T \in O$ one has $LF(T) = F(T)$ since there is $\underset{=}{\text{a}}$ terminal element in the above diagram namely coming from the identity $T \rightarrow T$.
Now the following lemma will be clear:

4.26 Lemma: *Let $O = O(X) \rightarrow O_1$ be an inclusion of orbit categories. Then the functor $X^{O_1} : O_1^{op} \rightarrow S.$ is the left Kan extension $L\, X^O$ of the corresponding functor for $O(X) = O$.*

Proof: A typical element in $X_n^{O_1}(T_1)$ is a D-map $\sigma_1 : T_1 \times \Delta[n] \rightarrow X$ with $T_1 \in O_1$. But this is a map $T_1 \rightarrow hom(\Delta[n], X)$. Now since T_1 is O_1-orbit this last map upon taking direct limits gives us the outer square in the diagram

$$\begin{array}{ccc} T_1 & \xrightarrow{\;O_1\;} & hom\,(\Delta[n],X)=X^{\Delta[n]} \\ \downarrow & \searrow \quad T \quad \nearrow & | \\ & \swarrow & | \\ \underset{\to}{\lim} T=\Delta[0] & \underset{\bar{\sigma}_1}{\to} & \underset{\to}{\lim}\, hom\,(\Delta[n],X) \end{array}$$

Let T be the indicated pullback. Then T is a D-orbit in $X^{\Delta[n]}$, so $T \in O(X)$. But clearly the map σ_1 factors *uniquely* through the map $T \to X^{\Delta[n]}$. This shows that the function complex $hom(T_1,X)$ is a colimits as in 4.25 above. From this lemma one gets the invariance of chains under extension of the orbit category $O(X)$:

4.27 Proposition: *Let $M : O(X) \to (R\text{-modules})$ be a coefficient system over $O(X)$, and let $M_1 : O_1 \to (R\text{-modules})$ be any extension of M to a larger category of D-orbits $O_1 \supseteq O = O(X)$. Then there is a canonical isomorphism for all $n \geq 0$*

$$C_n^O(X,M) \cong C_n^{O_1}(X,M_1).$$

Proof: From (4.26) above one gets immediately that the local chains on $C_n^{O_1}(X) : O_1^{op} \to (R\text{-modules})$ is the left Kan extension of the local chains $C_n^O(X)$. But it is immediate from the definition of Kan extension L that $A \otimes_C M \cong KA \otimes_D M'$ for any diagram M' with extension M along a functor $C \to C'$, since tensor product commutes with the colimit defining the Kan extension. Similar arguments work for the invariance of co-chains: Here one uses $hom(\underset{\to\alpha}{\lim}\ F\,|_\alpha, B) = \underset{\leftarrow}{\lim}\, hom(A_\alpha, B)$.

4.28 Remark: It is clear that taking $O = O(X)$ is crucial for (4.27). If one takes a smaller category of orbits the two chain complexes will be very different. However, if M_1 in (4.27) is LM then the conclusion will still hold by the same proof for any orbit category extension $O \subseteq O_1$. This allows us to define a coefficient system and obtain $C^D(X,M)$ where M is given on any $O \subseteq O_D$. One simply extends M to all of O_D by taking $LM : O_D \to (R\text{-modules})$.

4.29 Naturality of chains. It follows from (4.27) that one can associate for any homotopy functor $M : O_D \to R$-modules on R- module $C_*^D(X,M)$ defined as $C_*^O(X,M)$ for some orbit category that contains $O(X)$. Given a D-map $f : X \to Y$ between two D-spaces take $O = O(X \cup Y)$, to get a map

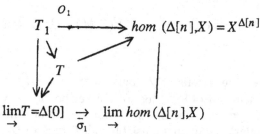

$$C_*^D(X,M) \cong C_*^O(X,M) \xrightarrow{f_*^O} C_*^O(Y,M) \cong C_*^D(Y,M).$$

Similarly for cochains.

4.30 Verification of the axioms. We must show that $H_*^D(-,M)$ is a classical D-homology in the sense of (4.8). Most of the verification is routine. We will spell out certain points where the concepts of homotopy functor $M : hO_D \to (R\text{-modules})$ and of an admissible pair $L \subseteq K$ of D-spaces comes in.

4.31 The homotopy axioms. This is done by a straightforward extension of the usual proof for homology or G-equivariant homology see [Bro]. A homotopy $f \simeq g : X \to Y$ between two D-maps gives rise to a $(n+1)$-chain $T \times \Delta[1] \times \Delta[n] \to Y$ connecting any two n-chain $T \times \Delta[n] \to Y$.

4.32 The dimension axiom. (compare [Andre] p.5) The task is to compute $H_n^D(T,M)$ for any orbit $T \in O_D$ and show that it vanishes in positive dimensions while being naturally isomorphic to $M(T)$ for $n = 0$. It is here that the *homotopy invariance* of M is crucial.

Claim: $H_0^D(T,M) \cong M(T)$; $H_D^0(T,M) \cong M(T)$ *where M is a homotopy functor (or cofunctor) on O_D.*

In fact we show that the natural maps $M(T) \xrightarrow{i_0} C_0^D(T,M) \xrightarrow{i_1} H_0^D(T,M)$ are isomorphisms. The map on the left is induced by the natural inclusion into the chains that by definition are certain quotients of $\sum_{P \to T \in O(T)} M(P)$ namely the map into the summand corresponding to the identity $T \to T$. But notice that, by the definition of $\otimes_D$, an element $a \in M(P)$ in this sum corresponding to a map $p : P \to T$ is to be identified with $M(P)(a) \in M(T)$ in the construction of the tensor product O_D as a quotient. Therefore i_0 above is an isomorphism. For the second isomorphism we use the homotopy invariance of M. We must show that the boundary map $\partial_1 : C_1^O(T,M) \to C_0^O(T,M)$ is the zero map. This map is the quotient under the $\otimes_D$ - identification of the map

$$\sum_{P \times I \to T} M(P) \xrightarrow{\partial_1} \sum_{P \to T} M(P) \to C_0^O(X,M) \cong M(T),$$

where the second map is the evaluation. Now ∂_1 into $M(T)$ is clearly the difference between the two maps induced on the boundary of $\Delta[1]$. But these two maps $P \rightrightarrows T$ are homotopic in O_D, so that the difference is zero. Similar arguments show that in fact $\partial_k : C_{k+1}^O(T,M) \to C_k^O(T,M)$ is zero for k even and an isomorphism for k odd. Notice that any map $\Delta[n] \times P \to T$ factors of course through the projection $pr_2 : \Delta[n] \times T \to T$ for $P, T \in O_D$. Therefore $C_k^o(T, M) = M(T)$ for all $T \in O_D$. So the situation here is the simplest possible generalization of the usual computation of the chains of a single point in singular or simplicial homology. The arguments for the cochains and cohomology are the exact duals of the above arguments.

4.33 The excision axiom. In fact here there is an isomorphism of local chains (4.18) i.e. $C_*^O(K, K \cap L) \to C_*^O(K \cup L, L)$ is an isomorphism, for K, L admissible pair. This isomorphism then must be verified for each $T \in O$. But in this case since $C_n^O(X, A)(T)$ is by definition equal to $C_n(X^T, A^T)$ where C_n is the usual chain complex, the desired chain isomorphism will follow from the usual one by showing that $K^T \cup L^T = (K \cup L)^T$ and $(K \cap L)^T = K^T \cap L^T$. The second being obvious, we observe that for the first follows from admissibility (4.7). One must show that any map $T \times \Delta[n] \xrightarrow{\sigma} K \cup L$ can be factored through $K \amalg L \to K \cup L$, where $\amalg$ is the disjoint union of D- spaces. But upon taking the direct limit of σ over D we get $\underline{\sigma} = dir\ \lim \sigma : \Delta[n] \to dir\ \lim(K \cup L)$ equal to $dir\ \lim K \cup dir\ \lim L$. Since σ is a simplex it must be in one of the direct limits (or both). But then by the admissibility conditions $T \times \Delta[n] \to K \cup L$ factor either through $K \to K \cup L$ or through $L \to K \cup L$. This proves excision. The rest of the axioms follow by similar techniques.

4.34 Other exact sequences. It is not hard to deduce from 4.23 above using the usual techniques that each admissible pair (X, A) gives rise to a long exact sequence in homology with coefficients in $M : O(X) \to R$-modules and each exact sequence of homotopy functors on hO_D, $0 \to M_1 \to M_2 \to M_2 \to 0$ gives a long exact sequence on homology with these coefficient modules.

4.35 Remark: If $M : O_D \to (R\text{-modules})$ is not a homotopy invariant functor we still can make the construction and get a D- homology theory satisfying all the axioms now $H_0(T, M)$ will not be in general isomorphic to $M(T)$ but to some quotient thereof.

5. APPLICATIONS

In this section we discuss several constructions and applications related to the concepts discussed above. We start with a spectral sequence that converges to the D-homology or D-cohomology. This is a local to global spectral sequence starting with the diagram of R-homology groups $\{H_* X(d)\}_d$. A special case of this spectral sequence yields a spectral sequence that converges to the R-homology of the direct limit of a given diagram of spaces. Notice that no such spectral sequence exists in the literature since it seems to depend heavily on the concept of D-homology defined above. We then discuss a related spectral sequence of Atiyah-Hirzebruch type that "computes" a generalized D-homology or cohomology theory in terms of an associated classical D-homology. In the case where $D-G$ is a finite p-group we show how to get a rather quick and conceptual proof of the well known Smith isomorphism: The proof shows that similar result holds for general abelian p-group. We then turn back to example (4.1) and show how classical D-cohomology is the proper domain for obstruction co-cycles for extending maps of D- spaces: Here we use directly ideas of [Bredon] and [Dwyer-Kan-2]. The last and potentially fruitful application concerns the function complex $hom_D(X,Y)$ of two $D.C.W.$-complex - we show how this space admits an important simplification that leads to a resolution of $\pi_* \, hom_D(X,Y)$ in the spirit of [Bousfield-Kan]

5.1 Local-global spectral sequence. In (4.18) above we have associated with any D-space X a category of D-orbits O a *local chain functor* $C_*^O(X)$ given by $C_*^O(X)(T) := C_*(X^T)$. Take $O = O(X)$ the orbit category of all orbits in $X^{\Delta[n]}$ $n \geq 0$ (4.25). Define the *local homology* of X with respect to O to be

$$H_*(X) = H_*^O(X) : O^{op} \to R\text{–modules}; \quad H_*^O(X)(T) = H_*(X^O(T))$$

given by the homology of C_*^O. Let M be a coefficient system on O thus $M : O \to R$-modules. It is well known that one can take a projective resolution of such diagram of modules [Bredon], [Watts]. Let $M_* \to M$ be such a resolution. Then one can consider the double complex $C_*^O(X) \otimes_O M_*$. This is the tensor product over O between a cofunctor and a functor on O as used in (4.18). The double complex gives rise, as usual, to a spectral sequence

$$E^2_{p,q} = Tor_p^O(H_q^O(X), M) \Rightarrow H^D_{p+q}(X, M)$$

where Tor_*^O is the left derived of the tensor product over O in the category of O-modules. Notice that for $O = O(X)$, $C_n^O(X)$ is a projective object in the category of O-modules, because the O-space X^O is free (see 4.23). Similarly, for cohomology one uses injective resolution for the cofunctor M on O get a spectral sequence

$$E_2^{p,q} = Ext_O^p(H_O^q(X),M) \Rightarrow H_O^{p+q}(X,M).$$

It is clear from (4.26), (4.27) that the functors Tor_p^O, Ext_O^p do not depend on O as long as it contains $O(X)$. Therefore it is legitimate to denote these functor by Tor^D, Ext_D meaning the common value for any orbit category over D containing $O(X)$.

5.2 The homology of a direct limit. Let $Z : O_D \to R$-module be the constant functor $Z(T) = Z =$ (the integers) for all $T \in O_D$. It is immediate from the definition of the tensor product that in that case $C_*^D(X) \otimes_D Z \cong C_*(\lim_{\to D} X)$ the usual chain-complex of the direct limit over D of the diagram X. Therefore the local-global spectral sequence above becomes

$$Tor_p^D(H_q^D(X),Z) \Rightarrow H_{p+q}(\lim_{\to D} X).$$

Similar spectral sequence arises for cohomology. For several examples of computing such $Ext_O^p(-,-)$ see [Bredon p. 25].

As a special case of the direct limit spectral sequence one gets a Leray spectral sequence for any map $K \to B$. To see that consider K as the direct limit over the category of simplices of B of the diagrams of the pull back spaces $\{K_\sigma\}$ over these simplices $\{\sigma\}$. (see 1.11). The E^2-term then becomes the usual homology of B with system of local coefficient $\{H_*(K_\sigma,R)\}$, this local system is not necessarily locally trivial. More generally, if $X = \bigcup_I X_\alpha$ is covering of X by subspaces one can consider X as the direct limit over I of the spaces $\{X_\alpha\}$. In this case the relevant orbit category is precisely the nerve of the covering with typical element $\alpha = \{i_1, \ldots, i_k\}$ so that $\bigcap_\alpha X_i \neq \emptyset$, and maps begin inclusions. Again the E^2-term is the homology of this nerve with local coefficients systems that assign to a simplex α the homology of the $H_*(\bigcap_\alpha X_i,R)$.

Furthermore, it is not hard to see that for the cohomology of a direct limit of a telescope one recovers the exact sequence of Milnor involving $\lim^1$: In that case the only orbits that appear in the telescope are *free* so the diagram of orbit-points is the original telescope.

5.3 Generalized equivariant (co)homology and its spectral sequence A generalized equivariant D-(co)homology is a theory that satisfies the usual axioms except the dimension axiom. That is, it is allowed to assign non-trivial values to D-orbits in nonzero dimension. A well-known example is Borel (co) homology for G-spaces. Here we defined $h_G^*(X) \equiv H^*(EG \times_G X,R)$, i.e. as the usual cohomology of the homotopy direct limit of X. Thus $h_G^*(pt) = H^*(BG,R)$. In a similar vein one can define the Borel cohomology of an arbitrary D-space to be the usual cohomology of its homotopy direct limit. This will then be a generalized D-cohomology theory. Another example is equivariant K-theory [Segal].

Let h_D^* any generalized homology or cohomology theory satisfying in addition an axiom for infinite disjoint union. One can associate to a finite dimensional D-space $X = X^{(N)}$ a spectral sequence converging to that cohomology whose E_2-term depends on the classical D-cohomology of the space with coefficient in $h_D^*(T)$ where $T \in O_D$. The construction is a direct generalization to D-cohomology of Bredon's construction [Bredon]. We get: $E_2^{p,q} \approx H_D^p(X, \underline{h}^q)$, where $\underline{h}^q : O_D \to$ (graded R-modules) assigns to an orbit T its cohomology $h_D^*(T)$. Using this spectral sequences one can prove the usual uniqueness theorem regarding cohomology theories.

5.4 Smith theory. In the case $D = G = Z/pZ$ the orbit category is very simple, having two objects with one map between them in addition to the self maps of the free orbit G/e: So schematically the category O_G^{op} appear as follows

$$[G/G] \to [G] \circlearrowleft$$

We would like to use this simple structure in order to illustrate how the spectral sequence for generalized cohomology yields a very simple proof of an isomorphism due to Smith:

5.5 Theorem (Smith): *Let $f : X \to Y$ be a G-map between two finite G-spaces with $G = Z/pZ$. Assume that f is a homotopy equivalence then when restricted to the fixed points $f^G : X^G \to Y^G$ it indices an isomorphism on the Z_p-cohomology.*

Proof: In the proof we employ the standard localization approach: one notices that the Borel cohomology of X for any G-space X is a module over the cohomology of BG. By inverting a suitable polynomial generator u in $H^* BZ/pZ$ we show using (5.1) above:

5.6 Lemma: *The inclusion $X^G \to X$ induces an isomorphism on the u-localized Borel cohomology.*

Since it follows directly from the equivalence $f : X \to Y$ that f induces an isomorphism on homotopy colimits and thus on Borel cohomology, we get that $f^G : X^G \to Y^G$ induces an isomorphism on u-localized Borel cohomology. But for finite spaces with a trivial G- action this implies Z/pZ cohomology isomorphism as needed.

Recall (5.3) above that Borel cohomology denoted here by $H_B^*(X)$ is defined to be the cohomology of $EG \times_G X$. This cohomology group is a module over $H_B^*(pt) = H^*(BG) \cong H^*(BZ_p)$. Let u be a polynomial generator of $H^2(BG)$, consider the $[\frac{1}{u}]H^*(BG)$- module $[\frac{1}{u}]H_B^*(X)$. Every element in $H_B^*(X)$ that is killed by some power of u becomes the zero element in $[\frac{1}{u}]H_B^*(X)$. Now the latter is a generalized G- equivariant cohomology theory. We now use the spectral sequence (5.3) above, converging to that cohomology. We claim that the map $X^G \to X$ induces an isomorphism on the E_2-term of the spectral sequence.
Let us denote by $[\frac{1}{u}]B^q$ the coefficient system gotten by applying Borel cohomology, localized with respect to u, to the orbit category of $G = Z/pZ$. Thus $[\frac{1}{u}]B^*$ consists of the map $[\frac{1}{u}]H_B^*(G/G) \to [\frac{1}{u}]H_B^*(G/\{e\})$, together with the self maps of the range. Then our spectral sequence has $E_2^{p,q} = H_G^p(X,[\frac{1}{u}]B^q) \Rightarrow H_B^{p+q}(X)$, where H_G^p denotes the Bredon cohomology (4.22) above. Notice however that $H_B^*(G/\{e\}) \cong H^*(EG \times_G G) \cong H^*(pt, Z/pZ) = Z/pZ$. Since u raises dimension by 2 it acts trivially on this group and so $[\frac{1}{u}]H_B^*(G/\{e\}) \cong 0$. Thus the E_2-terms of the spectral sequence for the generalized equivariant cohomology $[\frac{1}{u}]H_B^*$ is just the classical cohomology group $H^p(X^G,[\frac{1}{u}]H_B^q(G/G))$. Now clearly the same E_2-terms appears in the spectral sequence converging to the $[\frac{1}{u}]H_B^*$- cohomology of the trivial G-space X^G. Therefore the map $X^G \to X$ induces an isomorphism on the corresponding E^2-term. The lemma now follows since for *finite dimensional* spaces, the spectral sequence converges to the localized Borel cohomology (5.3).

Since the given equivariant map $f : X \to Y$ is a homotopy equivalence it follows (2.14 above) that it induces a homotopy equivalence on homotopy limits (direct and inverse) so that f induces an isomorphism on the u-localized Borel cohomology. Therefore we get from the lemma that $f^G : X^G \to Y^G$ induces an isomorphism $[\frac{1}{u}]H_B^*(f) : [\frac{1}{u}]H_B^*(X^G) \to [\frac{1}{u}]H_B^*(Y^G)$.

But X^G, Y^G have trivial G-action so that $EG \times_G X^G = BG \times X^G$ and $[\frac{1}{u}]H_B^*(X^G) = [\frac{1}{u}]H^*(BG) \otimes H^*(X^G)$, and the same for Y. Now recall that

X^G, Y^G are finite spaces, a simple counting argument in these finite dimensional spaces over Z/pZ shows a map $f^* : H^*(X^G) \to H(Y^G)$ is an isomorphism if and only if the map $[\frac{1}{u}]H^*(BG) \otimes f^*$ is an isomorphism. This completes the proof of (5.5).

5.7 Generalized Smith isomorphism. In fact the argument in 5.5 and 5.6 generalizes to get:

Proposition. *The Smith isomorphism theorem (5.5) holds for G an arbitrary abelian p - group.*

Proof: The crucial step in the proof is lemma 5.6. The isomorphism follows formally from it. To prove 5.6 all we used about G is that there is a polynomial generator on $u \in H^*(BG)$, and that since $|u| > 0$, this generator acts trivially on $H^*(pt) = H_B^*(G/\{e\})$. For a general abelian p-group we must find a polynomial generator u in $H^*(BG)$ that acts as the zero element on all $H^*(BH)$ for *all* the subgroup $H \subseteq G, H \neq G$. In that case again $[\frac{1}{u}]H_B^*(G/H) = 0$ for all $G \neq H \subseteq G$. So that the coefficient system in the spectral sequence (5.3 above) converging to $H_B^*(X)$ and $H_B^*(X^G)$ reduces to $[\frac{1}{u}]H^*(BG)$ in both cases, implying the desired isomorphism. But such a class u always exists: For each $G \neq H_\sigma \subseteq G$ there exist a polynomial generator $u_\sigma \in H^*(BG)$ that pulls back to zero in $H^*(BH_\sigma)$. Therefore $u = \prod_\sigma u_\sigma$ pulls back to zero in all the subgroups $H \neq G$ of G. This completes the proof.

5.8 A non-representable cohomology theory for J-spaces. We would like to observe here that in almost all cases a cohomology theory on D-spaces is not representable by a classifying D-space. The ultimate reason is that there are too many homotopy types of D-orbits - together they do not form a set: Recall example (1.7). We claim that the large class of non-equivalent J-orbits gives rise to a classical cohomology theory on J-spaces that is not representable in the sense of E. Brown. First notice that for each set S in the category of all sets, there is in Top^J an orbit P_s namely, $S \to \{*\}$. For every cardinal number we get exactly one orbit type. Let $\{P_s\}$ denote the full subcategory of O_J consisting of the orbits P_s. The category $\{P_s\}$ is isomorphic to the category of all sets. We say that a functor $F : C \to C'$ is small if there exist a small subcategory $C'' \subseteq C$ so that F is a Kan extension of some functor $F'' : C'' \to C'$.

Now the existence of a non-representable functor follows from the fact that representable functors *are always small* in the above sense - see exercise X.3.2 in [MacLane]. Therefore we have:

5.9 Proposition: *If B is any J-space, s is any set then the functor $P_s \to [P_s, B]_J = \pi_0 \, hom_J(P_s, B)$ is small.*

To get a non representable theory use the reference above and take M in 4.17 to be any non-small functor.

5.10 Obstruction theory. We now turn briefly to explain in what sense classical D-cohomology $H_D^*(-, M)$ is the correct domain for the obstruction to extension of a D-map $A \to Y$ over a D-cofibration $A \to X$ of $D-C.W.$-spaces. In fact using the discussion in 3.8 above one can simply say that by taking O to be the small orbit category containing $O(X) \cup O(A)$ we can apply $hom(T,-)$ for all $T \in O$ to transform the obstruction problem

5.11

$$\begin{array}{ccc} A & \xrightarrow{f} & Y \\ \downarrow & \nearrow_{\tilde{f}} & \\ X & & \end{array}$$

To an equivalent obstruction problem

$$\begin{array}{ccc} A^O & \xrightarrow{f^O} & Y^O \\ \downarrow & \nearrow & \\ X^O & & \tilde{f}^O \end{array}$$

for O-spaces. The advantage is that the spaces X^O, A^O are *free.* Therefore the treatment of Dwyer and Kan in [D-K-2] applies. In fact assuming, for the sake of simplicity, that $hom(T, Y)$ is a simple space (and $Y(d)$ is a Kan complex if we work in $S.^D$) then the obstruction to extend the restriction of $\tilde{f}^{(n)}$ to $(n-1)$- skeleton in the $C.W.$-structure of X, A over the $(n+1)$-skeleton lie in $H_D^{n+1}(X, A; \underline{\pi}_n Y)$ where $\underline{\pi}_n Y$ is the coefficient system $O^{op} \to$ (abelian groups) given by $(\underline{\pi}_n Y)(T) = \pi_n \, hom_D(T, Y)$. For any discrete group G we thus capture the case given in [Bredon]. In this framework one can therefore develop Postnikov decomposition, R-localization and other common techniques of homotopy theory, for general $D-C.W.$- complexes [May].

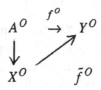

For example a D-Eilenberg-MacLane space $\underline{K}$ is a D-space with $hom_D(T,\underline{K})$ on Eilenberg-MacLane space for *every* $T \in O_D$. Taking $D = J$ from (1.1) one sees that the algebraic classification of say 2-stage Postnikov tower using the k-invariant gives an algebraic classification of *maps* of usual 2-stages not possible without these D-cohomology groups.

5.12 Function complexes. Given two D-spaces X,Y we saw that the space $hom_D(X,Y)$ is of central importance in equivariant homotopy theory. It is unlikely that an algebraic machinery can be developed to approximate its homotopy groups $\pi_i\, hom_D(X,Y)$ directly for a non-free X assuming X is D–$C.W$-complex and Y is Kan. Therefore it may be of interest to observe the following consequence of 3.11 above: Compare [El] for $D = G$.

5.13 Proposition: *Let X,Y be D-spaces. Then there exist free O-spaces X^O, Y^O and a weak homotopy equivalence $hom_D(X,Y) \simeq hom_O(X^O,Y^O)$ for some small orbit category $O \subseteq O_D$.*

Proof: [D-Z-2] Let O be the union of $O(X)$ and $O(Y)$, and X^O, Y^O as in (3.7) above. It was shown in [DF] the construction $X \to X^O$ gives an equivalence between two model categories: That of O- spaces with weak equivalences being *local* weak equivalence and co-fibrant diagrams being retracts of *free* ones; and that of D-spaces of type O where cofibrant spaces include all retracts of D–$C.W.$-complexs. But it is shown in [D-K-1] that an equivalence of simplicial model categories induces a weak equivalence on function complexes.

5.14 Filtration of $\pi_*\, hom_D(X,Y)$: Proposition 5.13 opens the way to a spectral sequence converging to $\pi_*\, hom_D^*(X,Y)$ of pointed D-spaces. Let R denote here the Bousfield-Kan free R-module functor on simplicial sets. If X,Y are in $S.^D$ and $Y(d)$ is R-nilpotent for all d, then clearly due to functoriality the map $Y \to R_\infty Y = tot\, R^* Y$ is a local weak equivalence where RY, $R_\infty Y$ are the said functors applied to each space in the diagram separately: $(RY)(d) = R(Y(d))$. Then using (5.13) above one can assume that X is free and therefore (2.9)

$$hom_D(X,Y) \xrightarrow{\simeq} hom_D(X,R_\infty Y)$$

is a weak equivalence of spaces. But $R_\infty Y$ is the homotopy inverse limit of the cosimplicial resolution of Y by means of $R^n Y$: $R_\infty Y = \underleftarrow{holim}\, R^* Y$; since $\underleftarrow{holim}$ is a function complex $hom_C(E,R^* Y)$ over the appropriate category C, it commutes with function complexes into $R^* Y$ and one has:

$$hom_D(X, \underleftarrow{holim}\, R^* Y) \simeq \underleftarrow{holim}\, hom_D(X,R^* Y).$$

Thus the usual spectral sequence associated with $\underleftarrow{holim}$ [B-K] gives us a spectral sequence as needed. Notice that for each n the D-space $R^n Y$ depends, though in a complicated way, only on the homology groups of Y with coefficient in R and the maps between these groups coming from Y. To get around this difficulty a different filtration of the function complex $hom_D(X, R_\infty Y)$ was developed in [D-Z-2]. It uses heavily the present techniques combined with those of [B-K] to get a D-equivariant version of the Bousfield-Kan homotopy spectral sequence that converges in favorable cases to the desired homotopy groups.

One gets a spectral sequence associated to the equivariant pointed function complex $hom_D^*(X,Y)$ for any small simplicial category D whose E_2-term is an unstable Ext-group.

REFERENCES

1. [A] M. Andre, *Methode Simpliciale en Algebra homologique et Algebra Commutative.* Lecture Notes in Math. 32 (1967) Springer-Verlag.
2. [B-K] A. Bousfield, D.Kan, *"Homotopy limits, completions and localizations.* Lecture notes in Math. No. 304 (1972), Springer-Verlag.
3. [Br] G. Bredon, *"Equivariant cohomology theories".* Lecture Notes in Math. No. 34 (1967), Springer-Verlag.
4. [Bro] T. Brocker, *Singulare Definition der Aquivarianten Bredon homology".* Manuscripta Math. 5 (1971), p. 91-102. Springer-Verlag.
5. [D-Z] E. Dror Farjoun, A. Zabrodsky, *"Homotopy equivalences between diagrams of spaces:.* J. of pure and applied algebra, 41 (1986), pp. 169-182.
6. [D-Z-2] --------: *The equivariant homotopy spectral sequence.* Proceeding Barcelona Conference. Springer Verlag (to appear).
7. [DF] E. Dror Farjoun: *"Homotopy theories for diagrams of spaces".* Proceeding A.M.S. (to appear).
8. [D-K-1] W. G. Dwyer and D. M. Kan, *"Function complexes in homotopical algebra".* Topology 19 (1980) pp. 427-440.
9. [D-K-2] --------: *"An obstruction theory for diagrams of simplicial sets".* Proc. Kon. Akad. van Wetensch. A87=Ind. Math. 46 (1984), pp. 139-146.
10. [D-K-3] --------: *"Singular functors and realization functors".* Proc. Kon. Akad. van Wetensch. A87=Ind. Math.46 (1984).
11. [D-K-4] --------: *Equivariant homotopy classification,* J.of Pure and Appl. Alg. 35 (1985), pp. 269-285.
12. [D-K-5] --------: *Function complexes for diagrams of simplicial sets.* Proc. Kan. Akad. van Wetensch. A86-Ind. Math. 45 (1983).

13. [El] A.D.Elmendorf, *"Systems of fixed point sets"*, Trans. A.M.S. 277, (1983), pp. 275-284.

14. [H] U. Hommel, *"Singulare algebraische topologie for D-raume"* Thesis, Univ. of Heidelberg, 1985.

15. [Ill] S. Illman, *"Equivariant singular homology and cohomology I"*, Memoirs A.M.S.No. 156, (1975).

16. [Mac] MacLane, *"Categories for the working Mathematician"*. Grad. Text in Math. #5 (1971) Springer-Verlag.

17. [May] J. P. May, *"Equivariant homotopy and cohomology theory"*, Contemporary Mathematics 12 (1981), pp. 209-217, A.M.S.

18. [M] H. Miller, *"The Sullivan fixed point conjecture on maps from classifying spaces,"* Ann. of Math. 120 (1984), pp. 39-87.

19. [SE] G. Segal, *"Equivariant stable homotopy theory."* Actes Congress Intern. Math. 1970 Tom 2 pp. 59-63.

20. [Qu] D. G. Quillen, *"Homotopical algebra"*. Lecture Notes in Math. 43 (1967), Springer Verlag.

21. [W] C. Watts, *"A homology theory for small categories"*. Proceedings of Conference on Categorical algebra, La Jolla, (1965), Springer Verlag.

THE KERVAIRE INVARIANT AND THE HOPF INVARIANT

M.G. Barratt, J.D.S. Jones and M.E. Mahowald

INTRODUCTION In this paper we look at the Kervaire invariant problem from the point of view of unstable homotopy theory and show that it is intimately connected with properties of the image of J. For some general discussion and an idea of the context in which to fit this work we refer to [3].

In the first part of this paper we describe a formulation of the Kervaire invariant problem in terms of the classical EHP sequence

$$\ldots \longrightarrow \pi_i S^n \xrightarrow{E} \pi_{i+1} S^{n+1} \xrightarrow{H} \pi_{i+1} S^{2n+1} \xrightarrow{P} \pi_{i-1} S^n \longrightarrow \ldots$$

Here E, H and P are the suspension, Hopf invariant and Whitehead product homomorphisms respectively. This sequence is exact in the metastable range; it is always exact on 2-primary components.

Write E^r for the r-th iterate of the suspension homorphism and $E_n : \pi_{i+n} S^n \longrightarrow \pi_i{}^S$ for the stabilisation homorphism. If $\alpha \in \pi_i{}^S$ and n is the minimum integer such that α is in the image of E_n then we say that S^n is the sphere of origin of α. If S^n is the sphere of origin of α then we define the stable Hopf invariant of α to be the set

$$H^S(\alpha) = \{ H(x) \mid x \in \pi_{i+n} S^n \text{ and } E_n(x) = \alpha \}.$$

First we study the sphere of origin and stable Hopf invariant of elements of $\pi^S{}_*$ with Kervaire invariant one. To state the results precisely we need to introduce some notation.

Let k be an integer, then define integers N = N(k), t = t(k) and a = a(k) by the following formulas:

$$N = 2^{k+1} - 2$$

$t(0) = 0, t(1) = 1, t(2) = 3, t(3) = 7$ and $t(4b+c) = 8b+t(c)$

$a = N - t + 1 = 2^{k+1} - t(k) - 1.$

Note that the k-th nontrivial homotopy group of SO occurs in dimension $t(k)$. We pick a generator of $\pi_{t(k)}SO$ and denote by β_K the generator of the image of J determined by applying the J homorphism to this chosen element. Now recall from [2], [30] and [20] that Adams' e-invariant, $e : \pi_*{}^S \longrightarrow \mathbb{Q}/\mathbb{Z}$ detects the image of J. We write $K : \pi_*{}^S \longrightarrow \mathbb{Z}/2$ for the Kervaire invariant, so from Browder's theorem, [6], $K = 0$ in dimensions not of the form $N(k)$.

It sometimes happens that $k = 1,2,3$ are special cases, but all the results one wants are true in these special cases. So in order to avoid a number of special arguments we implicitly exclude the cases $k = 1,2,3$ when they do not follow the general pattern

THEOREM A. Fix k and let α be any element of $\pi_{N+a}S^a$. Suppose that $eH(\alpha)$ is an odd multiple of $e(\beta_K)$, then $K(E_a\alpha) = 1$.

Note that in the notation of the theorem $H\alpha$ is in $\pi_{N+a}S^{2a-1} = \pi_{t(k)}{}^S$ and so the statement does in fact make sense.

COROLLARY. Fix k and suppose α is any element of $\pi_N{}^S$ such that:

(i) The sphere of origin of α is a.

(ii) There is an element γ in $H^S(\alpha)$ with $e(\gamma)$ an odd multiple of $e(\beta_K)$.

Then α has Kervaire invariant one.

We conjecture that the converse is true, that is if $\alpha \in \pi_N{}^S$ is any element with Kervaire invariant one then α satisfies (i) and (ii) of the above corollary. We discuss some partial results, under the additional hypothesis that α has order 2 in §5. The following conjectures also seem reasonable (and are independent of the existence or non-existence of elements with Kervaire invariant one):

(i) $E_a\pi_{N+a}S^a = \pi_N{}^S$.

(ii) $E_{a-1}\pi_{N+a-1}S^{a-1} = \ker(K : \pi_N{}^S \longrightarrow \mathbb{Z}/2)$.

Now assume that $k \geq 4$ so we may identify $\pi_{a+t(k)-1}S^{a-1}$ with $\pi_{t(k)}^S$. In view of Theorem A it is natural to ask if there is an element of $\pi_{N+a}S^a$ whose Hopf invariant is actually β_k. From the EHP sequence, the obstruction to the existence of such an element is the Whitehead product

$$P(\beta_k) = [\iota_{a-1}, \beta_k] \in \pi_{N+a-2}S^{a-1}.$$

Assume the stability hypothesis $3t(k) - 2 \leq 2n$, then a lot is known about Whitehead products of the form $[\iota_n, \beta_k]$:

$$[\iota_n, \beta_k] = 0 \quad \text{if } 3 \leq \nu(n+t(k)+2) \leq k$$
$$[\iota_n, \beta_k] \neq 0 \quad \text{if } \nu(n+t(k)+2) \geq k+1 \text{ but } n+t(k)+2 \neq 2^{k+1}.$$

Here $\nu(m)$ is the exponent of 2 occuring in the factorisation of the integer m into prime powers. These results and many related results are discussed in [19]. A complete proof of the first is given in [12]. The second is implicit in [22]; we give an explicit proof here. Of course the case we are interested in is the only case not covered by the above results.

In the second part of this paper we study Whitehead products

$$[\iota_{a(k+1)-1}, \gamma] \in \pi_{N(k+1)+a(k+1)-2}(S^{a(k+1)-1}),$$

where $\gamma \in \pi_{a(k+1)+t(k+1)-1}S^{a(k+1)-1}$ is such that $e(\gamma)$ is an odd multiple of $e(\beta_{k+1})$, under the assumption that there exists an element $\theta_k \in \pi_{N(k)}^S$ with Kervaire invariant one and certain other properties, the precise result is theorem B below.

Fix k, then from [1] we know that there is an element $w \in \pi_{2N-r+1}S^{N-r+1}$ with $E^r w = [\iota_{N+1}, \iota_{N+1}]$ where $r = t(k+1)$ and that for any choice of w, $eH(w)$ is an odd multiple of $e(\beta_{k+1})$.

THEOREM B. Let w be an element of $\pi_{2N-r+1}S^{N-r+1}$ with $E^r w = [\iota_{N+1}, \iota_{N+1}]$. Suppose that there is an element $\theta \in \pi_{2N-p+1}S^{N-p+1}$ with $2E^{p-q}\theta = E^{r-q}w \in \pi_{2N-q+1}S^{N-q+1}$ where

$$r \geq p \geq q, \quad p+q \geq r, \quad N-p+1 \equiv 0 \bmod 4$$

Then it follows that

$$E^{2p-r}((\theta\wedge\theta)\circ\eta) = [\iota_{e(k+1)-1}, \tilde{\delta}]$$

where $e(\tilde{\delta}) = e(\beta_{k+1})$ mod 2.

Here, of course, $\theta\wedge\theta$ is the element of $\pi_{4N-2p+2}S^{2N-2p+2}$ given by the smash product of $\theta \in \pi_{2N-p+1}S^{N-p+1}$ with itself. We will use a similar notation throughout the paper.

We have not tried to make the hypotheses on p and q best possible. A few numerical experiments may be necessary to reveal the meaning of this theorem. We explain the content when we assume that we are in the nontrivial cases $k \geq 4$ and the sphere of origin of θ is, as conjectured, S^{N-t+1} where $t = t(k)$. Then $w-2\theta$ is zero on S^{N+1} and to use the theorem we need to be able to prove that $w-2\theta$ is zero on S^{N+1-q} where

$$q \geq 2 \text{ if } k \equiv 0 \bmod 4$$
$$q \geq 3 \text{ if } k \equiv 1 \bmod 4$$
$$q \geq 4 \text{ if } k \equiv 2 \bmod 4$$
$$q \geq 1 \text{ if } k \equiv 3 \bmod 4$$

We can use weaker hypotheses on the sphere of origin of θ at the expense of having to prove that $w-2\theta$ is zero on a lower dimensional sphere.

Very roughly this theorem says that if we assume there is an element $\theta_k \in \pi_N{}^S$ of order 2 and Kervaire invariant one then

$$\eta\theta_k{}^2 = 0 \quad \text{stably,}$$
$$\eta\theta_k{}^2 = [\iota_{e(k+1)-1}, \tilde{\delta}] \quad \text{on } S^{e(k+1)-1}$$

where $e(\tilde{\delta})$ is an odd multiple of $e(\beta_{k+1})$, provided we make suitable sphere of origin assumptions. This result is discussed in [3] in the section entitled "The relation between the unstable and inductive approaches". The point is that the usual inductive approach to the Kervaire invariant problem starts from the assumption that there exists an elment $\theta_k \in \pi_*{}^S$ with $K(\theta_k) = 1$ and $2\theta_k = 0$. One deduces that $\eta\theta_k{}^2 = 0$ in $\pi_*{}^S$. To construct θ_{k+1} it is necessary to construct a null homotopy of $\eta\theta_k{}^2$ with a specific property, see 2.3 of [3]. In Theorem B we make the same assumptions, plus extra sphere of origin assumptions and this time we deduce that it is sufficient to construct a null homotopy of $\eta\theta_k{}^2$ on $S^{e(k+1)-1}$.

We believe that the two conditions on a null homotopy of $\eta\theta_k{}^2$ we arrive at are the same but we leave it to the reader to attach any meaning to this statement.

In rough outline the proof of theorem A goes as follows. First we go through a standard proceedure in suspension theory to convert the problem to one concerning the stable homotopy of real projective space. This is done in §1. We analyse the resulting problem using the bo-resolution, that is the Adams spectral sequence based on connective real K theory. The basic structure of this argument is described in §2 and in §§ 3,4 the details are filled in.

In §5 we determine the sphere of origin and Hopf invariant of elements of Kervaire invariant one which also have order two.

The proof of theorem B is based on the cup-1-construction in unstable homotopy theory. In §6 we state the properties of the cup-1-construction we need and deduce theorem B from these properties. In §§ 7,8 we fill in the details, that is we give the precise definition of the cup-1-construction and prove the required properties.

§1 THE EHP SEQUENCE AND PROJECTIVE SPACE

We begin this section by describing a basic technique in suspension theory, due in essence to Toda [32]. This consists of converting sphere of origin and Hopf invariant problems into problems concerning projective space. We then apply this technique to the problem of studying Whitehead products of the form $[\iota_n,\gamma]$ where $e(\gamma)$ is non zero.

We write P^n for the real n dimensional projective space and $P_k{}^n$ for the truncated projective space P^n/P^{k-1}. We write u for an inclusion and c for a collapsing map between truncated projective spaces. The basic idea is to compare the filtration of the space $\Omega^\infty S^\infty$ by the spaces $\Omega^n S^n$ with the filtration of the space QP^∞ by the spaces QP^n, where, as usual, Q denotes the functor $\Omega^\infty S^\infty$. We will assume all spaces and spectra are localised at 2.

There are several ways of describing the relation between the EHP sequence and projective space [32], [14]. We choose to start from the commutative diagram of fibrations, [9], [16],

$$
\begin{array}{ccccc}
\Omega^n S^n & \xrightarrow{E} & \Omega^{n+1} S^{n+1} & \xrightarrow{H} & \Omega^{n+1} S^{2n+1} \\
s_n \downarrow & & s_{n+1} \downarrow & & \downarrow \\
QP^{n-1} & \xrightarrow[Qu]{} & QP^n & \xrightarrow[Qc]{} & QS^n.
\end{array}
\tag{1.1}
$$

Here E is the suspension map, H is the James Hopf invariant map and $s_n : \Omega^n S^n \longrightarrow QP^{n-1}$ is an appropriate version of the Snaith map [31]. To ensure that the above diagram commutes it is necessary to use the Snaith maps constructed in [9] or [16].

This diagram gives a commutative diagram of exact sequences where the top row is the EHP sequence.

$$
\begin{array}{ccccccc}
\ldots \longrightarrow \pi_{i+n} S^n & \xrightarrow{E} & \pi_{i+n+1} S^{n+1} & \xrightarrow{H} & \pi_{i+n+1} S^{2n+1} \longrightarrow \ldots \\
\downarrow & & \downarrow & & \downarrow \\
\ldots \longrightarrow \pi_i^S P^{n-1} & \xrightarrow[u_*]{} & \pi_i^S P^n & \xrightarrow[c_*]{} & \pi_i^S S^n \longrightarrow \ldots
\end{array}
\tag{1.2}
$$

In particular we conclude the following:

(1.3) If $\gamma \in \pi_{i+n+1} S^{2n+1}$ is such that $P(\gamma) = 0$ then there exists an element $\alpha \in \pi_i^S(P^n)$ such that in $\pi_i^S(S^n)$, $c_*\alpha = E_{2n+1}\gamma$, that is $c_*\alpha$ is the stable element determined by γ.

The next step is to assume that $P(\gamma) = 0$ and to translate conditions on $e(\gamma)$ into conditions on the homotopy class $\alpha \in \pi_i^S(P^n)$. This is most conveniently done using J theory and we now summarise the relevant facts.

Denote -1 connected and 3 connected real K-theory by bo and bspin respectively. The Adams operation $\psi^3 - 1$ extends to an operation $\psi : bo \longrightarrow bspin$. There is a homology theory J which fits into the following exact sequence

$$\ldots \longrightarrow J_n(X) \longrightarrow bo_n(X) \longrightarrow bspin_n(X) \longrightarrow J_{n-1}(X) \longrightarrow \ldots$$
$$\psi$$

We define the d invariant and the e invariant to be the Hurewicz homomorphisms in the theories bo and J respectively

$$d : \pi_*{}^S(X) \longrightarrow bo_*(X)$$
$$e : \pi_*{}^S(X) \longrightarrow J_*(X).$$

We will also write d,e for the unstable d and e invariants that is the composites

$$\pi_*(X) \longrightarrow \pi_*{}^S(X) \longrightarrow bo_*(X)$$
$$\pi_*(X) \longrightarrow \pi_*{}^S(X) \longrightarrow J_*(X).$$

When $X = S^0$ we recover, essentially, Adams' d and e invariants [2] so in particular we know, from [2], [30], [20], that $e : \pi_*{}^S(S^0) \longrightarrow J_*(S^0)$ defines an isomorphism of $J_*(S^0)$ with the subgroup of $\pi_*{}^S(S^0)$ generated by the image of J and the μ family.

Now suppose that γ and α satisfy the conditions in 1.3, and that $e(\gamma) \neq 0$. Then it certainly follows that $e(\alpha) \neq 0$ in $J_j(P^n)$. So to proceed we need some information on the groups $J_*(P^n)$, more accurately $J_*(P^\infty)$.

LEMMA 1.4 (a) $J_{8q-2}(P^\infty) = \mathbb{Z}/2^{\nu(2q)}$.

(b) Let $g \in J_{8q-2}(P^\infty)$ be a generator of this cyclic group. Then $2^b g$ is in the image of $J_{8q-2}(P^m)$ if and only if $m \geq 8q-2-t(b+2)$.

(c) Let $m = 8q-2-t(b+2)$ and let x be any element of $J_{8q-2}(P^m)$. Then in $J_{8q-2}(P^\infty)$, $u_*x = 2^b g$ where g is a generator of $J_{8q-2}(P^\infty)$ if and only if $ec_*x = e(\beta_{b+2})$ mod $2J_{8q-2}(S^m)$.

This lemma is well known, [22], [23].

At this point we will establish some conventions, which will be in use throughout this paper, concerning the classical mod 2 Adams spectral sequence. Let A be the mod 2

Steenrod algebra. If Y is a space then $\mathrm{Ext}_A^{s,t}(H^*(Y;\ \mathbb{Z}/2),\ \mathbb{Z}/2)$ will be abbreviated to $\mathrm{Ext}^{s,t}(Y)$. If $\alpha \in \pi_n^S(Y)$ then we write $AF(\alpha) = s$ to mean that α has filtration precisely s in the mod 2 Adams spectral sequence. So $AF(\alpha) = s$ means that α is non-zero and is detected by a non-zero element in $\mathrm{Ext}^{s,n+s}(Y)$. Similar conventions will be used when Y is a spectrum.

Next we state a fact about the groups $\mathrm{Ext}(P^\infty)$:

$$\mathrm{Ext}^{1,t}(P^\infty) = \mathbb{Z}/2 \quad \text{if } t = 2^{k+1} - 1.$$

This is verified in [34]. There is a little more discussion of $\mathrm{Ext}(P^\infty)$ in §4.

It is not too difficult to check that if $\alpha \in \pi_N^S(P^\infty)$ with $N = 2^{k+1} - 2$ then $AF(\alpha) = 1$ if and only if the following condition is satisfied. Put $m = 2^k$, then in the stable complex $X = P^\infty \cup_\alpha e^{N+1}$, the cohomology operation φ is non zero on $H^{m-1}(X;\mathbb{Z}/2)$ if and only if $\varphi = Sq^m$ modulo decomposables. This condition is sometimes rather carelessly phrased as "α is detected by Sq^m". In the terms of §4 it is the statement that the non-zero element of $\mathrm{Ext}^{1,t}(P^\infty)$, where $t = 2^{k+1} - 1$, is $h_k e_k$.

We can now state the main technical theorem which describes the possible values the e invariant $\pi_{8q-2}^S(P^\infty) \longrightarrow J_{8q-2}(P^\infty)$ may assume.

THEOREM 1.5. (a) If $q \neq 2^r$, then $e : \pi_{8q-2}^S(P^\infty) \longrightarrow J_{8q-2}(P^\infty)$ is zero.

(b) If $q = 2^r$, $r \geq 1$ and $\alpha \in \pi_{8q-2}^S(P^\infty)$, then $2e(\alpha) = 0$; and

$$e(\alpha) \neq 0 \Rightarrow AF(\alpha) = 1.$$

This result is stated as 7.11 in [22]; unfortunately the proof given there has a gap in it, as explained in the correction to [22]. Indeed it requires quite a delicate comparison of mod 2 Adams filtration and bo-Adams filtration to prove the result.

We believe that the converse to 1.5 (b) is true, that is if $q = 2^r$, $r \geq 1$ and $\alpha \in \pi_{8q-2}^S(P^\infty)$, then

$$AF(\alpha) = 1 \Rightarrow e(\alpha) \neq 0.$$

but at the time of writing this paper we have not written down a complete proof. A partial converse, where we assume in addition that $2\alpha = 0$, is given in 5.2.

COROLLARY 1.6. Assume the stability hypothesis $3t(k) - 2 \leq 2n$ and let $\gamma \in \pi_{n+t(k)}S^n$ be any element such that $e(\gamma) = e(\beta_k) \bmod 2$. Then $[\iota_n, \gamma] \neq 0$ if $\nu(n + t(k) + 2) \geq k+1 \geq 3$ but $n + t(k) + 2 \neq 2^{k+1}$.

Proof. Suppose $[\iota_n, \gamma] = 0$, then by 1.3 there is an element $\alpha \in \pi_{n+t(k)}{}^S(P^n)$ such that, in $\pi_{n+t(k)}{}^S(S^n)$, $c_*\alpha = E_{2n+1}\gamma$ so in particular $ec_*(\alpha) = e(\beta_k) \bmod 2$. Assume that $\nu(n+t(k)+2) \geq k+1 \geq 3$, then from part (c) of 1.4 we can compute the e invariant of the element $u_*(\alpha) \in \pi_{n+t(k)}{}^S(P^\infty)$

$$eu_*(\alpha) = 2^{k-2}g, \quad \text{where g is a generator of } J_{n+t(k)}(P^\infty).$$

From part (a) of 1.4 the group $J_{n+t(k)}(P^\infty)$ is cyclic of order 2^r where $r = \nu(n+t(k)+2) - 2 \geq k-1$ and so

$$eu_*(\alpha) \neq 0.$$

If $n+t(k)+2 \neq 2^{k+1}$ this contradicts 1.5.

Next we explain how to prove theorem A assuming 1.5. We will need to use the following form of the Kahn-Priddy theorem [17], [18]. There is a map $s : QS^0 \longrightarrow QP^\infty$ which is compatible with the Snaith maps s_n occuring in 1.1 in the sense that the following diagram commutes

$$\begin{array}{ccc} \Omega^n S^n & \longrightarrow & QS^0 \\ s_n \downarrow & & \downarrow s \\ QP^{n-1} & \longrightarrow & QP^\infty. \end{array}$$

Here the horizontal arrows are the obvious inclusions. Now let $\tau : P^\infty \longrightarrow S^0$ be the stable transfer map, then $(\Omega^\infty\tau)s \simeq 1$. To find the result in this precise form it is necessary to look in [9] and [16].

Proof of Theorem A assuming 1.5. Fix k and suppose $\theta \in \pi_{N+a}S^a$ is such that $eH(\theta)$ is an odd multiple of $e(\beta_k)$, then from 1.3 with $\gamma = H\theta$, we deduce that there is an element $\alpha' \in \pi_N{}^S(P^{a-1})$ such that $ec_*(\alpha') = e(\beta_k) \bmod 2J_N(S^{a-1})$. From 1.4 and a little exercise with the numbers involved we conclude that the element $\alpha = u_*(\alpha') \in \pi_N{}^S(P^\infty)$ has the property that $e(\alpha) \neq 0$ and therefore by 1.5, $AF(\alpha) = 1$.

We now use the stable transfer map $\tau : P^\infty \longrightarrow S^0$; we check that $K\tau_*(\alpha) \neq 0$ if and only if $AF(\alpha) = 1$. The stable map τ has Adams filtration 1 so it defines a homorphism τ_* : $Ext^{s,t}(P^\infty) \longrightarrow Ext^{s+1,t+1}(S^0)$ and furthermore this map is an isomorphism if $s = 1$ and $t = 2^{k+1} - 1$, compare [34]. This homorphism in Ext groups is compatible with the induced homorphism in homotopy groups in the sense that if $x \in \pi_*{}^S(P^\infty)$ has Adams filtration s, then $AF(\tau_* x) \geq s + 1$; if x is detected by $a \in Ext^{s,t}(P^\infty)$ and $AF(\tau_* x) = s + 1$ then $\tau_*(x)$ is detected by $\tau_* a \in Ext^{s+1,t+1}(S^0)$.

Next we use Browder's theorem [6] in the following form; if $y \in \pi_N{}^S$ then

$$K(y) = 1 \Leftrightarrow AF(y) = 2.$$

We conclude that

$$AF(\alpha) = 1 \Leftrightarrow AF(\tau_*\alpha) = 2 \Leftrightarrow K(\tau_*\alpha) = 1.$$

Finally by the Kahn-Priddy theorem $\tau_*(\alpha) = E_a(\theta)$ and so we have shown that if $eH(\theta)$ is an odd multiple of $e(\beta_k)$ then $E_a(\theta)$ has Kervaire invariant one.

§2 THE bo-RESOLUTION AND THE e-INVARIANT

We work stably in the category of CW spectra. We do not use any special notation to distinguish between a CW complex and its suspension spectrum. All spaces and spectra are to be localised at 2. We write π_* for homotopy groups in the category of spectra.

Let X be a spectrum, then $E_2{}^{s,t}(X;bo)$, the E_2 term of the bo-Adams spectral sequence of X, is the homology of the chain complex

(2.1) $$\pi_t bo\wedge X \longrightarrow \pi_t bo\wedge Ibo\wedge X \longrightarrow \ldots \longrightarrow \pi_t bo\wedge I^s bo\wedge X \longrightarrow \ldots .$$

Here Ibo is the cofibre of the unit map $S^0 \longrightarrow bo$ and $I^s bo$ is the smash product $Ibo\wedge\ldots\wedge Ibo$ with s factors. The boundary homorphisms $\partial : \pi_t bo\wedge I^s bo\wedge X \longrightarrow \pi_t bo\wedge I^{s+1} bo\wedge X$ are induced by the maps

$$bo\wedge I^s bo\wedge X \longrightarrow Ibo\wedge I^s bo\wedge X = S^0\wedge I^{s+1} bo\wedge X \longrightarrow bo\wedge I^{s+1} bo\wedge X.$$

The homotopy type of $bo\wedge Ibo$ is analysed in [21]. There are spectra B_n and inverse equivalences

$$\lambda : bo\wedge Ibo \longrightarrow \bigvee_{n\geq 1} bo\wedge B_n$$

(2.2)

$$\mu : \bigvee_{n\geq 1} bo\wedge B_n \longrightarrow bo\wedge Ibo.$$

The spectra B_n are the so-called integral Brown-Gitler spectra [21], [13]. The main ingredient in our proofs is complete information on the action of these maps λ and μ in integral homology modulo torsion and we now recall this information from [28] and [24].

The torsion free quotient of $H_*(bo; \mathbb{Z}_{(2)})$ is $\mathbb{Z}_{(2)}$ in every degree of the form $4n$ with $n \geq 0$. We pick generators $a_n \in H_{4n}(bo; \mathbb{Z}_{(2)})$ for the torsion free summands. The torsion subgroup of $H_*(bo; \mathbb{Z}_{(2)})$ is finite and of exponent 2. The spectra B_n are $4n-1$ connected and the groups $H_i(B_n; \mathbb{Z}_{(2)})$ are finite groups of exponent 2 if $i \geq 4n+1$; $H_{4n}(B_n; \mathbb{Z}_{(2)}) = \mathbb{Z}_{(2)}$ and we pick a generator u_n. The maps λ and μ have the property that

$$\lambda_*(a_s\otimes a_t) = \sum_{1\leq n\leq t} (n,t-n)k_n a_{s+t-n}\otimes u_n \quad \text{modulo torsion}$$

(2.3)

$$\mu_*(a_s\otimes u_t) = \sum_{1\leq n\leq t} (-1)^{t-n}(n,t-n)m_n a_{s+t-n}\otimes a_n \quad \text{modulo torsion.}$$

Here (c,d) is the binomial coefficient $(c+d)!/c!\,d!$ and k_n, m_n are odd integers.

We use the notation λ_n and μ_n for the n'th components of λ and μ. There is an equivalence $bo\wedge B_1 \longrightarrow bspin$ and under this equivalence the composite map

$$bo \longrightarrow Ibo \longrightarrow bo \wedge Ibo \xrightarrow{\lambda_1} bo \wedge B_1 \longrightarrow bspin$$

is the Adams operation ψ used in §1, [28], [23], [24]. Therefore there is a commutative diagram

$$\begin{array}{ccc} \pi_t bo \wedge X & \xrightarrow{\partial} & \pi_t bo \wedge Ibo \wedge X \\ \psi_* \downarrow & & \downarrow \lambda_{1*} \\ \pi_t bspin \wedge X & \xrightarrow[\cong]{} & \pi_t bo \wedge B_1 \wedge X. \end{array} \tag{2.4}$$

Here and at various places we simplify our notation by writing ψ_*, λ_{1*} for $\psi \wedge 1_*$ and $\lambda_1 \wedge 1_*$. We get a homorphism, which will still be denoted by λ_{1*}

$$\lambda_{1*} : E_2^{1,t}(X;bo) \longrightarrow J_{t-1}(X).$$

Let $F^1\pi_n(X)$ be the kernel of the homorphism $d : \pi_n(X) \longrightarrow bo_n(X)$, then there is a homorphism $F^1\pi_n(X) \longrightarrow E_2^{1,n+1}(X;bo)$ and a commutative diagram

$$\begin{array}{ccc} F^1\pi_n X & \longrightarrow & E_2^{1,n+1}(X;bo) \\ \downarrow & & \downarrow \lambda_{1*} \\ \pi_n X & \xrightarrow[e]{} & J_n(X). \end{array} \tag{2.5}$$

This diagram puts definite restrictions on the values the e invariant may assume.

Our aim is to exploit this diagram in the case $X = P^\infty$ and to do this we clearly need some information on the various groups when $X = P^\infty$. We will need the following three lemmas. If n is an integer then let $\varphi(n)$ be the number of elements in the set $\{t \mid 0 < t \leq n,\ t \equiv 0, 1, 2, 4 \text{ modulo } 8\}$.

LEMMA 2.6. (1) $\pi_{4p-1} bo \wedge P^\infty = \mathbb{Z}/(2^{\varphi(4p)})$.

(2) $\pi_{4p-1} bspin \wedge P^\infty = \mathbb{Z}/(2^{\varphi(4p)-3})$.

(3) Let v_p be a generator of $\pi_{4p-1} bo \wedge P^\infty$ and let w_p be a generator of $\pi_{4p-1} bspin \wedge P^\infty$. Then $\psi_* v_p = kpw_p$ where k is odd.

(4) $J_{4p-2} P^\infty = \mathbb{Z}/(2^{\nu(p)})$.

This lemma is well known, see [10] and [23]. The next lemma gives information on the homorphism $\partial : \pi_{4p-1} bo \wedge P^\infty \longrightarrow \pi_{4p-1} bo \wedge \bar{I} bo \wedge P^\infty$. We retain the notation v_p for a generator of $\pi_{4p-1} bo \wedge P^\infty$.

LEMMA 2.7 (1) If $p \neq 2^r$ then $AF(4\partial v_p) = 2$.

(2) If $p = 2^r$ with $r \geq 2$ then $AF(4\partial v_p) = 4$.

The significance of (1), for example, is that it tells us $AF(\partial v_p) = 0$ and that $4\partial v_p$ does not jump in filtration. This is the lemma which ultimately gives the restrictions on the values the e invariant in P^∞ may attain. One can understand the role of these lemmas in the proof of 1.5 as follows; ∂v_p is in the group $\pi_* bo \wedge \bar{I} bo \wedge P^\infty \cong \oplus_{n \geq 1} \pi_* bo \wedge B_n \wedge P^\infty$ and 2.4 and 2.6 tell us that $\lambda_{1*} \partial v_p \in \pi_{4p-1} bo \wedge B_1 \wedge P^\infty$, is divisible by $2^{\nu(p)}$. However 2.7 tells us that ∂v_p itself is not divisible by 2 if $p \neq 2^r$ and if $p = 2^r$ with $r \geq 3$ then ∂v_p is divisible by at most 4. Our aim is to show that if we start with an element

$$\alpha \in \pi_* bo \wedge \bar{I} bo \wedge P^\infty = E_1^{1,*}(P^\infty; bo)$$

which is an infinite cycle in the bo-Adams spectral sequence and has the property that in $\pi_* bo \wedge B_1 \wedge P^\infty$, $2\lambda_{1*}\alpha = \lambda_{1*}(\partial v_p)$ then in fact we can construct a new element $\beta \in \pi_* bo \wedge \bar{I} bo \wedge P^\infty$, which is also an infinite cycle in the bo-Adams spectral sequence, with the property that $2\beta = \partial v_p$. If $p \neq 2^r$ this contradicts 2.7. If $p = 2^r$ with $r \geq 2$ then 2.7 gives us information about β and a careful analysis leads to a proof that the element of $\pi_* P^\infty$ corresponding to the infinite cycle β has Adams filtration 1.

In order to construct this element β we will need the following lemma.

LEMMA 2.8. If $x \in \pi_* bo \wedge Ibo \wedge P^\infty$ is such that $AF(x) \geq 2$ and $\lambda_{1*}(x) \in \pi_* bo \wedge B_1 \wedge P^\infty$ is zero then

$$\partial x = 0 \Leftrightarrow x = 0.$$

For a proof see [21], [11] and the bounded torsion theorem of [24].

Proof of 1.5 (a) assuming 2.7. First we describe in more detail the homorphism $F^1\pi_n P^\infty \longrightarrow E_2^{1,n+1}(P^\infty; bo)$ mentioned above. The cofibration sequence

$$P^\infty \xrightarrow{i} bo \wedge P^\infty \xrightarrow{\rho} Ibo \wedge P^\infty \xrightarrow{\delta} SP^\infty$$

gives a homorphism $\delta_* : \pi_{n+1}\, Ibo \wedge P^\infty \longrightarrow \pi_n\, P^\infty$ whose image is $F^1\pi_n\, P^\infty$. Write

$$j : Ibo \wedge P^\infty = Ibo \wedge S^0 \wedge P^\infty \longrightarrow Ibo \wedge bo \wedge P^\infty$$

for the inclusion. If $\alpha \in F^1\pi_n\, P^\infty$, pick $\beta \in \pi_{n+1}\, Ibo \wedge P^\infty$ such that $\delta_*\beta = \alpha$. Then $j_*\beta \in \pi_{n+1}$ $Ibo \wedge bo \wedge P^\infty$ is a cycle in the chain complex 2.1, whose homology class in $E_2^{1,n+1}(P^\infty; bo)$ will be denoted $[j_*\beta]$; the function referred to above is $\alpha \longrightarrow [j_*\beta]$.

From [10] and [23] $\pi_{8q-2}\, bo \wedge P^\infty = 0$ and so $F^1\pi_{8q-2}\, P^\infty = \pi_{8q-2}\, P^\infty$. Pick $\alpha \in \pi_{8q-2}\, P^\infty$ and suppose that $e(\alpha) \in J_{8q-2}\, P^\infty = \mathbb{Z}/(2^{\nu(2q)})$ has order precisely 2^a. Pick β as in the previous paragraph; we use the information on $e(\alpha)$ to determine $\lambda_{1*} j_*(\beta) \in \pi_{8q-1} bo \wedge B_1 \wedge P^\infty$. In view of the diagram 2.4, the computations of 2.6 and the fact that $e(\alpha)$ has order precisely 2^a it must follow that $2^a \lambda_{1*} j_*(\beta) = m\lambda_{1*}\partial v_{2q}$ where m is odd, or equivalently, since we are working over $\mathbb{Z}_{(2)}$, $n2^a\lambda_{1*}j_*(\beta) = \lambda_{1*}\partial v_{2q}$ where n is odd. Let $\gamma = nj_*\beta$ and $y = \partial v_{2q} - 2^a\gamma$, so $AF(4y) \geq 2$. By construction, $\lambda_{1*}(4y) = 0$ and $4y$ is an infinite cycle in the bo-Adams spectral sequence. Therefore it follows from 2.8 that $4y = 0$; so $4\partial v_{2q} = 2^{a+2}\gamma$ and therefore $AF(4\partial v_{2q}) \geq a+2$. It now follows from 2.7 that $a = 0$ and therefore $e(\alpha) = 0$.

§3 THE bo-RESOLUTION OF P^{∞}

The aim of this section is to do some explicit calculations with the bo-resolution for P^{∞}; more precisely to prove 2.7. The conventions described at the beginning of §2 remain in force.

Let $\tau : P^{\infty} \longrightarrow S^0$ be the stable transfer map and let R be the fibre of τ. Let e^{-1} be a generator of $H^{-1}(R; \mathbb{Z}_{(2)})$ and let $\alpha_n \in H^{4n}(bo; \mathbb{Z}_{(2)})$ be a generator for the torsion free part. Let $\sigma : bo\wedge R \longrightarrow \bigvee_{n\geq 0} K(\mathbb{Z}_{(2)}, 4n-1)$ be the map defined by the cohomology classes $\alpha_n \otimes e^{-1}$. (Note that $\bigvee_{n\geq 0} K(\mathbb{Z}_{(2)}, 4n-1)$ is equivalent to $\prod_{n\geq 0} K(\mathbb{Z}_{(2)}, 4n-1)$ so that this construction of σ does make sense.) Then a cohomology computation, see [23], proves the following result:

(3.1) This map σ is an equivalence.

From 3.1, we see that the computation of the bo resolution for R amounts to computations in $\mathbb{Z}_{(2)}$ homology and these are, in principal, straightforward. We use the cofibration sequence $S^{-1} \longrightarrow R \longrightarrow P^{\infty} \longrightarrow S^0$ to deduce information about the bo-resolution for P^{∞}.

First we do the following computations, at least modulo torsion: We explicitly describe the homotopy of $bo\wedge R$ and the homotopy of $bo\wedge \overline{bo}\wedge R$. Then we compute the homorphism ∂ : $\pi_* bo\wedge R \longrightarrow \pi_* bo\wedge \overline{bo}\wedge R$.

Let $h : \pi_* X \longrightarrow H_*(X; \mathbb{Z}_{(2)})$ be the Hurewicz homomorphism. If X is $bo\wedge R$ or $bo\wedge \overline{bo}\wedge R$ then we know from 3.1 that h is injective. In view of the precise construction of σ and the equivalences of 2.2 it is possible to pick elements of the groups $\pi_* bo\wedge R$ and $\pi_* bo\wedge \overline{bo}\wedge R$ as follows:

$$x_r \in \pi_{4r-1} bo\wedge R,\ r \geq 0;\quad h(x_r) = a_r \otimes e_{-1} \quad \text{modulo torsion,}$$

(3.2)

$$y_{s,t} \in \pi_{4s+4t-1} bo\wedge \overline{bo}\wedge R,\ s \geq 0,\ t \geq 1;\quad h(y_{s,t}) = \mu_*(a_s \otimes u_t)\otimes e_{-1} \quad \text{modulo torsion.}$$

Here $e_{-1} \in H_{-1}(R; \mathbb{Z}_{(2)})$ is the generator dual to $e^{-1} \in H^{-1}(R; \mathbb{Z}_{(2)})$. In view of 3.1 and 2.2 these elements generate the torsion free part of $\pi_* bo\wedge R$ and $\pi_* bo\wedge \overline{bo}\wedge R$ respectively. The choice of $\mu_*(a_s \otimes u_t)\otimes e_{-1}$ modulo torsion as the Hurewicz image of the generators of the torsion free part of $\pi_* bo\wedge \overline{bo}\wedge R$ may seem slightly unnatural but it helps with the computations, see 3.4

Let C_p be the sub $\mathbb{Z}_{(2)}$ module of $\pi_{4p-1}\mathrm{bo}\wedge\mathrm{Ibo}\wedge R$ generated by the $y_{s,t}$ with $s+t = p$, $t \geq 1$. Let S_p be the torsion subgroup of $\pi_{4p-1}\mathrm{bo}\wedge\mathrm{Ibo}\wedge R$. Write $\partial : \pi_{4p-1}\mathrm{bo}\wedge R \longrightarrow \pi_{4p-1}\mathrm{bo}\wedge\mathrm{Ibo}\wedge R = C_p \oplus S_p$ as (∂_C, ∂_S) where $\partial_C : \pi_{4p-1}\mathrm{bo}\wedge R \longrightarrow C_p$ and $\partial_S : \pi_{4p-1}\mathrm{bo}\wedge R \longrightarrow S_p$.

(3.3) $\partial_C(x_p) = \sum_{1\leq n\leq p} (p-n,n)k_n\, y_{p-n,n}$ where k_n is a unit in $\mathbb{Z}_{(2)}$.

To show this one simply checks that

$$h\partial(x_p) = h(\textstyle\sum_{1\leq n\leq p} (p-n,n)k_n\, y_{p-n,n}) \text{ modulo torsion}$$

so it then follows that $\partial x_p - \sum_{1\leq t\leq p} m_t(p-t,t)\, y_{p-t,t}$ is in S_p; but this is the statement in 3.3. To do the computations with the Hurewicz homorphism we apply λ_* to each side. Then since $h\partial(x_p) = 1\otimes a_p \otimes e_{-1}$ and $\lambda_*\mu_* = 1$ we must check that

$$\lambda_*(1\otimes a_p) = \textstyle\sum_{1\leq n\leq p}(n,p-n)k_n a_{p-n}\otimes u_n.$$

But this follows immediately from 2.3.

We now have the information we require on the bo-resolution for R namely $\partial x_p \in \pi_{4p-1}\mathrm{bo}\wedge\mathrm{Ibo}\wedge R$ modulo torsion. The next step in the argument is to get information on $\pi_*\mathrm{bo}\wedge\mathrm{Ibo}$, with the ultimate aim of obtaining some information on the homorphism $\pi_* S^{-1}\mathrm{bo}\wedge\mathrm{Ibo} \longrightarrow \pi_*\mathrm{bo}\wedge\mathrm{Ibo}\wedge R$.

From 3.6.2 of [21] or 3.9 of [11] we can compute the homotopy of $\mathrm{bo}\wedge B_n$ modulo torsion and the Hurewicz homorphism $\pi_*(\mathrm{bo}\wedge B_n) \longrightarrow H_*(\mathrm{bo}\wedge B_n; \mathbb{Z}_{(2)})$ modulo torsion. To state the result introduce the numerical function $\omega(s,t)$ defined as follows:

$$\begin{aligned} \omega(s,t) &= \max(0, \varphi(4s)-2t+\alpha(t)) && \text{if } t \equiv 0 \text{ modulo } 2 \\ &= \max(0, \varphi(4s)-2t+1+\alpha(t)) && \text{if } t \equiv 1 \text{ modulo } 2. \end{aligned}$$

Here $\alpha(s)$ is, as usual, the number of ones in the dyadic expansion of s. Then the torsion free subgroup of $\pi_*\mathrm{bo}\wedge B_n$ is generated by elements $w_{i,n}$ where $h(w_{i,n}) = 2^{\omega(i,n)}a_i\otimes u_n \in H_*(\mathrm{bo}\wedge B_n; \mathbb{Z}_{(2)})$. Now define $z_{s,t}$ to be $\mu_* w_{s,t}$ then:

(3.4) $\quad z_{s,t} \in \pi_{4s+4t}bo\wedge bo, s \geq 0, t \geq 1; \quad h(z_{s,t}) = 2^{\omega(s,t)}\mu_*(a_s \otimes u_t)$ modulo torsion.

In view of the equivalences in 2.2 and the choice of $w_{s,t}$ we know that the $z_{s,t}$ generate $\pi_* bo\wedge bo$ modulo torsion. Let D_p be the sub $\mathbb{Z}_{(2)}$ of $\pi_{4p}bo\wedge bo$ generated by the $z_{s,t}$ with $s+t = p$, $t \geq 1$ and let T_p be the torsion subgroup of $\pi_{4p}bo\wedge bo$.

Now let $u : bo\wedge bo\wedge S^{-1} \longrightarrow bo\wedge bo\wedge R$ be the inclusion and decompose the homorphism u_* : $\pi_{4p}bo\wedge bo\wedge S^{-1} = D_p \oplus T_p \longrightarrow \pi_{4p}bo\wedge bo\wedge R = C_p \oplus S_p$ into its components u_{DC}, u_{DS}, u_{TS}, $u_{TC} = 0$; we are using an obvious notation.

(3.5) $$u_{DC}(z_{s,t}) = 2^{\omega(s,t)}y_{s,t}.$$

This follows since from 3.4 and 3.2; $h(u_*(z_{s,t}) - 2^{\omega(s,t)}y_{s,t})$ is zero modulo torsion and since h is injective it follows that $u_*(z_{s,t}) - 2^{\omega(s,t)}y_{s,t}$ is torsion and this is the statement in 3.5.

Let $E_p = \oplus_{1\leq n\leq p} \mathbb{Z}/(2^{\omega(p-n,n)})$ be the cokernel of u_{DC} and let γ_p be the element of E_p determined by $\partial_C x_p$. We next prove the following statement concerning γ_p:

LEMMA 3.6. (a) If $p \neq 2^r$ then $4\gamma_p$ is not divisible by 8.

(b) If $p = 2^r$ with $r \geq 2$ then γ_p is divisible by 4 but $4\gamma_p$ is not divisible by 32.

From 3.3 the components of γ_p are $k_n(p-n,n) \in \mathbb{Z}/(2^{\omega(p-n,n)})$ where k_n is odd. For (a) it is sufficient to exhibit one value of n for which $4(p-n,n)$ is not divisible by 8 in $\mathbb{Z}/(2^{\omega(p-t,t)})$.

Case 1: p is odd, p > 1 Take $n = 1$, so $(p-1,1) = p$ is odd and the only way the result could be false is if $4(p-1,1)$ is zero in $\mathbb{Z}/(2^{\omega(p-1,1)})$. So we need to prove that $\omega(p-1,1) \geq 3$; however

$$\omega(p-1,1) = \max(0, \varphi(4(p-1)) \geq \varphi(4) = 3$$

and therefore $4(p-1,1)$ is not divisible by 8 in $\mathbb{Z}/(2^{\omega(p-1,1)})$.

Case 2: p is even but $p \neq 2^r$ Take n to be any integer with $n < p/2$ and $(p-n,n) = 1$

modulo 2. Then since $(p-n,n)$ is odd we only need to exclude the possibility that $4(p-n,n)$ is zero in $\mathbb{Z}/(2^{\omega(p-n,n)})$; we need to prove that $\omega(p-n,n) \geq 3$. Since $p-n \geq n+1$ it follows that $\omega(p-n,n) \geq \omega(n+1,n)$ and we leave it to the reader to check that $\omega(n+1,n) \geq 3$ for all n.

This completes the proof of part (a); we now prove part (b). To prove the first statement in part (b) it is sufficient to prove that if $p = 2^r$ then $(p-n,n)$ is divisible by 4 in $\mathbb{Z}/(2^{\omega(p-n,n)})$.

Case 3: $p = 2^r$. The formula $\nu(a,b) = \alpha(a) + \alpha(b) - \alpha(a+b)$ shows that $\nu(p-n,n) \geq 2$ if $n \neq 2^{r-1}$ and therefore $(p-n,n)$ is divisible by 4 in $\mathbb{Z}/(2^{\omega(p-n,n)})$ if $n \neq 2^r$. If $n = 2^{r-1}$ then $\varphi(4(p-n)) = 2(p-n) = 2n$ and $\alpha(n) = 1$ so that $\omega(p-n,n) = 1$; $(p-n,n)$ is even and therefore zero in $\mathbb{Z}/(2^{\omega(p-n,n)}) = \mathbb{Z}/2$. In all cases $(p-n,n)$ is divisible by 4 in $\mathbb{Z}/2^{\omega(p-n,n)})$ and therefore γ_p is divisible by 4.

To prove the second statement in part (b) we assume that $r \geq 2$ and then exhibit one value of n for which $4(p-n,n)$ is not divisible by 32 in $\mathbb{Z}/(2^{\omega(p-n,n)})$; we take $n = 2^{r-2}$ so that $\nu(p-n,n) = 2$. If $r \geq 3$ then $\omega(p-n,n) = \varphi(2^{r+2}-2^r)-2^{r-1}+1 = (2^{r+1}-2^{r-1})-2^{r-1}+1 = 2^r+1$. In this case $4(p-n,n)$ is not divisible by 32 in $\mathbb{Z}/(2^{\omega(p-n,n)})$. If $r = 2$, then $\omega(p-n,n) = \varphi(12)-2+2 = 7$ and once more $4(p-n,n)$ is not divisible by 32 in $\mathbb{Z}/(2^{\omega(p-n,n)})$.

The proof of 2.7. We will use the cofibration sequence

$$S^{-1} \xrightarrow{u} R \xrightarrow{c} P^{\infty} \xrightarrow{\tau} S^0$$

and apply $bo\wedge-$ or $bo\wedge Ibo\wedge-$ to this cofibration. Let $v_p = c_*(x_p) \in \pi_{4p-1}bo\wedge P^{\infty}$ where x_p is as in 3.2 so $v_p \in \pi_{4p-1}bo\wedge P^{\infty}$ is a generator of this cyclic group.

From 3.1 and the structure of $H_*(Ibo;\mathbb{Z}_{(2)})$ we know that $bo\wedge Ibo\wedge R$ is equivalent to a product of $\mathbb{Z}/2$ and $\mathbb{Z}_{(2)}$ Ellenberg-Maclane spectra. We deduce the following facts concerning $\pi_*bo\wedge Ibo\wedge R$: The torsion in $\pi_*bo\wedge Ibo\wedge R$ has exponent 2 and Adams filtration 0; $AF(2^k y_{s,t}) = k$ where the $y_{s,t}$ are the elements of $\pi_*bo\wedge Ibo\wedge R$ chosen in 3.2. If α is any element of $\pi_*bo\wedge Ibo\wedge R$ and $AF(\alpha) = k \geq 1$ then α is divisible by 2^k. We will have a lot of use for these facts.

First we deal with the case $p \neq 2^r$. From 3.3 we know that

$$4\partial(x_p) = \sum_{1\le n\le p} 4(p-n,n)k_n\, y_{p-n,n} \quad \text{where } k_n \text{ is a unit in } \mathbb{Z}_{(2)}$$

and by examining the binomial coefficients we conclude that since $p \neq 2^r$, $AF(4\partial x_p) = 2$. Since $\partial v_p = c_*\partial(x_p)$ we deduce that $AF(4\partial v_p) \geq 2$.

Now suppose that $AF(4\partial v_p) \geq 3$. Let $a \in Ext^{2,4p+1}(bo\wedge Ibo\wedge R)$ be the representative for $4\partial x_p$ in the classical mod 2 Adams spectral sequence. Since $AF(c_*4\partial x_p) \geq 3$ it must follow that in $Ext^{2,4p+1}(bo\wedge Ibo\wedge P^\infty)$, $c_*a = 0$. Now an Ext exact sequence shows that there must be an element $b \in Ext^{2,4p+1}(bo\wedge Ibo\wedge S^{-1})$ such that $u_*b = a$. It is known that the Adams spectral sequence for $bo\wedge Ibo$ collapses at E_2 so b is an infinite cycle; let $\beta \in \pi_{4p-1}bo\wedge Ibo\wedge R$ be a homotopy class detected by b. Since $u_*b = a$ in $Ext^{2,4p}(bo\wedge Ibo\wedge R)$ it follows that $u_*\beta$ and $4\partial x_p$ are both detected by a and therefore $AF(4\partial x_p - u_*\beta) \geq 3$. From our remarks concerning $\pi_*bo\wedge Ibo\wedge R$ this means that there exists $e \in \pi_*bo\wedge Ibo\wedge R$ such that $4\partial x_p - u_*\beta = 8e$; therefore $4\partial x_p$ is divisible by 8 modulo the image of u_*. In the notation of 3.6 this says that $4\partial_p$ is divisible by 8 which is a contradiction to 3.6 (a). Therefore we must have $AF(4\partial v_p) = 2$.

If $p = 2^r$ with $r \geq 2$ then set $m = 2^{r-1}$. Now using 3.5 it follows that $4y_{m,m}$ is in the image of $u_* : \pi_{4p-1}bo\wedge Ibo\wedge S^{-1} \longrightarrow \pi_{4p-1}bo\wedge bo\wedge R$ so that $4\partial v_p = c_*4\partial x_p = c_*4(\partial x_p - y_{m,m})$. Now a check on binomial coefficients shows that $AF(4(\partial x_p - y_{m,m})) = 4$ so that $AF(4\partial v_p) \geq 4$. Now repeat the argument above to dispose of the possibility that $AF(4\partial v_p) \geq 5$.

§4 THE FINAL STEPS IN THE PROOF OF THEOREM A

We now complete the proof of Theorem A by proving Theorem 1.5 (b). We assume $p = 2^r$ with $r \geq 2$. We will use the following two cofibrations:

$$S^0 \xrightarrow{i} bo \xrightarrow{\rho} Ibo \xrightarrow{\delta} S^1$$

$$S^{-1} \xrightarrow{u} R \xrightarrow{c} P^\infty \xrightarrow{\tau} S^0$$

We will also use the notation

$$j : \mathbb{I}bo \wedge P^{\infty} \longrightarrow bo \wedge \mathbb{I}bo \wedge P^{\infty}$$

for the inclusion. We need two lemmas concerning Ext groups and Adams filtration.

LEMMA 4.1. (1) $Ext^{0,4p-1}(\mathbb{I}bo \wedge P^{\infty}) = \mathbb{Z}/2.$

(2) $Ext^{1,4p}(\mathbb{I}bo \wedge P^{\infty}) = \mathbb{Z}/2.$

(3) Multiplication by $h_0 \in Ext^{1,1}(S^0)$ gives an isomorphism of $Ext^{0,4p-1}(\mathbb{I}bo \wedge P^{\infty})$ with $Ext^{1,4p}(\mathbb{I}bo \wedge P^{\infty})$.

LEMMA 4.2. (1) Let v_p be a generator of $\pi_{4p-1} bo \wedge P^{\infty}$. Then in $\pi_{4p-1} \mathbb{I}bo \wedge P^{\infty}$,

$$AF(\rho_* v_p) = 1.$$

(2) Let $x \in \pi_{4p-2} P^{\infty}$ and $y \in \pi_{4p-1} \mathbb{I}bo \wedge P^{\infty}$ be such that $\delta_* y = x$; then

$$AF(x) = 1 \Leftrightarrow AF(y) = 0.$$

In the statement of the lemma we have abbreviated $(\rho \wedge 1)_*$ to ρ_* and $(\delta \wedge 1)_*$ to δ_*. We will often use a similar abbreviation.

Proof of Theorem 1.5 (b) assuming 4.1 and 4.2. As in the proof of 1.5 (a) given in §2 we see that if $\alpha \in \pi_{4p-2} P^{\infty}$ is such that $e(\alpha)$ has order precisely 2^a then there is an element $\gamma \in \pi_{4p-1} \mathbb{I}bo \wedge P^{\infty}$ such that $\delta_* \gamma = \alpha$ and

$$2^{a+2} j_* \gamma = 4\lambda v_p.$$

Now, using 2.7 (2), we compute Adams filtrations

$$4 = AF(4\lambda v_p) = AF(2^{a+2} j_* \gamma) \geq a + 2 + AF(\gamma).$$

There are four cases to consider:

Case (1): $a = 0$, $e(\alpha) = 0$.
Case (2): $a = 1$, $AF(\gamma) = 0$.

Case (3): $a = 1$, $AF(\gamma) = 1$.

Case (4): $a = 2$, $AF(\gamma) = 0$.

We are assuming $e(\alpha) \neq 0$ so this eliminates case (1). Suppose case (3) occurs, then $AF(\gamma) = 1$ and $8j_*\gamma = 4\partial v_p$. From 4.2 (1) and 4.1 (2) it must now follow that γ and $\rho_*(v_p)$ are detected by the same element in $Ext^{1,4p}(Ibo \wedge P^{\infty})$ so that $AF(\gamma - \rho_*(v_p)) \geq 2$. Multiply $\gamma - \rho_*(v_p)$ by 4 and apply j_*; since $j_*\rho_* = \partial$ we conclude that

(4.3) $$AF(4j_*\gamma - 4\partial v_p) \geq 4.$$

From 2.7 (2) we know that

(4.4) $$AF(4\partial v_p) = 4.$$

By construction $8j_*\gamma = 4\partial v_p$ and therefore from 4.4 it follows that

(4.5) $$AF(4j_*\gamma) \leq 3.$$

But 4.4 and 4.5 together show that $AF(4j_*\gamma - 4\partial v_p) \leq 3$ which contradicts 4.3. Therefore we conclude that case (3) cannot occur.

Now assume case (4) occurs and let $\varepsilon = 2\gamma$ so that by construction $8j_*\varepsilon = 4\partial v_p$ and $AF(\varepsilon) \geq 1$. In fact $AF(\varepsilon) = 1$ or we would have a contradiction to 2.7 (2). Now repeat the argument used to eliminate case (3) with γ replaced by ε to eliminate case (4).

We conclude that if $e(\alpha) \neq 0$ then only case (2) can occur so that $2e(\alpha) = 0$ and since $\delta_*(\gamma) = \alpha$, 4.2 (2) shows that $AF(\alpha) = 1$. This proves 1.5 (b).

It remains to prove 4.1 and 4.2. We assume the following facts about the groups $Ext(P^{\infty})$.

(4.6) (1) $$Ext^{0,t}(P^{\infty}) = 0 \quad \text{if } t \neq 2^n - 1$$
$$= \mathbb{Z}/2 \quad \text{if } t = 2^n - 1 \text{ with nonzero element } e_n.$$

(2) $Ext^{1,t}(P^{\infty}) = \mathbb{Z}/2$ if $t = 2^{n+1} - 1$ with non zero element $h_n e_n$, where h_n is the non zero element in $Ext^{1,t}(S^0)$ with $t = 2^n$.

(3) $\mathrm{Ext}^{2,t}(P^\infty) = \mathbb{Z}/2$ if $t = 2^{n+1}$, $n \geq 3$, with nonzero element $h_0h_ne_n$.

These computations can be verified using the Λ algebra complex [5] with homology $\mathrm{Ext}(P^\infty)$; see [34] for a brief discussion and some calculations. The calculation is very similar to the calculation done in [33]; it is carried out explicitly in [8].

The groups $\mathrm{Ext}(bo \wedge P^\infty)$ are well known, see [10] and [23].

Proof of 4.1. Since $p = 2^r$, $i_* : \mathrm{Ext}^{0,4p-1}(P^\infty) \longrightarrow \mathrm{Ext}^{0,4p-1}(bo\wedge P^\infty)$ is an isomorphism. Further from [10] and [23] we know that $\mathrm{Ext}^{s,t}(bo\wedge P^\infty) = 0$ if $(s,t) = (0,4p)$, $(1,4p-1)$ and $(2,4p)$. The cofibration sequence $P^\infty \longrightarrow bo\wedge P^\infty \longrightarrow \overline{bo}\wedge P^\infty$ induces a short exact sequence of mod 2 homology groups and therefore a long exact sequence in Ext groups. If we put the above information into this exact sequence we deduce that if $s = 0$ or 1 then the boundary homorphism

$$\delta_* : \mathrm{Ext}^{s,4p-1}(\overline{bo}\wedge P^\infty) \longrightarrow \mathrm{Ext}^{s+1,4p-1}(P^\infty)$$

is an isomorphism of $\mathrm{Ext}(S^0)$ modules. Lemma 4.1 follows from 4.6.

Proof of 4.2. Part (1) is more or less well known but we cannot find an explicit proof in the literature so we will outline an argument. Consider the commutative diagram

$$\begin{array}{ccccc}
\pi_{4p-1}bo\wedge P^{4p-2} & \xrightarrow{\rho_*} & \pi_{4p-1}\overline{bo}\wedge P^{4p-2} & \xrightarrow{\delta_*} & \pi_{4p-2}P^{4p-2} \\
u_1 \downarrow & & u_2 \downarrow & & u_3 \downarrow \\
\pi_{4p-1}bo\wedge P^\infty & \xrightarrow[\rho_*]{} & \pi_{4p-1}\overline{bo}\wedge P^\infty & \xrightarrow[\delta_*]{} & \pi_{4p-2}P^\infty
\end{array}$$

Each vertical arrow is induced by the inclusion $P^{4p-2} \longrightarrow P^\infty$. From [10] and [23] we know that u_1 is injective and its image has index 2. For dimensional reasons u_2 is an isomorphism. Let $y = u_2^{-1}\rho_*(v_p) \in \pi_{4p-1}\overline{bo}\wedge P^{4p-2}$; an easy argument shows that $\delta_* y \neq 0$ in $\pi_{4p-2}P^{4p-2}$ but $u_3\delta_* y = 0$ in $\pi_{4p-2}P^\infty$. Therefore $\varphi_* y$ is the unique nontrivial element in the kernel of u_3, that is the attaching map of the $4p-1$ cell in P^{4p-1}. A straightforward modification of the argument of [7] with the Whitehead square $[\iota_{2p-1}, \iota_{2p-1}]$ replaced by $\delta_* y$, the attaching map of the 4p-1 cell in P^{4p-1}, shows that $AF(\delta_* y) = 2$. Since δ is zero in mod 2 homology it follows that $AF(y) \leq 1$ and therefore $AF(u_2 y) \leq 1$ since u_2 is an isomorphism. But $u_2 y = \rho_*(v_p)$ so we have shown that $AF(\rho_* v_p) \leq 1$. However $AF(v_p) = 0$ and we checked in the proof of 4.1 that $\rho_* : Ext^{0,4p-1}(bo \wedge P^\infty) \longrightarrow Ext^{0,4p-1}(Ibo \wedge P^\infty)$ is zero so $AF(\rho_* v_p) \geq 1$; this proves that $AF(\rho_* v_p) = 1$.

We now prove part (2) of 4.2. An Ext exact sequence shows that $Ext^{0,4p-2}(Ibo \wedge P^\infty) = \mathbb{Z}/2$. Since δ is zero in homology δ_* increases Adams filtration and gives an isomorphism of $Ext^{0,4p-2}(Ibo \wedge P^\infty)$ with $Ext^{1,4p-1}(P^\infty)$. It follows that $\delta_* \beta$ is detected by the unique non-zero element of $Ext^{1,4p-1}(P^\infty)$ if and only if β is detected by the unique non-zero element of $Ext^{0,4p-2}(Ibo \wedge P^\infty)$. This proves 4.2 (2).

§5 ON THE SPHERE OF ORIGIN AND HOPF INVARIANT OF ELEMENTS OF ORDER TWO AND KERVAIRE INVARIANT ONE.

In this section we continue to assume that all spaces and spectra are localised at two. We revert to the notation π_*^S and π_* to distinguish between stable homotopy groups and ordinary homotopy groups. We use the notation $N = N(k)$, $t = t(k)$, $a = a(k)$ established in the introduction. The object of this section is to prove the following theorem.

THEOREM 5.1. Fix k and suppose that $\theta_k \in \pi_N^S$ is an element of order two with Kervaire invariant one.

(1) θ_k is not in the image of $E_{a-1} : \pi_{N+a-1} S^{a-1} \longrightarrow \pi_N^S$.

(2) Suppose that there exists an element $x \in \pi_{N+a} S^a$ such that $E_a(x) = \theta_k$. Then $eH(x)$ is an odd multiple of $e(\beta_k)$.

This result says that the sphere of origin of an element θ_k of order two and Kervaire invariant one cannot be less than S^a. Further if the sphere of origin of this element θ_k is S^a then every element of the stable Hopf invariant of θ_k has e invariant an odd multiple of $e(\beta_k)$. This of course supports the conjecture on the sphere of origin and Hopf invariant of elements of Kervaire invariant made in the introduction.

The main point in the proof is to obtain a partial converse to theorem 1.5 (b).

THEOREM 5.2. Suppose that $\alpha \in \pi_N{}^S P^\infty$ has order 2 and $AF(\alpha) = 1$; then $e(\alpha) \neq 0$ in $J_N(P^\infty)$.

Proof of 5.1 assuming 5.2. Suppose $\alpha \in \pi_N{}^S P^\infty$ has order 2 and $AF(\alpha) = 1$, then by 5.2, $e(\alpha)$ is non zero. We know from theorem 1.5 (b) that $e(\alpha)$ has order two and a little exercise with the numbers in 1.4 shows that $e(\alpha)$ is not in the image of $J_N(P^{a-2})$. Therefore α is not in the image of $\pi_N{}^S P^{a-2}$.

Now assume the hypotheses of 5.1 but suppose that $x \in \pi_{N+a-1}S^{a-1}$ is such that $E_{a-1}x = \theta_k \in \pi_N{}^S$. Then take $\alpha' = s_{a-1*}(x) \in \pi_N{}^S P^{a-2}$ where $s_{a-1}: \Omega^{a-1}S^{a-1} \longrightarrow QP^{a-2}$ is the Snaith map used in §1. Let $u : P^{a-2} \longrightarrow P^\infty$ be the inclusion. Then $u_*(\alpha') = s_*(E_{a-1}x) = s_*(\theta_k)$, where again $s : \Omega^\infty S^\infty \longrightarrow QP^\infty$ is the Snaith map used in §1, and so $\alpha = u_*(\alpha')$ has order 2. From the Kahn-Priddy theorem $\tau_*(\alpha) = \theta_k$ and so, compare the argument given in the proof of theorem A at the end of §1, $AF(u_*\alpha) = 1$. Therefore $\alpha \in \pi_N{}^S P^\infty$, $AF(\alpha) = 1$, α has order 2 and α is in the image of $\pi_N{}^S P^{a-2}$. But this contradicts the previous paragraph. We have now established 5.1 (1).

Now suppose that $x \in \pi_{N+a}S^a$ is such that $E_a(x) = \theta_k$; let $\alpha' = s_{a*}(x) \in \pi_N{}^S(P^{a+1})$ and let α be the image of α' in $\pi_N{}^S(P^\infty)$. Then, as above, we know that $AF(\alpha) = 1$ and so from 5.2, $e(\alpha) \neq 0$ and therefore $e(\alpha') \neq 0$. Virtuosity with the numbers in 1.4 shows that $e(c_*\alpha')$ is an odd multiple of $e(\beta_k)$. Now, from 1.2, $c_*s_{a*}(x) = E_{2n+1}H(x)$ so it follows that $eH(x) = e(\beta_k)$ modulo 2.

This proves 5.1; it remains to prove 5.2.

Proof of 5.2. Let $\alpha \in \pi_N{}^S(P^\infty)$ be an element of order 2 with $AF(\alpha) = 1$. Let α' be any element of $\pi_N{}^S(P^N)$ such that α' maps to α in $\pi_N{}^S(P^\infty)$. The first step is to prove the following:

(5.3) $2\alpha'$ is non zero in $\pi_N{}^S(P^N)$.

We use the notation and results in 4.6. The homotopy class α' is detected by the element $h_k e_k \in \mathrm{Ext}^{1,N+1}(P^N)$, and if $k \geq 2$, $h_0 h_k e_k \in \mathrm{Ext}^{2,N+2}(P^N)$ is non zero. Therefore if $2\alpha' = 0$ it must follow that, in the classical mod 2 Adams spectral sequence, $h_0 h_k e_k$ is the boundary of an element in $Ext^{0,N+1}(P^N)$. It is easy to check that the group $Ext^{0,N+1}(P^N)$ is zero so $2\alpha'$ cannot be zero. Note that if we replace P^N by P^∞ throughout this argument we find that $Ext^{0,N+1}(P^\infty) = \mathbb{Z}/2$ with generator e_{k+1} and $\partial_2 e_k = h_0 h_k e_k$.

The kernel of the homorphism $\pi_N{}^S(P^N) \longrightarrow \pi_N{}^S(P^\infty)$ is $\mathbb{Z}/2$ generated by w_{k+1}, the attaching map of the N+1 cell of P^{N+1}. From 5.3 we know that $2\alpha'$ is non zero and by assumption $2\alpha'$ maps to zero in $\pi_N{}^S(P^\infty)$; therefore $2\alpha' = w_k$. We now recall some more facts concerning the J theory of projective spaces, see [22], [23] for details:

(5.4). (1) $J_N(P^N) = \mathbb{Z}/(2^k)$.

(2) $J_N(P^\infty) = \mathbb{Z}/(2^{k-1})$.

(3) The homorphism $J_N(P^N) \longrightarrow J_N(P^\infty)$ is onto.

(4) The nontrivial element in the kernel of $J_N(P^N) \longrightarrow J_N(P^\infty)$ is $e(w_k)$.

We have established that $2\alpha' = w_k$ and therefore $2e(\alpha') = e(w_k)$. We easily deduce that $e(\alpha')$ maps to the nontrivial element of order two in $J_N(P^\infty)$ and therefore $e(\alpha)$ is this nontrivial element. This completes the proof of 5.2.

§6 THE CUP-1-CONSTRUCTION

We now begin the second main task of this paper, that is the proof of Theorem B. The methods used belong entirely to unstable homotopy theory, rather than the stable homotopy theory of the previous sections; the main tool is the cup-1-construction. A great deal is known about the cup-1-construction but by no means all of it appears in the literature. In an effort to be self contained we will start at the beginning so we will inevitably repeat some of the material covered in [25], [26], [27] and [29].

There is a function, the cup-1-construction,

$$Sq_1 : \pi_{2k}S^{2n} \longrightarrow \pi_{4k+1}S^{4n}$$

with the following properties.

(6.1) (1) $Sq_1(E^2x) = E^4Sq_1(x)$

(2) $Sq_1(\iota_{2n}) = 0$

(3) $Sq_1(x+y) = Sq_1(x) + Sq_1(y) + b(k,n)(x \wedge y) \cdot \eta$ where $b = b(k,n)$ is an integer which depends only k and n.

A more careful argument shows that $b = k+n+1$ but we will not give this argument here.

If $\alpha \in \pi_p(S^n)$ then let $C(\alpha)$ be the two cell complex $S^n \cup_\alpha e^{p+1}$.

THEOREM 6.2 If $\alpha \in \pi_{2k}S^{2n}$ let $\alpha^* = Sq_1\alpha \in \pi_{4k+1}S^{2n}$; then there is a homotopy equivalence of $C(\alpha) \wedge C(\alpha)$ with a complex with the following cell diagram.

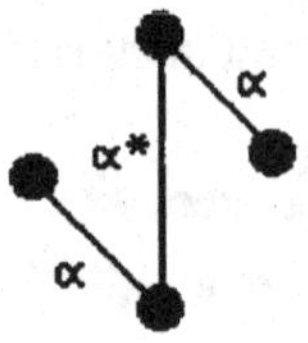

For the reader unfamiliar with the use of cell diagrams a brief explanation is given in [4]. It will considerably simplify notation if we use the following convention: Given elements $x \in \pi_{p+n}S^n$ and $y \in \pi_{q+m}S^m$, an integer k with $k-n \geq 0$ and $p+k-m \geq 0$ then $x \cdot y \in \pi_{p+q+k}S^k$ will denote the composition $(E^{k-n}x) \cdot (E^{p+k-m}y)$. A similar convention will apply to secondary compositions, that is Toda brackets.

We usually assume we are working in the metastable range, that is we are working with π_qS^m where $q < 3m-1$.

THEOREM 6.3. Suppose k and n are integers such that $2k < 6n-1$. Let $x \in \pi_{2k}S^{2n}$ be such that $E^rx = 0$, then:

(1) $E^rSq^1(x) = 0$ in $\pi_{4k+r+1}S^{4n+r}$.

(2) If $E^{r-1}x = P(\varphi)$ and $H(x) = \psi$, then

$$E^{r-1}Sq_1(x) = P(E\varphi \wedge \psi).$$

The first point to make is that the statement in (2) makes sense, that is in the notation of (2):

$$\varphi \in \pi_{2k+r+1}(S^{4n+2r-1}),\quad \psi \in \pi_{2k}(S^{4n-1}),\quad E\varphi\wedge\psi \in \pi_{4k+r+2}(S^{8n+2r-1})$$

$$P(E\varphi\wedge\psi) \in \pi_{4k+r}(S^{4n+r-1}) \text{ and } E^{r-1}Sq_1(x) \in \pi_{4k+r}(S^{4n+r-1}).$$

There is an analogous result for elements x in the homotopy of an odd sphere.

THEOREM 6.4. Assume that $2k-1 < 6n-4$ and let $x \in \pi_{2k-1}S^{2n-1}$ be such that $E^r x = 0$, then:

(1) $E^{r-2}Sq^1(Ex) = 0 \quad \text{in } \pi_{4k+r+1}S^{4n+r-1}$.

(2) If $E^{r-1}x = P(\varphi)$ and $H(x) = \psi$, then

$$E^{r-3}Sq_1(Ex) = P(E\varphi\wedge\psi).$$

The final result we need on the cup-1-construction is a formula for $2Sq_1(x)$ valid when $x \in \pi_{2k}S^{2n}$ with k odd and n even.

THEOREM 6.5. If $x \in \pi_{2k}S^{2n}$ where k is odd and n is even then

$$2Sq_1(x) = (x\wedge x)\circ\eta.$$

Given this collection of properties of the cup-1-construction we can give the proof of Theorem B. We use the notation in the statement of theorem B in the introduction.

Proof of Theorem B. Let w be an element of $\pi_{2N-r+1}S^{N-r+1}$ with $E^r w = [\iota_{N+1},\iota_{N+1}]$ so that $e(Hw) = e(\beta_{k+1})$ modulo two. For simplicity we will assume that $N-r+1$ is even. Our hypotheses are that there is an element $\theta \in \pi_{2N-p+1}S^{N-p+1}$ with $E^{r-q}w = 2E^{p-q}\theta$ in $\pi_{2N-q+1}S^{N-q+1}$ where $r \geq p \geq q$ and $p+q \geq r$ and $N-p+1$ is divisible by 4.

Let $\gamma = Hw$; since $E^{r+1}w = 0$, we conclude from 6.3 (2) that

$$E^r Sq_1(w) = P(\gamma).$$

Now set $x = E^{r-p}w - 2\theta \in \pi_{2N-p+1}(S^{N-p+1})$ so that $E^{p-q}x = 0$, and therefore $E^{p-q}Sq_1(x) = 0$ by 6.3 (1). From 6.1 (3)

$$Sq_1(x) = Sq_1(E^{r-p}w - 2\theta) = Sq_1(E^{r-p}w) + Sq_1(-2\theta) + b(E^{r-p}w)\wedge(-2\theta)\circ\eta$$

$$= Sq_1(E^{r-p}w) - Sq_1(2\theta).$$

Two straightforward arguments are required here, the first is to check that

$$(E^{r-p}w)\wedge(-2\theta)\circ\eta = 0$$

and the second to check, using 6.1, that $Sq_1(-2\theta) = -Sq_1(2\theta)$.

Therefore since $E^{p-q}Sq_1(x) = 0$

$$E^{p-q}Sq_1(E^{r-p}w) = E^{p-q}Sq_1(2\theta)$$

$$E^{2r-p-q}Sq_1(w) = E^{p-q}Sq_1(2\theta)$$

and since $p+q \geq r$ it follows, by applying E^{p+q-r}, that

$$P(\gamma) = E^r Sq_1(w) = E^{2p-r}Sq_1(2\theta).$$

From 6.1 (3) we know that

$$Sq_1(2\theta) = 2Sq_1(\theta).$$

Finally we have assumed that $N-p+1 \equiv 0 \bmod 4$ and it follows that $2N-p+1 \equiv 2 \bmod 4$ so we can use 6.5 with $x = \theta \in \pi_{2N-p+1}S^{N-p+1}$ to deduce that

$$2Sq_1(\theta) = (\theta\wedge\theta)\circ\eta.$$

Combining the last three formulas with the fact that $e(\gamma) = e(Hw) = e(\beta_{K+1})$ modulo two gives the result.

The case where N-t+1 is odd is similar but using 6.4 in place of 6.3. The details are left to the reader.

§7 THE DEFINITION AND PROPERTIES OF THE CUP-1-CONSTRUCTION

Let X be a space with base point x_0. Let T be the involution defined on $X \wedge X$ by switching factors; we let T act on S^1 by the antipodal map. If Y is a space then Y_+ denotes the space obtained from Y by adjoining a disjoint base point. We will use the notation D(X) for the following form of the quadratic construction on X, [25], [26], [27], [29] and [15];

$$D(X) = S^1{}_+ \wedge_T X \wedge X = S^1 \times_T X \wedge X / S^1 \times_T x_0 \wedge x_0.$$

There are natural homeomorphisms

$$h_1 : D(S^2 X) \longrightarrow S^4 D(X)$$

defined in [15 pp 476-477] (in the notation of [15], $D = D_2{}^1$). We define natural homeomorphisms

$$h_n : D(S^{2n} X) \longrightarrow S^{4n} D(X)$$

by iterating h_1 in the obvious manner. In particular the following diagram commutes

$$\begin{array}{ccc} D(S^2(S^{2n}X)) & \longrightarrow & S^4 D(S^{2n}X) \\ h_{n+1} \downarrow & & \downarrow S^4 h_n \\ S^{4n+4} D(X) & = & S^{4n+4} D(X). \end{array}$$

Evidently $D(S^0) = P^1{}_+$ and so there is a map $P^1{}_+ \longrightarrow S^0$ given by mapping the disjoint base point to the base point of S^0 and P^1 to the other point; this map will be denoted $\rho : D(S^0) \longrightarrow S^0$. Define

$$\rho_k : D(S^{2k}) \longrightarrow S^{4k}$$

to be the composite $S^{4k}\rho\cdot h_k$. Let $i_k : S^{4k} \longrightarrow D(S^{2k})$ be the inclusion of $S^{2k}\wedge S^{2k}$ and let q_k : $D(S^{2k}) \longrightarrow S^{4k+1}$ be the map which collapses $S^{2k}\wedge S^{2k}$ to a point. Then the map

$$e_k : D(S^{2k}) \longrightarrow S^{4k}\vee S^{4k+1}$$

with components ρ_k and q_k is an equivalence. Note that to define a map $D(S^{2k}) \longrightarrow S^{4k}\vee S^{4k+1}$ it is sufficient, for dimension and connectivity reasons, to give its components. We have arranged our choices so that $S^4 e_k\cdot h_k = e_{k+1}$. Let

$$f_k : S^{4k}\vee S^{4k+1} \longrightarrow D(S^{2k})$$

be a homotopy inverse of e_k, then, up to homotopy, f_k restricted to S^{4k} is i_k. We define

$$\varsigma_k : S^{4k+1} \longrightarrow D(S^{2k})$$

to be f_k restricted to S^{4k+1}; so ς_k is well-defined up to homotopy.

7.1 The definition of the cup-1-construction. Given $\alpha : S^{2k} \longrightarrow S^{2n}$ we define $Sq_1(\alpha)$ to be the composite

$$S^{4k+1} \xrightarrow{\varsigma_k} D(S^{2k}) \xrightarrow{D(\alpha)} D(S^{2n}) \xrightarrow{\rho_n} S^{4n}.$$

It is simple to check that $Sq_1(E^2x) = E^4Sq_1(x)$ and $Sq_1(\iota_{2n}) = 0$, that is 6.1 (1) and (2), directly from the definition.

Now we start on the proof of 6.1 (3) that is the formula for $Sq_1(x+y)$. The map $x+y$ factors as

$$S^{2k} \xrightarrow{\xi} S^{2k}\vee S^{2k} \xrightarrow{x\vee y} S^{2n}\vee S^{2n} \xrightarrow{F} S^{2n}$$

where ξ is the map with both components the identity and F is the folding map, (the identity on each of the wedge summands S^{2n}). Therefore $Sq_1(x+y)$ factors as

$$S^{4k+1} \xrightarrow{\zeta_k} D(S^{2k}) \xrightarrow{D(\xi)} D(S^{2k} \vee S^{2k}) \xrightarrow{D(x \vee y)} D(S^{2n} \vee S^{2n}) \xrightarrow{D(F)} D(S^{2n}) \xrightarrow{\rho_n} S^{4n}$$

Following Nishida [29], compare [15] §4, there is a natural map $\tau : D(X \vee Y) \longrightarrow S^1{}_+ \wedge X \wedge Y$ and a natural equivalence

$$g : D(X \vee Y) \longrightarrow D(X) \vee D(Y) \vee S^1{}_+ \wedge X \wedge Y.$$

The components of this equivalence are $D(q_1)$, $D(q_2)$ and τ where q_1 and q_2 are the projections of $X \vee Y$ onto X and Y respectively. Note that since we are mapping into a wedge it is not possible to define a map simply by giving its components. Rather the maps τ and f have to be constructed directly (by analysing the T equivariant homotopy type of $S^1{}_+ \wedge (X \vee Y) \wedge (X \vee Y)$) and then checking that the components of f are as asserted. This is done in [29] and in essence in §2 of [15].

Next we use the equivalence $S^1{}_+ \wedge S^{2k} \wedge S^{2k} \longrightarrow S^{4k} \vee S^{4k+1}$ given by the base point preserving maps of $S^1{}_+$ to S^0 and S^1 respectively. Combining these equivalences we find that $Sq_1(x+y)$ factors as

$$S^{4k+1} \longrightarrow D(S^{2k}) \vee D(S^{2k}) \vee S^{4k} \vee S^{4k+1} \longrightarrow D(S^{2n}) \vee D(S^{2n}) \vee S^{4n} \vee S^{4n+1} \longrightarrow S^{4n}$$

where the various maps are as follows. For dimension and connectivity reasons the first map is determined by its components and these are

$$\zeta_k \, , \; \zeta_k \, , \; b'(k)\eta \, , \; 1$$

for some integer b'(k) which depends only on k. This is easily verified from the components of g and, in the case of the fourth component, the explicit definition of τ given in [29], see also [15]. In fact, since we do not need to determine the precise value of the constant b in 6.1 (3), the precise degree of the fourth component is of no relevance.

The second map is given by wedging together the following four maps:

$$D(x),\ D(y),\ x\wedge y,\ Ex\wedge y.$$

This follows from the naturality of the equivalence g and the precise choice of the equivalence $S^1{}_+\wedge S^{2k}\wedge S^{2k} \longrightarrow S^{4k}\vee S^{4k+1}$.

The third map is given by the following maps on the respective wedge factors

$$\rho_n,\ \rho_n,\ 1\ ,\ b''(n)\eta$$

where $b''(n)$ is an integer which depends only on n.

It now follows that

$$Sq_1(x+y) = Sq_1(x) + Sq_1(y) + b(k,n)(x\wedge y)\cdot\eta$$

where $b(k,n) = b'(k)+b''(n)$. This proves 6.1 (3).

A more delicate analysis shows that $b'(k) = k+1$ and $b''(n) = n$; we do not need these facts so we omit the proofs.

We move on now to the proof of 6.2; it is proved by an analysis of the cell structure of the following spaces and the maps between them

$$C(\alpha)\wedge C(\alpha) \longrightarrow D(C(\alpha)) \longleftarrow C(D(\alpha)).$$

The first map is the usual inclusion and the second is given by the obvious null-homotopy of the composite

$$D(S^{2k}) \longrightarrow D(S^{2n}) \longrightarrow D(C(\alpha)).$$

The point is that all the spaces have exactly one 4k+2 cell and the two maps into $D(C(\alpha))$ both have degree one in this dimension. This gives us two ways to determine the attaching map of the 4k+1 cell in $D(C(\alpha))$; this leads to a relation, that is 6.2. The proof is, in essence, routine but in full detail quite long so we content ourselves with the following summary.

First we analyse the cell structure of $C(D(\alpha))$. Using the equivalences e_k and e_n above we see that $C(D(\alpha))$ is the mapping cone of the map $S^{4k} \vee S^{4k+1} \to S^{4n} \vee S^{4n+1}$ with matrix

$$\begin{matrix} \alpha\wedge\alpha & \alpha^* \\ 0 & \alpha\wedge\alpha \end{matrix}$$

where as in the statement of 6.2 we have written α^* for $Sq_1(\alpha)$.

Now we analyse the 4k+2 skeleton of $D(C(\alpha))$. We need the following facts whose proof is left to the reader:

(a) The 2n+2k+2 skeleton is

$$S^{4n} \cup_\alpha e^{2n+2k+1} \vee S^{4n+1} \cup_\alpha e^{2n+2k+2}.$$

All the ingredients required are contained in §2 of [15]. We have slipped into the convention of using the notation α for any suspension of $\alpha \in \pi_{2k}S^{2n}$.

There is one 4k+2 cell in $D(C(\alpha))$ and we need to determine its attaching map.

(b) The map $C(D(\alpha)) \to D(C(\alpha))$ has degree one in dimension 4k+2.

It now follows that:

(c) The attaching map of the 4k+2 cell in $D(C(\alpha))$ is $j_*(\alpha^*)$ where $j : S^{4n} \to D(C(\alpha))$ is the inclusion of the bottom cell.

Next we use Toda's analysis of the cell structure of $C(\alpha)\wedge C(\alpha)$. In [32] chapter 3, Toda shows that $C(\alpha)\wedge C(\alpha)$ has the following cell structure:

$$((S^{4n} \cup_\alpha e^{4n+2k+1}) \vee S^{4n+2k+1}) \cup_{i_*\beta} e^{4k+2}$$

where $\beta : S^{4k+1} \to S^{4n}$ and $i : S^{4n} \to ((S^{4n} \cup_\alpha e^{4n+2k+1}) \vee S^{4n+2k+1})$ is the inclusion of the bottom cell.

Finally we need:

(d) The map $C(\alpha)\wedge C(\alpha) \longrightarrow D(C(\alpha))$ has degree one in dimension $4k+2$ and therefore the attaching map of the $4k+2$ cell in $D(C(\alpha))$ is $j_*\beta$.

The final step in the argument is to verify that (c) and (e) together show that:

(f) $\beta - \alpha^* = x\alpha$ for some element $x \in \pi_{4k+1}S^{2k+2n}$.

This of course completes the proof of 6.2.

Finally we give the proof of 6.5. Since $2n \equiv 0 \bmod 4$, $\rho_n : S^1{}_+\wedge_T S^{2n}\wedge S^{2n} \longrightarrow S^{4n}$ extends over $S^2{}_+\wedge_T S^{2n}\wedge S^{2n}$. Therefore $Sq_1(x)$ factors as

$$S^{4k+1} \xrightarrow{\zeta_k} S^2{}_+\wedge_T S^{2k}\wedge S^{2k} \xrightarrow{1\wedge_T x\wedge x} S^2{}_+\wedge_T S^{2n}\wedge S^{2n} \longrightarrow S^{4n}.$$

where ζ_k has degree one in integral homology. However

$$S^2{}_+\wedge_T S^{2k}\wedge S^{2k} \simeq S^{2k}P_{2k}{}^{2k+2}$$

and since k is odd a standard relation in the homotopy of $S^{2k}P_{2k}{}^{2k+2}$ shows that $2\zeta_k$ is $i\cdot\eta$ where $i : S^{2k}\wedge S^{2k} \longrightarrow S^2{}_+\wedge_T S^{2k}\wedge S^{2k}$ is the inclusion. Now

$$S^{4k} \xrightarrow{i} S^2{}_+\wedge_T S^{2k}\wedge S^{2k} \xrightarrow{1\wedge_T x\wedge x} S^2{}_+\wedge_T S^{2n}\wedge S^{2n} \longrightarrow S^{4n}.$$

is just $x\wedge x$ and so 6.5 follows.

§8 THE PROOFS OF 6.3 AND 6.4

The proof of 6.3 and 6.4 is based on the following geometrical observation concerning vector bundles. Let $V_j(\mathbb{R}^n)$ be the Stiefel manifold of j-frames in $\mathbb{R}^n$. We will identify a map $f : X \longrightarrow V_j(\mathbb{R}^n)$ with a vector bundle $\xi = \xi(f)$ of dimension $n-j$ over X and an isomorphism $t = t(f) : \xi(f) + \varepsilon^j \longrightarrow \varepsilon^n$ where ε^m denotes the trivial m dimensional bundle. In [14] James constructs the "intrinsic map" $h : V_j(\mathbb{R}^n)*V_k(\mathbb{R}^m) \longrightarrow V_k(\mathbb{R}^{n+m})$ when $j \geq k$. Of particular interest to us is the case where $j = n \geq k$ so that $V_n(\mathbb{R}^n)$ is the orthogonal group $O(n)$.

Let $f : X \longrightarrow O(n)$ be a map with corresponding vector bundle ζ over SX. So ζ is the bundle obtained by the "clutching construction" using the map f; equivalently ζ thought of as a map $SX \longrightarrow BO(n)$ is adjoint to f. Let $g : Y \longrightarrow V_k(\mathbb{R}^m)$ be a map with corresponding vector bundle $\xi = \xi(g)$ over Y. The product bundle $\zeta \times \xi$ over $SX \times Y$ restricted to Y (embedded as the product of the base point with Y) is $\varepsilon^n + \xi$ and so is provided with a trivialisation. Using this trivialisation we get a well defined bundle γ over $SX \times Y/Y$ such that $c^*\gamma = \zeta \times \xi$. Let η be the bundle over $X*Y$ defined by the map

$$h(f*g) : X*Y \longrightarrow V_k(\mathbb{R}^{n+m}).$$

There are natural equivalences $SX \times Y/Y = SX \wedge Y_+ \longrightarrow SX \wedge Y \vee SX$ and $X*Y \longrightarrow SX \wedge Y$ and using these equivalences we can identify γ and η with bundles over $SX \wedge Y \vee SX$ and $SX \wedge Y$.

LEMMA 8.1. The restrictions of γ to the factors $SX \wedge Y$ and SX are, respectively, η and $\zeta + \varepsilon^{m-k}$.

The proof is a more-or-less direct geometric argument from the definition of the map h. It is however quite laborious and is omitted in the interests of brevity.

Suppose that $X = S^p$, $Y = S^q$ and $k = 1$ so

$$f : S^p \longrightarrow O(n), \quad g : S^q \longrightarrow V_1(\mathbb{R}^m) = S^{m-1}.$$

Then $h(f*g)$ is the following composite

$$S^p*S^q \longrightarrow O(n)*V_1(\mathbb{R}^m) \longrightarrow V_1(\mathbb{R}^{n+m}) = S^{n+m-1}.$$

From the properties of the intrinsic join operation this map is the join of the two maps

$$pf : S^p \longrightarrow O(n) \longrightarrow S^{n-1}, \quad g : S^q \longrightarrow V_1(\mathbb{R}^m) = S^{m-1}$$

where $p : O(n) \longrightarrow S^{n-1}$ is the usual projection. Call the first of these maps φ and the second ψ.

Let $\Delta : \pi_q(V_1(\mathbb{R}^m)) \longrightarrow \pi_{q-1}(O(m-1))$ be the boundary homorphism in the homotopy exact sequence of the fibration $O(m-1) \longrightarrow O(m) \longrightarrow V_1(\mathbb{R}^m)$ and let $J : \pi_i O(r) \longrightarrow \pi_{i+r}(S^r)$ be the J homorphism. We know that

$$J\Delta(x) = P(E^m x) \in \pi_{q+m-2}(S^{m-1}), \quad x \in \pi_q(S^{m-1})$$

$$E^n \rho_*(x) = HJ(x) \in \pi_{p+n}(S^{2n-1}), \quad x \in \pi_p O(n).$$

Now put $\alpha = J(f) \in \pi_{p+n} S^n$ and $\beta = J\Delta(g) \in \pi_{q+m-2} S^{m-1}$ so that

$$\lambda = J\Delta(h(f * g)) = P(E^{m+n}\varphi * \psi) = P(E^{m+n+1}\varphi \wedge \psi) \in \pi_{p+q+n+m-1}(S^{n+m-1})$$

where $E^n\varphi = H(\alpha)$ and $P(E^m\psi) = \beta$.

Next we use the well-known fact that if $d : A \longrightarrow O(r)$ is a map then the mapping cone of $Jd : S^r A \longrightarrow S^r$ is equivalent to the Thom complex of the bundle δ over SA defined by the map d. Now use the notation introduced before 8.1 in our special case; so we get a bundle γ over $(S^{p+1} \times S^q)/S^q \simeq S^{p+q+1} \vee S^{p+1}$ such that $c^*\gamma = \xi \times \xi$ where c is the collapsing map $S^{p+1} \times S^q \longrightarrow S^{p+q+1} \vee S^{p+1}$. This gives us a map of Thom complexes

$$C(\alpha) \wedge C(\beta) \longrightarrow T(\gamma).$$

However using 8.1 and the above remark on Thom complexes over suspensions we see that

$$T(\gamma) \simeq (S^{n+m-1} \cup_{\alpha'} e^{n+m+p}) \cup_{\lambda'} e^{p+q+n+m}.$$

Here α' is $E^{m-1}\alpha$ and λ' is the composite

$$S^{p+q+n+m-1} \xrightarrow{\lambda} S^{n+m-1} \xrightarrow{i} (S^{n+m-1} \cup_{\alpha} e^{n+m+p}),$$

where i is the inclusion.

We apply this argument in the following case. Suppose $\alpha \in \pi_{k+n}(S^n)$, $H(\alpha) = E^n\varphi$, $E^r\alpha = 0$ and $E^{r-1}\alpha = \beta = P(E^{n+r}\psi) \in \pi_{k+n+r-1}(S^{n+r-1})$ where $\psi \in \pi_{k+n+r+1} S^{n+r+1}$. As we are in the metastable range and α and β are stably trivial we know that there are elements f and g such that $\alpha = J(f)$ and $\beta = J\Delta(g)$. So we may use the above reasoning with $p = k$, $m = n+r$ and $q = k+1$ in the above argument. Then an easy argument shows that

$$C(\alpha)\wedge C(\beta) \simeq (S^{2n+r-1} \cup_\sigma e^{2k+2n+r+1}) \vee S^{k+2n+r} \vee S^{k+2n+r}.$$

Furthermore if n and k are even then from 6.2

$$\sigma = E^{r-1}Sq_1(\alpha).$$

However

$$T(\gamma) \simeq (S^{2n+r-1} \cup_\lambda e^{2k+2n+r+1}) \vee S^{2n+r+k},$$

where as above

$$\lambda = P(E^{2n+r+1}\varphi\wedge\psi) \in \pi_{2k+2n+r}(S^{2n+r-1}).$$

But we have a degree one map from $C(\alpha)\wedge C(\beta) \longrightarrow T(\gamma)$ and so we must have $\lambda = \sigma$. This proves 6.3; 6.4 requires a similar argument which will be left to the reader.

REFERENCES.

[1] J.F. ADAMS, Vector fields on spheres. Ann. Math. 75 (1962), 603-632.

[2] J.F. ADAMS, On the groups J(X)-IV. Topology 5 (1965), 21-71.

[3] M.G. BARRATT, J.D.S. JONES and M.E. MAHOWALD, The Kervaire invariant problem. Proceedings of the Northwestern homotopy theory conference Contemporary Mathematics 19, AMS 1983, 9-23.

[4] M.G. BARRATT, J.D.S. JONES and M.E. MAHOWALD, Relations amongst Toda brackets and the Kervaire invariant in dimension 62. J. London Math. Soc. (Series 2) 30 (1984), 533-550.

[5] A.K. BOUSFIELD, E.B. CURTIS, D.M. KAN, D.G. QUILLEN, D.L. RECTOR and J.W. SCHLESINGER, The mod p lower central series and the Adams spectral sequence. Topology 5 (1966), 331-342.

[6] W.E. BROWDER, The Kervaire invariant of framed manifolds and its generalisations. Ann.

of Math. 90 (1969), 157–186.

[7] E.H. BROWN and F.P. PETERSON, Whitehead products and cohomology operations. Oxford Quart. J. Math. (Series 2) 15 (1964), 116–120.

[8] R.L. COHEN and M.E. MAHOWALD, On the Adams spectral sequence of projective spaces. To appear.

[9] F.R. COHEN, J.P. MAY and L.R. TAYLOR, Splitting certain spaces CX. Math. Proc. Camb. Phil. Soc. 84 (1978), 465–496.

[10] D.M. DAVIS, Generalized homology and the generalised vector field problem. Oxford Quart. J. Math. (Series 2) 25 (1974), 169–193.

[11] D.M. DAVIS, S. GITLER and M.E. MAHOWALD, The stable geometric dimension of vector bundles over real projective spaces. Trans. A.M.S. 268 (1981), 39–61.

[12] S. FEDER, S. GITLER and K.Y. LAM, Composition properties of projective homotopy classes. Pacific J. Math. 68 (1977) 47–61.

[13] P.G. GOERSS, J.D.S. JONES and M.E. MAHOWALD, Generalised Brown-Gitler spectra. To appear in Trans. A.M.S.

[14] I.M. JAMES, The topology of Stiefel manifolds. L.M.S. Lecture note series 24, C.U.P. (1976)

[15] J.D.S. JONES and S.A. WEGMANN, Limits of stable homotopy and cohomotopy groups. Math. Proc. Camb. Phil. Soc. 94 (1983) 473–482.

[16] N.J. KUHN, The geometry of James-Hopf maps. Pacific Journal of Math. 102 (1982), 397–412.

[17] D.S. KAHN and S.B. PRIDDY, Applications of the transfer to stable homotopy theory. Bull. A.M.S. 78 (1972), 981–987.

[18] D.S. KAHN and S.B. PRIDDY, The transfer and stable homotopy theory. Math. Proc. Camb. Phil. Soc. 83 (1978) 103–112.

[19] M.E. MAHOWALD, The metastable homotopy of S^n. AMS memoirs 72 (1967).

[20] M.E. MAHOWALD, The order of the image of the J-homorphism. Bull. A.M.S. 76 (1970), 1310–1313.

[21] M.E. MAHOWALD, bo-resolutions. Pacific J. Math. 92 (1981), 365–383.

[22] M.E. MAHOWALD, The image of J in the EHP sequence. Ann. of Math. 116 (1982) 65–112 also correction Ann. of Math. 120 (1984), 97–99.

[23] M.E. MAHOWALD and R.J. MILGRAM, Operations which detect Sq^4 in connective K-theory and their applications. Oxford Quart. J. Math. (Series 2) 27 (1976), 415–432.

[24] M.E. MAHOWALD and W. LELLMAN, The bo-Adams spectral sequence. To appear.

[25] R.J. MILGRAM, A construction for s-parallelizable manifolds and primary homotopy

operations. Topology of manifolds edited by J.C. Cantrell and C.H. Edwards, Markham (1970), 483-489.

[26] R.J. MILGRAM, Symmetries and operations in homotopy theory. Proc. Symp. Pure Math. 22, AMS (1971), 203-210.

[27] R.J. MILGRAM, Group representations and the Adams spectral sequence. Pacific J. Math. 41 (1972), 157-182.

[28] R.J. MILGRAM, The Steenrod algebra and its dual for connective K-theory. Reunion sobre teoria de homotopia Universidad de Northwestern, Agosto 1974. Mexican Math. Soc. (1975), 127-158.

[29] G. NISHIDA, The nilpotency of elements of the stable homotopy of spheres, J. Math. Soc. Japan 25 (1973) 301-347.

[30] D.G. QUILLEN, The Adams conjecture. Topology 10 (1971), 29-56.

[31] V.P. SNAITH, A stable decomposition for $\Omega^n S^n X$. J. London Math. Soc. (Series 2) (1974), 577-583.

[32] H. TODA, Composition methods in the homotopy groups of spheres. Annals of Math. Studies 49, P.U.P. (1962).

[33] J.S.P. WANG, On the cohomology of the mod 2 Steenrod algebra. Ill. J. Math. 11 (1967), 480-490.

[34] G.W. WHITEHEAD, Recent advances in homotopy theory. Regional conference series, AMS conference board (1972).

Addresses.

Barratt, Mahowald: Mathematics Department, Northwestern University, Evanston, Illinois 60201, USA.

Jones: Mathematics Institute, University of Warwick, Coventry CV4 7AL, England.

STABLE SPLITTINGS OF MAPPING SPACES

C.-F. Bödigheimer

0. Introduction.

In this note we elaborate on two observations concerning configuration spaces; they will lead to a stable splitting of certain mapping spaces into infinite bouquets of simpler spaces.

Let K be a finite complex, K_o a subcomplex, and X a connected CW-complex. Then choose a smooth, compact and parallelizable m-manifold M with a submanifold M_o such that the pairs (K,K_o) and (M,M_o) are homotopy equivalent. For the space $\mathrm{map}(K,K_o;S^mX)$ of based maps from K/K_o to S^mX we prove

PROPOSITION 1.

There is a stable equivalence

$$\mathrm{map}(K,K_o;S^mX) \simeq_s \bigvee_{k=1}^{\infty} \mathcal{D}_k \quad ;$$

the spaces $\mathcal{D}_k$ *depend on* M,M_o *and* X, *in particular* $\mathcal{D}_1 = (M \setminus M_o, \partial M \setminus M_o) \wedge X$.

Several special cases of this proposition are well-known.

EXAMPLE 1. $K = M = [0,1]$, $K_o = M_o = \{0,1\}$.
The proposition gives a splitting of the suspension spectrum $S^\infty \Omega SX$; a refinement of the proof would yield the splitting of $S\Omega SX$ found by Milnor [17], see Remark 3.

EXAMPLE 2. $K = M = D^m$, $K_o = M_o = \partial D^m$.
This is the stable splitting of $\Omega^m S^m X$ found by Snaith [20].

EXAMPLE 3. $K = M = S^1$, $K_o = M_o = \emptyset$.
A stable splitting of the free loop space ΛSX of SX has recently been obtained by Goodwillie (unpublished).

EXAMPLE 4. $K = M = S^{m-1} \times [0,1]$, $K_o = M_o = S^{m-1} \times \{0,1\}$.
This example gives a stable splitting of $\bar{\Omega}^m S^m X$, the space of maps $f : S^m \longrightarrow S^m X$ such that $f(s_o) = f(-s_o) = *$, where s_o and $*$ are the basepoints; it is particularly interesting for $\mathbb{Z}/2$- and S^1-equivariant homotopy theory, (N.B. $\bar{\Omega}^m S^m X \simeq \Omega^m S^m X \times \Omega S^m X$.)

EXAMPLE 5. $K = \mathbb{D}^m$, $K_o = \partial\mathbb{D}^m$, $M = \mathbb{D}^m \times [0,1]$, $M_o = \partial\mathbb{D}^m \times [0,1]$.
In this case we obtain - also for non-connected X - a stable splitting of $\Omega^m S^{m+1} X$; it is different but equivalent to the corresponding one replacing X by SX in Example 2.

EXAMPLE 6. $K = M = G$ a compact Lie group of dimension m, $K_o = M_o = \emptyset$.
Here the mapping space is the space of all unbased maps from G to $S^m X$.

EXAMPLE 7. K = point, $K_o = \emptyset$, $M = \mathbb{D}^m$, $M_o = \emptyset$.
We have $\mathrm{map}(K,K_o;S^m X) = S^m X = \mathcal{D}_1$, all other $\mathcal{D}_k$ are contractible.

EXAMPLE 8. In general one can choose an embedding $K \subset \mathbb{R}^m$ of K, a regular neighbourhood M, a submanifold M_o with $K_o \subset M_o$ and a deformation retraction of pairs $r_t : (M,M_o) \longrightarrow (K,K_o)$. Hence $\mathrm{map}(K,K_o;S^m X)$ always stably splits into a bouquet, if m is at least the embedding dimension of K.

Such splittings are usually obtaiend by splitting appropriate configuration space models for the mapping spaces. In Section 1 we will define these models. In Section 2 we observe that (under certain connectivity assumptions) they are equivalent to mapping spaces. In Section 3 we observe that these models split stably, and we conclude Proposition 1. In Section 4 we list some properties of the splittings.

We do not claim any originality. In fact, all the constructions and proofs either can be found in the literature (e.g. [2], [5], [16] and [20]) are well-known to the experts. Only the importance of such splittings may justify the publication of a unified approach.

The author is indepted to the referee and to F. Cohen for demanding more details to make the following pages more self-contained and readable.

1. The Configuration Spaces.

Let N be a smooth m-manifold, N_o a submanifold (closed as a subspace), and X a CW-complex with basepoint $*$. We denote by $C(N,N_o;X)$ the space of finite configurations of particles in N with parameters (or labels) in X, which are annihilated in N_o or for vanishing; more precisely, let $\tilde{C}(N,k) = \{(z_1,\ldots,z_k) \in N^k \mid z_i \neq z_j \text{ for } i \neq j\}$ be the space of ordered (unlabeled) configurations of k points in N; then $C(N,N_o;X)$ is the quotient of $\coprod_{k=1}^{\infty} \tilde{C}(N,k) \times X^k$ by the following identifications:

(1.1) actions of the symmetric groups Υ_k

$$(z_1,\ldots,z_k;x_1,\ldots,x_k) \sim (z_{s(1)},\ldots,z_{s(k)};x_{s(1)},\ldots,x_{s(k)}) \text{ for } s \in \Upsilon_k;$$

(1.2) annihilation of particles with parameter $*$

$$(z_1,\ldots,z_k;x_1,\ldots,x_k) \sim (z_1,\ldots,z_{k-1};x_1,\ldots,x_{k-1}) \text{ if } x_k = *;$$

(1.3) annihilation of particles in N_o

$$(z_1,\ldots,z_k;x_1,\ldots,x_k) \sim (z_1,\ldots,z_{k-1};x_1,\ldots,x_{k-1}) \text{ if } z_k \in N_o.$$

Because of (a) we will write a configuration $\xi \in C = C(N,N_o;X)$ as a formal sum $\xi = \Sigma z_i x_i$ bearing in mind that C is a subspace of the infinite symmetric product $SP_\infty((N/N_o) \wedge X)$; then (1.2) and (1.3) can be replaced by: $zx = 0$ if $x = *$ or $z \in N_o$, respectively, where 0 denotes the basepoint in C (which is represented by any $\xi = \Sigma z_i x_i$ such that for all i, $x_i = *$ or $z_i \in N_o$ holds).

Such configuration spaces have been extensively studied by Fadell-Neuwirth [8] for $N_o = \emptyset$ and $X = S^o$, by Mc Duff [16] for $N_o = \partial N$ and $X = S^o$, and by Cohen-Taylor [2] for $N_o = \emptyset$.

<u>EXAMPLE 9.</u> $C(\mathbb{R}^m;X) = C(\mathbb{R}^m,\emptyset;X)$ are the well known configuration spaces of May [13] and Segal [19]. $C(\mathbb{R}^1;X)$ is homotopy equivalent to the James construction [9].

The length k of a configuration $\xi = \sum_{i=1}^{k} z_i x_i$ induces a natural filtration of C by closed subspaces $C_k(N,N_o;X) = (\coprod_{i=1}^{k} \tilde{C}(N,k) \times X^k)/\sim$. The inclusion $C_{k-1} \longrightarrow C_k$ is a cofibration, because $N_o \longrightarrow N$ and $* \longrightarrow X$ are. C_o consists of 0 only, and C_1 is $(N,N_o) \wedge X$.

If the pair (N,N_o) or X is connected then each particle z_i of a configuration ξ can be moved to N_o or its parameter x_i can be moved to $*$; therefore ξ can be moved to 0, i.e. C is connected. If N is connected, $N_o = \emptyset$ and $X = S^o$, then the strata $C_k - C_{k-1} = \tilde{C}(N,k)/\Upsilon_k = C(N,k)$ of $C = C(N) = C(N,\emptyset;S^o)$ are the connected components of C.

So far we have not used that N is a manifold - indeed N might have been any space; in particular, $C(\mathbb{R}^\infty;X)$ will be of importance to us (see Example 13 and Section 3).

<u>EXAMPLE 10.</u> The connected components of $C(\mathbb{R}^\infty)$ are the classifying spaces of the symmetric groups; those of $C(\mathbb{R}^2)$ are the classifying spaces of Artin's braid groups.

<u>EXAMPLE 11.</u> $C(\mathbb{D}^m, \partial\mathbb{D}^m; X)$ is homotopy equivalent to $S^m X$, see [16; p. 95].

The construction C is a homotopy functor in X, but only an isotopy functor in (N, N_o).

So, for example, the inclusion $N \setminus \partial N \longrightarrow N$ induces a homotopy equivalence $C(N \setminus \partial N, N_o \setminus \partial N; X) \longrightarrow C(N, N_o; X)$. The excision property $C(N, N_o; X) \cong C(N \setminus U, N_o \setminus U; X)$ for $U \subset N_o$ and U open in N, and the product property $C(N, N_o; X) \cong C(N', N' \cap N_o; X) \times C(N'', N'' \cap N_o; X)$ for $N = N' \cup N''$ and $N' \cap N'' \subset N_o$ follow easily from the definition. The crucial property of C is contained in the following lemma.

<u>Lemma.</u>

<u>Let</u> $H \subset N$ <u>be an</u> m-<u>dimensional submanifold</u>. <u>Then the isotopy cofibration</u>

$$(H, H \cap N_o) \longrightarrow (N, N_o) \xrightarrow{q} (N, H \cup N_o)$$

<u>induces a quasifibration</u>

$$C(H, H \cap N_o; X) \longrightarrow C(N, N_o; X) \xrightarrow{Q} C(N, H \cup N_o; X)$$

<u>provided</u> $(H, H \cap N_o)$ <u>or</u> X <u>is connected</u>.

<u>Proof:</u> Except for the presence of a parameter space X the proof is that of [16; Proposition 3.1]; we list the various steps.

(1) We filter the base space $B = C(N, H \cup N_o; X)$ by $B_k = C_k(N, H \cup N_o; X)$, and the total space $E = C(N, N_o; X)$ by $E_k = Q^{-1}(B_k)$, and we denote the fibre by $F = C(H, H \cap N_o; X)$.

(2) Observe that for each k there is homeomorphism

$$h_k : E_k \setminus E_{k-1} \cong (B_k \setminus B_{k-1}) \times F \quad \text{over} \quad B_k \setminus B_{k-1} .$$

(3) A tubular neighbourhood U of H defines for each k a neighbourhood U_k of B_k in B_{k+1}, and an isotopy retraction $r : U \to H$ induces

retractions $r_k : U_k \to B_k$, and retractions $\bar{r}_k : Q^{-1}(U_k) \to Q^{-1}(B_k) = E_k$ lying over r_k.

(4) For every $b \in U_k$ the induced map

$$F \xleftarrow[h_{k+1}|]{\cong} Q^{-1}(b) \xrightarrow{\bar{r}_k|} Q^{-1}(r_k(b)) \xrightarrow[h_k|]{\cong} F$$

is a homotopy equivalence (precisely because $(H, H \cap N_o)$ or X is connected).

It follows from the Dold-Thom criterion [8 ; 2.10, 2.15, 5.2] that Q is a quasifibration. □

2. The Section Spaces

The space $C(N,N_o;X)$ is under certain connectivity conditions equivalent to the space of sections of a certain bundle with fibre S^mX, and whence sometimes equivalent to a space of maps into S^mX. To make this precise let W be any smooth m-manifold without boundary which contains N (for example, $W = N$ if $\partial N = \emptyset$, or $W = N \cup (\partial N \times [0,1[)$ otherwise); if $\hat{T}(W)$ denotes the fibrewise compactification of the tangent bundle $T(W)$ of W, then define $\hat{T}(W;X) = \hat{T}(W) \wedge_{\hat{\tau}} X$ to be fibrewise smash product of $\hat{T}(W)$ and X; this is a new bundle $\hat{\tau} : \hat{T}(W;X) \longrightarrow W$ with fibre S^mX.

The inclusion of the basepoint into each fibre yields a section g_∞ of $\hat{\tau}$. For $A_o \subset A \subset W$ let $\Gamma(A,A_o;X)$ denote the space of sections of $\hat{\tau}$ which are defined on A and agree with g_∞ on A_o; it is equipped with the (compactly generated topology induced by the) compact-open topology. (For example, if $X = S^o$ then $\hat{T}(W,S^o) = \hat{T}(W)$ and the sections are the vector fields with possible poles.)

The main theorem about configuration spaces on manifolds is the following duality.

PROPOSITION 2.

For compact N there is a map $\gamma : C(N,N_o;X) \longrightarrow \Gamma(W \setminus N_o, W \setminus N;X)$, *which is a (weak) homotopy equivalence provided* (N,N_o) *or X is connected.*

Proof: The proof is essentially contained in Mc Duff [16 ; Theorem 1.4] or [15]. For convenience we indicate the various steps.

(1) Following ideas of Gromov the map γ is defined as in [16 ; p. 95], or as in [15 ; p. 90] using Example 11, we have $\gamma(0) = g_\infty$.

(2) We start to prove the assertion with the case of (N,N_o) being a handle $(\mathbb{D}^m, \mathbb{D}^k \times S^{m-k-1})$ of index k. First, the assertion is true for k = 0 by Example 11. Consider for $k = 1,2,\ldots,m$ in $I^k = [0,1]^m$ the subspace I_k^m of all $y = (y^1,\ldots,y^m)$ such that $y^i = 0$ or $y^i = 1$ for some $i = k+1,\ldots,m$, or $y^k = 1$; set $H^k = [0,1]^{k-1} \times [0,\frac{1}{2}] \times [0,1]^{m-k}$. In the sequence

(3) $(H_k, H_k \cap I_k^m) \longrightarrow (I^m, I_k^m) \longrightarrow (I^m, H_k \cup I_k^m)$ the left hand pair is a handle of index k, the right hand pair is a handle of index k-1. We apply C(;X) to (3) and obtain by the above lemma a quasifibration for $k = 1,\ldots,m-1$ if X is arbitrary, and in addition for k = m if X is arbitrary, and in addition for k = m if X is connected. We apply $\Gamma($ $;X)$ to the complements in $W = \mathbb{R}^m$ of (3) and obtain a fibration; γ maps the quasifibration to the fibration. Notice that both total spaces are contractible. Hence we conclude by induction the assertion for all handles of index $k = 0,1,\ldots,m-1$ if X is arbitrary, and in addition for the handle of index m if X is connected.

(4) For the case $(N,\partial N)$ choose a handle decomposition of N, and if $(N,\partial N)$ is connected choose one without handles of index m. Attaching a new handle gives a quasifibration for C and a fibration for Γ, γ mapping

one to the other. Induction on the number of handles proves the assertion for $(N,\partial N)$.

(5) For the case (N,N_o) with $N_o \subset \partial N$ we choose a complementary submanifold $L \subset \partial N$, i.e. $L \cup N_o = \partial N$ and $L \cap N_o = \partial L = \partial N_o$. We attach a closed collar to N, $\bar{N} = N \cup (N \times [0,1])$, and consider the sequence

(6) $(\bar{L},\bar{L} \cap \bar{N}_o) \longrightarrow (\bar{N},\bar{N}_o) \longrightarrow (\bar{N},\bar{L} \cup \bar{N}_o)$ with $\bar{L} = L \times [0,1]$ and $\bar{N}_o = N_o \times [0,1]$. The assertion is true for the right hand pair by (4) since $(\bar{N},\bar{L} \cup \bar{N}_o) = (\bar{N},\partial\bar{N}) \cong (N,\partial N)$. As before, the assertion will follow for $(\bar{N},\bar{N}_o) \cong (N,N_o)$ if we can prove it for $(\bar{L},\bar{L} \cap \bar{N}_o) = (\bar{L},\partial\bar{L}) = (L,\partial L) \times [0,1]$.

(7) For this case we use the sequence

(8) $(L,\partial L) \times [0,1] \longrightarrow (L,\partial L) \times ([0,2],\{2\}) \longrightarrow (L \times [0,2],\partial(L \times [0,2]))$. The assertion is true for the right hand pair by (4); it is true for the middle pair, since this gives contractible spaces. Hence the assertion follows for the left hand pair.

(9) For the case of an arbitrary submanifold $N_o \subset N$ we replace N_o by closed tubular neighbourhood and then remove the interior of this neighbourhood. By isotopy invariance and excision property both manipulations leave the homotopy type of C unaltered. But now we are in case (5). □

<u>EXAMPLE 12</u>. (Example 8 continued). Under the assumptions of Proposition 1 set $N = M \setminus M_o$ and $N_o = \partial M \setminus M_o$, and $W = M \cup (\partial M \times [0,1[)$ if $\partial M \neq \emptyset$, or $W = M$ if $\partial M = \emptyset$. As a corollary we have

$$\begin{aligned}
C(M \setminus M_o,\partial M \setminus M_o;X) &\simeq \Gamma(W \setminus (\partial M \setminus M_o),W \setminus (M \setminus M_o);X) \text{ by Proposition 1} \\
&= \Gamma(W \setminus \partial M) \cup M_o,(W \setminus M) \cup M_o;X) \\
&= \Gamma(M \setminus \partial M,M_o \setminus \partial M;X) \text{ by excision} \\
&\simeq \Gamma(M,M_o;X) \text{ by extension over } \partial M \\
&\cong \operatorname{map}(M,M_o;S^m X) \text{ by parallelizability} \\
&\simeq \operatorname{map}(K,K_o;S^m X) \ ,
\end{aligned}$$

where we should replace M_o by an open tubular neighbourhood to ensure compactness of $M \setminus M_o$.

EXAMPLE 13 (Example 2, 9 and 10 continued). If $N = \mathbb{D}^m$, $N_o = \emptyset$ and $W = \mathbb{R}^m$, then γ is the well-known approximation $C(\mathbb{R}^m;X) \simeq C(\mathbb{D}^m;X) \longrightarrow \operatorname{map}(\mathbb{R}^m, \mathbb{R}^m \setminus \mathbb{D}^m; S^m X) \simeq \Omega^m S^m X$ of May [13] and Segal [19]. Passing to the limit over m yields $\gamma^\infty : C(\mathbb{R}^\infty;X) \longrightarrow \Omega^\infty S^\infty X =$ (X). See also Vogel [21].

Remark 1. For $C(M \setminus M_o, \partial M \setminus M_o;X)$ to be a model for $\operatorname{map}(K,K_o;S^m X)$ it is obviously enough that (M,M_o) is relatively compact and relatively parallelizable; but more important is that X need not be connected if $(M \setminus M_o, \partial M \setminus M_o)$ happens to be connected, see e.g. Example 5. In general, γ approximates the homology of the section space, see [16]; so in case $\partial N \neq \emptyset$, γ is a completion of homology modules over $H_*(\Omega \operatorname{map}(\partial N;S^m X))$. An interesting example is $C(\mathbb{R}P^m)$, since $\Gamma(\mathbb{R}P^m) = \Gamma(\mathbb{R}P^m;S^o)$ is the space of self-maps of S^m which are equivariant with respect to the antipodal action.

3. The Stable Splittings.

In [20] Snaith has obtained a stable splitting of $\Omega^m S^m X$ using the models $C(\mathbb{R}^m;X)$. Since then several authors have given very elegant proofs of this result, see F. Cohen [5], R. Cohen [6], Cohen-May-Taylor [3], May-Taylor [14], Vogt [22]. Our construction of a stable splitting of $C = C(N,N_o;X)$ is almost verbatim taken from [5].

Let $D_k = D_k(N,N_o;X)$ denote the filtration quotients C_k/C_{k-1} and consider the bouquet $V = V(N,N_o;X) = \bigvee_{k=1}^{\infty} D_k$ with the filtration given by $V_k = \bigvee_{j=1}^{k} D_j$.

Next we define the "power set map" $P : C \longrightarrow C(\mathbb{R}^\infty;V)$. Take some

$\xi = \sum_i z_i x_i \in C$ and a (non-empty) subset $\alpha = \{i_1,\ldots,i_k\}$ of the index set $I(\xi)$ of ξ. Define Z_α to be the (unlabeled) configuration $Z_\alpha = \sum_{j=1}^{k} z_{i_j}$ consisting of all z_i in ξ such that $i \in \alpha$; Z_α is in $\tilde{C}(N,k)/\Sigma_k = C(N,k)$ which is an km-manifold; we choose an embedding of their disjoint union $C(N) = \coprod_{k=1}^{\infty} C(N,k)$ into $\mathbb{R}^\infty$, and let $\bar{Z}_\alpha \in \mathbb{R}^\infty$ denote the image of Z_α under this embedding. Correspondingly, define ξ_α to be the subconfiguration $\xi_\alpha = \sum_{j=1}^{k} z_{i_j} x_{i_j}$ of ξ consisting of all labeled particles $z_i x_i$ of ξ such that $i \in \alpha$; ξ_α is in $C_k = C_k(N,N_o;X)$; using the quotient map $C_k \longrightarrow D_k$ and the inclusion $D_k \longrightarrow V$ we let $\bar{\xi}_\alpha \in V$ denote the image of ξ_α under the composition of these two maps. Finally, we define $P(\xi) = \sum_\alpha \bar{Z}_\alpha \bar{\xi}_\alpha$ in $C(\mathbb{R}^\infty;V)$ where the sum is over all subsets of $I(\xi)$.

Notice that the $\bar{Z}_\alpha$ are mutually different since two of the same length k have already different Z_α in $C(N,k)$, and the various $C(N,k)$ are disjointly embedded into $\mathbb{R}^\infty$. P is continous since it is well-defined: (1.1) is respected because a permutation of $I(\xi)$ only permutes the new indices α; (1.2) and (1.3) are respected because if $z_i \in N_o$ or $x_i = *$, then, for any α such that $i \in \alpha$, $\bar{\xi}_\alpha$ is the basepoint in D_k and in V, hence $\bar{Z}_\alpha \bar{\xi}_\alpha = 0$ in $C(\mathbb{R}^\infty;V)$.

Now let $\sigma : S^\infty C \longrightarrow S^\infty V$ denote the adjoint of the composition $\gamma^\infty \circ P : C \longrightarrow C(\mathbb{R}^\infty;V) \longrightarrow Q(V) = \Omega^\infty S^\infty V$ with γ^∞ as in Example 13.

PROPOSITION 3.

σ <u>is a stable equivalence</u> $C(N,N_o;X) \longrightarrow \bigvee_{k=1}^{\infty} D_k(N,N_o;X)$ <u>for any</u> (N,N_o) <u>and</u> X.

Proof: σ obviously preserves the filtration and we have a commutative lower square in the diagram

$$\begin{array}{ccc}
S^\infty(C_k/C_{k-1}) & = & S^\infty(V_k/V_{k-1}) \\
\uparrow & & \uparrow \\
S^\infty C_k & \xrightarrow{\sigma} & S^\infty V_k \\
\uparrow & & \uparrow \\
S^\infty C_{k-1} & \xrightarrow{\sigma} & S^\infty V_{k-1}
\end{array} ,$$

whereas the upper square is only homotopy commutative. Since the vertical sequences are cofibrations and since $C_1 = V_1$ the assertion follows by induction on k. □

<u>Proof of Proposition 1</u> (Example 7 and 11 continued). The stable splitting of $\mathrm{map}(K,K_o;S^mX)$ now follows from that of $C(M \setminus M_o, \partial M \setminus M_o;X)$. The spaces $\mathcal{D}_k$ are $D_k(M \setminus M_o, \partial M \setminus M_o;X)$, in particular we have
$\mathcal{D}_1 = C_1(M \setminus M_o, \partial M \setminus M_o;X) = (M \setminus M_o, \partial M \setminus M_o) \wedge X$. □

<u>EXAMPLE 14</u> (Example 2, 8 and 12 continued). The splitting we obtain for $K = M = \mathbb{D}^m$ and $K_o = M_o = \partial\mathbb{D}^m$ is the Snaith splitting of [20].

<u>Remark 2.</u> In the proof of Proposition 3 we did not use that N is a manifold; the proof covers also the case of $C(\mathbb{R}^\infty;X)$ which is equivalent to $\Omega^\infty S^\infty X$ if X is connected. A stable splitting of $\Omega^\infty S^\infty X$ was first obtained by Kahn, see [1], [10], [11] and [12]. Furthermore, we did not use that (N,N_o) or X is connected. This and Remark 1 shows that Proposition 1 is more generally true than stated, see e.g. Example 5.

<u>Remark 3.</u> A splitting of $S\Omega SX$ is achieved by refining the power set map to a map $P : C = C(\mathbb{R};X) \longrightarrow C(\mathbb{R};V(\mathbb{R};X))$; the order of the particles z_i on the real line induces a lexicographic order of the sets α, and the hereby induced order of the z_α is used to define particles $\bar{z}_\alpha$ in $\mathbb{R}$ instead of $\mathbb{R}^\infty$.

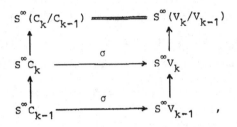

4. Naturality and Homology.

Assume we have two situations as in the introduction, a map $f : (K,K_o) \longrightarrow (K',K'_o)$ together with an embedding $F : (M,M_o) \longrightarrow (M',M'_o)$ making the obvious diagram commutative, $m = m'$ and $X = X'$. Then f induces $f^*: map(K',K'_o;S^mX) \longrightarrow map(K,K_o;S^mX)$, while F induces $F^* : C(M' \setminus M'_o, \partial M \setminus M'_o;X) \longrightarrow C(M \setminus M_o, \partial M \setminus M_o;X)$ and $F^*_k : D_k(M' \setminus M'_o, \partial M' \setminus M'_o;X) \longrightarrow D_k(M \setminus M_o, \partial M \setminus M_o;X)$. The approximation map γ of Proposition 2 and the splitting map σ of Proposition 3 commute with these induced maps.

Examples for such maps f are the inclusions $K_o \longrightarrow K$ and $K \longrightarrow (K,K_o)$, the inclusion of a bottom cell of K and the pinch map onto a top cell of K.

γ and σ are natural with respect to the suspension $map(K,K_o;S^mX) \longrightarrow map(S(K,K_o);S^{m+1}X)$, which for C and V is induced by the equatorial inclusion $(M,M_o) \longrightarrow (M,M_o) \times ([0,1],\{0,1\})$.

An analysis of the splitting map σ reveals that each of the spaces $\mathcal{D}_k$ is already after a finite number of suspensions a retract of C. An upper bound for the smallest number is given by the embedding dimension of C(N,k). In our standard situation of Example 7 we have $N = M \setminus M_o$ as a submanifold of $\mathbb{R}^m$, so $\mathcal{D}_1 = (M \setminus M_o, \partial M \setminus M_o) \wedge X$ is a retract of $map(K,K_o;S^mX)$ after at most m suspensions.

The (stable) projection onto this first summand $S^m map(K,K_o;S^mX) \simeq S^mC \longrightarrow S^m\mathcal{D}_1 = S^m(M \setminus M_o, \partial M \setminus M_o) \wedge X$ induces the homology slant product $H_q(map(K,K_o;S^mX)) \longrightarrow \bigoplus_j H^{j-q}(K,K_o;H_{j-m}(X))$.

For $X = S^o$ this homomorphism has been proved by Moore [18] to be an isomorphism if $q < 2(m - H\dim(K,K_o))$ which is twice the connectivity of the mapping space.

Studying the spaces $\mathcal{D}_k$ (which are Thom spaces for X a sphere) is a possible approach to the homology of the mapping spaces; we will return to this in a further article.

References.

[1] M. Barratt and P. Eccles: Γ^+-structures-III: the stable structure of $\Omega^\infty S^\infty A$. Topology 13 (1974), 199 - 207.

[2] F. Cohen and L. Taylor: Computations of Gelfand-Fuks cohomology, the cohomology of function spaces and the cohomology of configuration spaces. Geometric Applications of Homotopy Theory I (Evanston 1977), Springer LNM 657, 106 - 143.

[3] F. Cohen, P. May and L. Taylor: Splitting of certain spaces CX. Math. Proc. Cambridge Phil. Soc. 84 (1978), 465 - 496.

[4] F. Cohen, P. May and L. Taylor: Splitting of some more spaces. Math. Proc. Cambridge Phil. Soc. 86 (1979), 227 - 236.

[5] F. Cohen: The unstable decomposition of $\Omega^2 S^2 X$ and its applications. Math. Z. 182 (1983), 553 - 568.

[6] R. Cohen: Stable proofs of stable splittings. Math. Proc. Cambridge Phil. Soc. 88 (1980), 149 - 151.

[7] A. Dold und R. Thom: Quasifaserungen und unendliche symmetrische Produkte. Ann. of Math. 67 (1958), 239 - 281.

[8] E. Fadell and L. Neuwirth: Configuration spaces. Math. Scand. 10 (1962), 119 - 126.

[9] I. James: Reduced product spaces. Annals of Math. 62 (1955), 170 - 197.

[10] D. Kahn and S. Priddy: Applications of the transfer to stable homotopy theory. Bull. Math. Soc. 78 (1972), 981 - 987.

[11] D. Kahn: On the stable decomposition of $\Omega^\infty S^\infty A$. Geometric Applications of Homotopy Theory II (Evanston 1977), Springer LNM 658, 206 - 214.

[12] D. Kahn and S. Priddy: The transfer and stable homotopy theory. Math. Proc. Cambridge Phil. Soc. 83 (1978), 103 - 112.

[13] P. May: The Geometry of Iterated Loop Spaces. Springer LNM 271 (1972).

[14] P. May and L. Taylor: Generalized splitting theorem. Math. Proc. Cambridge Phil. Soc. 93 (1983), 73 - 86.

[15] D. Mc Duff: Configuration spaces. K-Theory and Operator Algebras (Athens, Georgia, 1975), Springer LNM 575, 88 - 95.

[16] D. Mc Duff: Configuration spaces of positive and negative particles. Topology 14 (1975), 91 - 107.

[17] J. Milnor: On the construction FK. Lecture Notes Princeton University 1956, repr. in J. F. Adams: Algebraic Topology - a Student's Guide. London Math. Soc. Lect. Notes Ser. 4 (1972), 119 - 136.

[18] J. Moore: On a theorem of Borsuk. Fund. Math. 43 (1956), 195 - 201.

[19] G. Segal: Configuration spaces and iterated loop-spaces. Invent. Math. 21 (1973), 213 - 221.

[20] V. Snaith: A stable decomposition of $\Omega^n S^n X$. J. London Math. Soc. 7 (1974), 577 - 583.

[21] P. Vogel: Cobordisme d'immersions. Ann. Scient. Ec. Norm. Sup. 7 317 - 358.

[22] R. Vogt: Splittings of spaces CX. Manuscr. Math. 38 (1982), 21 - 39.

Sonderforschungsbereich 170
"Geometrie und Analysis"
Bunsenstraße 3/5

3400 Göttingen, BRD

Added in proof:

R. Cohen has independently found a model and a stable splitting for ΛSX (see his "A Model for the Free Loop Space of a Suspension", to appear).

The splitting of $\Omega^2 S^{2n+1}$

FRED COHEN AND MARK MAHOWALD[1]

Snaith [S] showed that as a stable complex $\Omega^2 S^{2n+1} = \vee \Sigma^{(2n-1)k} D_k$ where the spectra D_k are the successive quotients of the May-Milgram filtration, [M]. Milgram [M2] showed that the spectra D_k are the pieces given by the Thom complexes of certain symmetric group bundles. We will give more details latter. The cohomology of the pieces D_k satisfy

$$H^*(D_k) = A/A\{\chi Sq^i | i > k/2\}, [M1].$$

In [CM] it is shown:

THEOREM 1. *[CM] If there is a map*

$$f : \Sigma^N \Omega^2 S^{2n+1} \to \Sigma^{N+2^i(2n-1)} D_{2^i}$$

with $H^*(f)$ non-zero in dimension $N + 2^i(2n-1)$ then $N \geq 2^{i+1}$.

In [C] Cohen constructed such maps for $N = 2^{i+1}$ and thus these are the best possible. Our purpose here is to give another proof of theorem 1. It uses a different construction which might be useful in other places. In some sense it calculates the Nisihida relations but does the complete calculation in a special case so the formulas seem to be much easier to handle.

First we must fix notation. Let $D_k = B([k/2])$. This is compatible with the usual Brown-Gitler notation. It is known [C2] that D_k is a Brown-Gitler spectrum although this fact will not be used here.

There is a map

$$g : \Omega^2 S^{2n+1} \to K(Z/2, 2^i(2n-1))$$

which is the adjoint to the stable composition

$$\Omega^2 S^{2n+1} \xrightarrow{s} \Sigma^{2^i(2n-1)} B(2^{i-1}) \xrightarrow{a} \Sigma^{2^i(2n-1)} K(Z/2)$$

[1]Both authors are supported by grants from the NSF.

where s is a piece of the Snaith map and a is the suspension of the unit. Suppose we have a map

$$f : \Sigma^N \Omega^2 S^{2n+1} \to \Sigma^{N+2^i(2n-1)} B(2^{i-1})$$

Adjointing gives a map

$$\Omega^2 S^{2n+1} \to \Omega^N \Sigma^{2^i(2n-1)} B(2^{i-1}) \to \Omega^N \Sigma^N K(Z/2, 2^i(2n-1))$$

Let $\tilde{g}$ be this composite. There is a second natural map α of $\Omega^2 S^{2n+1}$ into $\Omega^N \Sigma^N K(Z/2, 2^i(2n+1))$, namely the composite

$$\Omega^2 S^{2n+1} \xrightarrow{\beta} K(Z/2, 2^i(2n+1)) \to \Omega^N \Sigma^N K(Z/2, 2^i(2n+1))$$

where β_* is non-zero on the unique $2^i(2n+1)$-dimensional primitive. The difference $\alpha - \beta$ lifts to the fiber $U(N)$ of the non-trival map

$$\Omega^N \Sigma^N K(Z/2, 2^i(2n+1)) \to K(Z/2, 2^i(2n+1)).$$

After adjointing one obtains a map

$$h : F_{2^{i+1}}/F_{2^i} \to U(N)$$

where F_j denotes the jth May-Milgram filtration of $\Omega^2 S^{2n+1}$.

LEMMA 2. *The induced map, h_*, evaluated on the 2^{i+1}-st power of the fundamental class of $\Omega^2 S^{2n+1}$, is non-zero.*

PROOF: Restricting to $F_{2^{i+1}}$, we obtain a map

$$F_{2^{i+1}} \to D_2(\Sigma^{2^i(2n+1)} B(2^{i-1}))$$

where D_2 denotes the second summand in the Snaith decomposition for $\Omega^N \Sigma^N (\Sigma^{2^i(2n+1)} B(2^{i-1}))$. Since this map is non-zero on the 2^{i+1}-st power of the fundamental class, the lemma follows. More details are given in [CM].

Our proof of the main theorem follows from the calculations below where it is shown that $N \geq 2^{i+1}$.

First we recall some of the properties of the fibration defining $U(N)$.

1) The Serre long exact sequence is valid for

$$j \leq N + 2^i(2n-1) + 2^{i+1}(2n-1) - 1 + N - 1 = \text{conn } B + \text{ conn } F + 1.$$

Thus if $j \leq 2N + 3\ 2^i(2n-1) - 2$ we have

$$H^{j-1}(\Sigma^N K(Z/2, 2^i(2n-1)) \to H^{j-1}(U(N)) \to H^j(K(Z/2, N + 2^i(2n-1)).$$

2) This formula allows us to describe $H^*(U(N))$ as $\mu^*(\sigma^N(\alpha \cup \beta))$ and the vector space generated by $\overline{Sq}^I$ where $\delta\overline{Sq}^I = Sq^I\iota$, I is admissible and the excess of I is greater that $2^i(2n-1)$.

3) In this language the first non-zero class in $H^*(U(N))$ is

$$\overline{Sq}^{2^i(2n-1)+1} \in H^{N+2^{i+1}(2n-1)}(U(N)).$$

4) If $\Sigma Sq^{i'} Sq^{i''}\iota$ is a relation in $H^*(KZ/2, N+2^{i(2n-1)})$ and each $Sq^{i''} = \delta\overline{Sq}^{i''}$ then $\Sigma Sq^{i'}\overline{Sq}^{i''}$ is a relation in $H^*(U(N))$ modulo the image of μ^*.

LEMMA 3. *There is a map over the Steenrod algebra*

$$H^*(\Sigma^{2^{i+1}+1}(P^{-2}_{-2^{i+1}-1})) \leftarrow H^*(B(2^i))$$

which has degree 1 in dimension zero.

PROOF: This follows from the Brown-Gitler property. The A-module structure of $H^*(B(2^i))$ is that of the cohomology of a Brown-Gitler spectrum so the A modulo maps out of $H^*(B(2^i))$ are those which would be induced by appropriate spectrum maps.

The relation we wish to apply is one in a family constructed as follows. Let $D: Sq^j = Sq^{j-1}$ and let D act as a derivation.

THEOREM 4. *Let T be the free algebra over $Z/2$ generated by symbols*

$$Sq^i, \quad i = 1, \ldots,.$$

Let R be the vector space spanned by

$$\{D^t Sq^{2n-1} Sq^n\}$$

for all t and n. Then $A = T/R$.

This is a well-known theorem. A proof can be constructed by showing, by induction, that R acts trivially in repeated smash products of RP. Standard arguments, compare [MT], show that every element in R is a relation in A. Finally, the argument in [SE], Theorem 3.1, shows that T/R has a basis of Sq^I with I admissible.

To complete the proof of Theorem 1 we take

$$D^{2^{i+2}n+2^{i+1}-1}(Sq^{2^{i+2}n+2^{i+1}-1}Sq^{2^{i+1}n+2^i}) = 0$$

We will call this relation R. When we expand R, we get

$$\Sigma_{0\leq j\leq 2^{i+1}-1} Sq^j Sq^{2^{i+1}n+2^i-j} = 0$$

Thus in $H^*(U(N))$

$$\Sigma_{0\leq j\leq 2^{i+1}-1} Sq^j \overline{S}q^{2^{i+1}n+2^i-j} = 0$$

modulo μ^*.

There is a map $P^{-1}_{-2^{i+1}-1} \to CP^{-1}_{-2^i}$. Thus for all $j \geq 0$

$$Sq^{2j}\alpha^{-2-2j} = 0.$$

On the other hand

$$j \equiv 3(4) \quad j < 2^{i+1}-1 \qquad Sq^j\overline{S}q^{2^{i+1}n+2^i-j} = (Sq^2Sq^{j-1})Sq^{2^{i+1}n+2^i-j-1}$$

$$j \equiv 1(4) \quad j > 1 \qquad Sq^jSq^{2^{i+1}n+2^ij} = (Sq^2Sq^{j-1} + Sq^{j+1})Sq^{2^{i+1}n+2^i-j-1}$$

Since $Sq^1Sq^{2k+1} = 0$, the term corresponding to $j = 1$ can be dropped. Thus R can be written

$$Sq^{2^{i+1}n+2^i} + Sq^{2^{i+1}-1}Sq^{2^{i+1}n-2^i+1} + \text{ terms } a_j$$

where each a_j is an even Sq applied to a class in $H^*(U(N))$. Thus the composite

$$H^*(U(N)) \to H^*(\Sigma^{N+2^{i+1}(2n-1)}B(2^i)) \to H^*(\Sigma^{2^{i+1}+1+N+2^{i+1}(2n+1)}P^{-2}_{-2^{i+1}-1}),$$

which we call ψ, satisfies

$$0 = \psi(R) = Sq^{2+i+1-1}\psi(\overline{S}q^{2^{i+1}n-2^i+1}) + \psi(\overline{S}q^{2^{i+1}n+2^i})$$
$$= \alpha^{-2} + \psi(\overline{S}q^{2^{i+1}n+2^i}).$$

Thus the second term must be non-zero. This means that $Sq^{2^{i+1}n+2^i}\iota \neq 0$ in $H^*(K(Z/2, N + 2^i(2n-1))$ and this implies $N \geq 2^{i+1}$. This is what we wished to prove.

[CM] F. Cohen and M. Mahowald, Unstable properties of $\Omega^n S^{n+k}$, Contem. Maths. AMS vol 12(1982), 81-90.

[C] F. Cohen, The unstable decomposition of $\Omega^2\Sigma^2 X$ and its applications, Math. Zeit. 182(1983), 553-568.

[C2] Ralph Cohen, Geometry of $\Omega^2 S^3$ and braid orientations, Invent. Math. 54(1979), 53-67.

[M] J. P. May, *The Geometry of Iterated Loop Spaces*, Lecture Notes in Mathematics 271, Springer-Verlag.

[M1] M. E. Mahowald, A new infinite family in ${}_2\pi_*^S$ Topology 16(1977), 249-256.

[M2] R. J. Milgram, Group representations and the Adams spectral sequence, Pacific J. Math. 41(1972), 157-182.

R. Mosher and M. C. Tangora, *Cohomology operations and Applications in Homotopy Theory*, Harper and Row (1968).

[SE] N. Steenrod and D. Epstein, *Cohomology Operations*, Ann. of Math. Studies #50(1962).

[S] V. P. Snaith, A stable decomposition of $\Omega^n S^n X$, J. Lond. Math. Soc. (2), 7(1974), 577-583.

A Model for the Free Loop Space of a Suspension

by

Ralph L. Cohen

Combinatorial models of loop spaces have proved to be extremely useful in homotopy theory in the last thirty years. The first and most famous of these models is the James model of the (based) loop space of the suspension of a connected space, $\Omega\Sigma X$. This model has led to an understanding and generalization of Whitehead's Hopf invariant that has since been perhaps the most important tool in unstable homotopy theory. The James model is the naturally filtered space

$$J(x) = \coprod_{n\geq 1} X^n/\sim$$

where X^n is the n-fold cartesian product of X with itself and where the relation "$\sim$" is given by $(x_1,\dots,x_{i-1},*,x_{i+1},\dots,x_n) \sim (x_1,\dots,x_{i-1},x_{i+1},\dots,x_n)$, where $* \in X$ is the basepoint. James' theorem that $J(X)$ is weakly equivalent to $\Omega\Sigma X$ when X is connected [6], was shortly thereafter highlighted by Milnor's theorem [11] that once suspended, $J(X)$ splits into the wedge of the subquotients of its filtration. That is,

$$\Sigma(\Omega\Sigma X) \simeq \Sigma J(X) \simeq \bigvee_{n\geq 1} \Sigma X^{(n)}$$

where $X^{(n)}$ denotes the n-fold smash product.

Motivated by the work of Dyer and Lashof on homology operations, Milgram described a model we will call $C(X)$ of $Q(X) = \Omega^\infty\Sigma^\infty X = \varinjlim_N \Omega^N\Sigma^N X$ when X is connected. Similar models of $Q(X)$ were later constructed by Barratt [1] and by May [8]. $C(X)$ has the form

$$C(X) = \coprod_{n\geq 1} E\Sigma_n \times_{\Sigma_n} X^n/\sim$$

where $E\Sigma_n$ is a contractible space acted upon freely by the symmetric group

---

The research in this paper was partially supported by an N.S.F. grant and an N.S.F.-P.Y.I. award.

Σ_n, which acts on X^n by permuting coordinates. The constructions of Milgram and of May also yield models of the finitely iterated loop spaces $\Omega^N\Sigma^N X$, again when X is connected. A theorem of V. Snaith [12], which can be viewed as a generalization of Milnor's theorem, asserts that all of the above models stably split into the wedge of the subquotients of their natural filtrations. (That is, their associated suspension spectra split.)

Recently, the author together with Carlsson produced a complex $Z(X)$ which is like $C(X)$ except that it utilizes the cyclic groups $\mathbb{Z}_n$ instead of the symmetric groups. That is, $Z(X)$ is of the form

$$Z(X) = \coprod_{n\geq 1} E\mathbb{Z}_n \times_{\mathbb{Z}_n} X^n/\sim$$

and it was proved in [2] that for X connected, $Z(X)$ is homotopy equivalent to $ES^1_+\wedge_{S^1}\Lambda(\Sigma X)$. This notation needs some explanation. For any space Y, $\Lambda(Y) =$ $\mathrm{Maps}(S^1,Y)$ is the space of unbased, or free, maps from the circle S^1 to Y. $\Lambda(Y)$ is acted upon by the circle group by rotation of loops. The space $ES^1_+\wedge_{S^1}\Lambda(\Sigma X)$ can be viewed as the quotient of the homotopy orbit space of $\Lambda(\Sigma X)$ by BS^1:

$$ES^1_+\wedge_{S^1}\Lambda(\Sigma X) = ES^1\times_{S^1}\Lambda(\Sigma X)/(ES^1\times_{S^1}* = BS^1)$$

It was also proved in [2] that $Z(X)$ stably splits and so we get a splitting of suspension spectra

$$\Sigma^\infty(ES^1_+\wedge_{S^1}\Lambda(\Sigma X)) \simeq \Sigma^\infty Z(X) \simeq \bigvee_{n\geq 1} \Sigma^\infty(E\mathbb{Z}_{n_+}\wedge_{\mathbb{Z}_n} X^{(n)}).$$

This splitting proved very useful in the analysis of Waldhausen's algebraic K-theory of spaces done in [3].

In this note we will describe a combinational model $L(X)$ for the free loop space of the suspension of X, $\Lambda(\Sigma X)$. $L(X)$ will be of the form

$$L(X) = \coprod_{n\geq 1} S^1\times_{\mathbb{Z}_n} X^n/\sim$$

where $\mathbb{Z}_n$ acts on the circle S^1 by rotation by multiples of $2\pi/n$. $L(X)$ is naturally contained in $Z(X)$ as a subcomplex, and this inclusion realizes (up to homotopy) the map $\Lambda(\Sigma X) \hookrightarrow ES^1_+ \wedge_{S^1} \Lambda(\Sigma X)$ given by inclusion as the righthand coordinate. Similiarly, $L(X)$ contains the James construction $J(X)$ in such a way as to realize the inclusion of the basepoint preserving loops $\Omega\Sigma X$ inside of the free loops $\Lambda\Sigma X$. We will also prove that $L(X)$ stably splits and so we will obtain the following equivalences of suspension spectra: the second of which was first observed by Goodwillie:

$$\Sigma^\infty(\Lambda(\Sigma X)) \simeq \Sigma^\infty L(X) \simeq \bigvee_{n\geqslant 1} \Sigma^\infty(S^1_+ \wedge_{\mathbb{Z}_n} X^{(n)}).$$

This paper is divided into two sections. In section 1 we define $L(X)$ and prove the splitting mentioned above. In section 2 we define a map $h: L(X) \longrightarrow \Lambda(\Sigma X)$ and use homological calculations made in [2] to prove that h is an equivalence when X is connected. Throughout this paper X will denoted a connected, based space of the based homotopy type of a based C.W. complex.

§ 1. The Space L(X) and its Splitting.

We begin by defining the space $L(X)$. To do this we need to make some definitions.

Definitions 1.1. (i) Let S^1 be the unit circle in $\mathbb{R}^2$. A sequence of distinct points in S^1, $(x_1,\dots,x_n)$ is said to be *increasing* if there exists an increasing sequence of real numbers $\theta_1,\theta_2,\dots,\theta_n$ with

$$\theta_1 < \theta_2 <\dots< \theta_n < \theta_1 + 2\pi \quad \text{such that} \quad x_j = e^{i\theta_j},\ j = 1,\dots,n.$$

(ii) A sequence of nondegenerate closed arcs in S^1, $(a_1,\dots,a_n)$ is said to be *increasing* if and only if they satisfy the following properties:

a. The a_i's has disjoint interiors.

b. If $m(a_i) \in S^1$ is the midpoint of the arc a_i then the sequence of points $(m(a_1),\dots,m(a_n))$ is increasing.

Example: Consider the following diagram of arcs in S^1

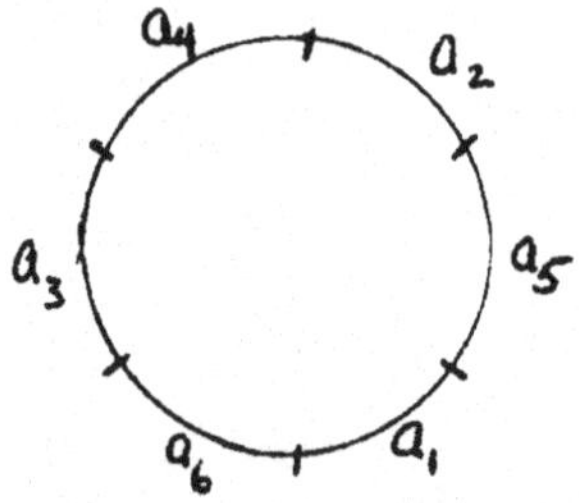

The sequences (a_2,a_4,a_6,a_1,a_5), (a_5,a_3,a_1), and (a_6,a_2) are increasing, but the sequence (a_4,a_5,a_3) is not.

Definition 1.2. Let $F_0(S^1,n)$ be the space of all of increasing sequences of nondegenerate closed arcs, $(a_1,\dots,a_n)$ in S^1. $F_0(S^1n)$ is topologized as a subquotient of the spaces of mappings, $\mathrm{Map}(\coprod_n I, S^1)$ which is endowed with the compact-open topology. By convention we let $F_0(S^1,0) =$ point.

Notice that $F_0(S^1,n)$ is acted upon freely by the cyclic group $\mathbb{Z}_n$, where the action is given by cyclically permuting the coordinates of an increasing sequence of arcs, $(a_1,\ldots,a_n)$.

We may now define the space $L(X)$.

Definition 1.3. Let X be a space with basepoint $* \in X$. We define $L(X)$ to be the complex

$$L(X) = \coprod_{n\geq 0} F_0(S^1,n) \times_{\mathbb{Z}_n} X^n/\sim$$

where the equivalence relation "$\sim$" is given by

$(a_1,\ldots,a_n)\times_{\mathbb{Z}_n}(x_1,\ldots,x_{i-1},*,x_{i+1},\ldots,x_n)\sim(a_1,\ldots,a_{i-1},a_{i+1},\ldots,a_n)\times_{\mathbb{Z}_{n-1}}(x_1,\ldots,x_{i-1},x_{i+1},\ldots,x_n)$ for $i = 1,\ldots,n$.

In the introduction we said that $L(X)$ is a complex built out of the spaces $S^1\times_{\mathbb{Z}_n} X^n$. The following lemma will explain this.

Lemma 1.4. The space $F_0(S^1,n)$ is $\mathbb{Z}_n$-equivariantly homotopy equivalent to S^1 given the $\mathbb{Z}_n$-action of rotation by $2\pi/n$.

Proof. Let $F_0(n)$ be the space of all n tuples of increasing sequences $(x_1,\ldots,x_n)$ of points in S^1. Consider the map $j\colon F_0(n) \longrightarrow F_0(S^1,n)$ given by mapping an n tuple of points $(x_1,\ldots,x_n)$ to the n-tuple of arcs $(c_1,\ldots,c_n)$ where for $k < n$ c_k is the arc from x_k to x_{k+1} going in a counterclockwise direction, and c_n is the arc from x_n to x_1 going in a counterclockwise direction. It is standard, and quite easy to see, that j is an equivariant homotopy equivalence. Thus it is enough to show that $F_0(n)$ is equivariantly homotopy equivalent to S^1.

To see this consider the fibration

$$p\colon F_0(n) \longrightarrow S^1$$

given by projecting an n-tuple $(x_1,\ldots,x_n)$ to the first coordinate, x_1. Notice that the fiber of p is homeomorphic to the space $F_0(I,n-1) = \{(t_1,\ldots,t_{n-1}) \in (0,1)^{n-1}$ such that $t_1 < t_2 <\ldots< t_{n-1}\}$, but this space is evidently homeomorphic to the open $n-1$ simplex, and is therfore contractible. Thus the fibration p is a homotopy equivalence.

Now consider the map

$$s: S^1 \longrightarrow F_0(n)$$

defined by the formula $s(x) = (x, x+\frac{2\pi}{n}, x+\frac{4\pi}{n}, \ldots, x+\frac{2(n-1)\pi}{n})$. Notice that s is a section of the fibration $p: F_0(n) \longrightarrow S^1$, and hence is a homotopy equivalence, because p is. Notice furthermore that s is a $\mathbb{Z}_n$-equivariant map. Now since the $\mathbb{Z}_n$-actions on both S^1 and $F_0(n)$ are free, the fact that s is both an equivalence and an equivariant map implies that s is in fact an equivariant homotopy equivalence. This completes the proof of lemma 1.4.

The goal of the rest of this paper is to prove the following two results.

__Theorem 1.5__. There is a natural map $h: L(X) \longrightarrow \Lambda\Sigma X$ which is a homotopy equivalence when X is connected.

__Theorem 1.6__. There is a stable splitting of suspension spectra

$$\Sigma^{\infty}L(X) \simeq \bigvee_{m\geq 1} \Sigma^{\infty}(S^1_+\wedge_{\mathbb{Z}_m} X^{(m)}),$$

where here, as above, S^1_+ denotes S^1 with a disjoint basepoint.

__Remark__. Notice that $L(X)$ is a naturally filtered space with m^{th} filtration

$$F_m(L(X)) = \coprod_{n=1}^{m} F_0(n)\times_{\mathbb{Z}_n} X^n/\sim.$$

Notice furthermore that we have an natural equivalence of the mth subquotient

$$F_m(L(X))/F_{m-1}(L(X)) = F_0(m)_+\wedge_{\mathbb{Z}_m} X^{(m)} \simeq S^1_+\wedge_{\mathbb{Z}_m} X^{(m)}$$

so that theorem 1.2 asserts that $L(X)$ stably splits into the wedge of the subquotients of this filtration.

We will prove theorem 1.5 in the next section. We now prove theorem 1.6.

Our proof will go along the lines of the proof of the splitting of the complex $Z(X)$ given in [2], which in turn was an adaptation of the author's proof of the Snaith splitting of the complex $C(X)$ given in [4].

These splittings begin with an equivalence of spectra

$$g\colon \Sigma^\infty(X_+) \longrightarrow \Sigma^\infty(X\vee S^0).$$

Notice that X_+ and $X\vee S^0$ are the same space but their basepoints are in different components. However their suspension spectra are clearly (based) homotopy equivalent. g is such an equivalence. g then induces an equivalence of the following equivariant smash product spectra

$$1\wedge g^{(n)}\colon S^1_+\wedge_{\mathbb{Z}_n}\Sigma^\infty(X_+)^{(n)} \xrightarrow{\simeq} S^1_+\wedge_{\mathbb{Z}_n}\Sigma^\infty(X\vee S^0).$$

Now as easily can be checked (as is done by May in [9]) if Y is any space there is an equivalence of spectra

$$S^1_+\wedge_{\mathbb{Z}_n}\Sigma^\infty Y^{(n)} \simeq \Sigma^\infty(S^1_+\wedge_{\mathbb{Z}_n}Y^{(n)}).$$

Putting these equivalences together we get an equivalence of suspension spectra

$$G_n\colon \Sigma^\infty(S^1_+\wedge_{\mathbb{Z}_n}(X_+)^{(n)}) \xrightarrow{\simeq} \Sigma^\infty(S^1_+\wedge_{\mathbb{Z}_n}(X\vee S^0)^{(n)}) .$$

Now the homeomorphism $(X_+)^{(n)} \cong (X^n)_+$ extends to give a homeomorphism

$$F_0(n)_+\wedge_{\mathbb{Z}_n}X_+^{(n)} \cong (F_0(n)\times_{\mathbb{Z}_n}X^n)_+.$$

Thus we can define a map $\sigma_n\colon \Sigma^\infty(F_0(n)_+\wedge_{\mathbb{Z}_n}X^{(n)}) \longrightarrow \Sigma^\infty(F_0(n)\times_{\mathbb{Z}_n}X^n)$ to be the composition

(1.7) $$\sigma_n\colon \Sigma^\infty(F_0(n)_+\wedge_{\mathbb{Z}_n}X^{(n)}) \xrightarrow[1\wedge i^{(n)}]{} \Sigma^\infty(F_0(n)_+\wedge_{\mathbb{Z}_n}(X\vee S^0)^{(n)}) \xrightarrow[G_n^{-1}]{}$$

$$\Sigma^\infty(F_0(n)_+\wedge_{\mathbb{Z}_n}(X_+)^{(n)}) \xrightarrow[\cong]{} \Sigma^\infty(F_0(n)\times_{\mathbb{Z}_n}X^n)_+) \xrightarrow[p]{} \Sigma^\infty(F_0(n)\times_{\mathbb{Z}_n}X^n)$$

where in this composition $i\colon X \longrightarrow X\vee S^0$ and $p\colon(F_0(n)\times_{\mathbb{Z}_n}X^n)_+ \longrightarrow F_0(n)\times_{\mathbb{Z}_n}X^n$ are the natural inclusion and projection maps respectively.

Notice that σ_n is a stable section of the projection map π_n: $F_0(n)\times_{\mathbb{Z}_n}X^n \longrightarrow F_0(n)_+\wedge_{\mathbb{Z}_n}X^{(n)}$. But by the remark after the statement of 1.6, π_n factors up to homotopy as the composition

$$F_0(n)\times_{\mathbb{Z}_n}X^n \underset{j}{\hookrightarrow} F_n(L(X)) = \coprod_{k=1}^{n} F_0(k)\times_{\mathbb{Z}_k}X^n/\sim \xrightarrow[\text{proj.}]{} F_n(L(X))/F_{n-1}(L(X)) =$$

$$F_0(n)_+\wedge_{\mathbb{Z}_n}X^{(n)}.$$

Thus if we let $s_n\colon \Sigma^\infty(F_0(n)_+\wedge_{\mathbb{Z}_n} X^{(n)}) \longrightarrow F_n(L(X))$ be the composition $j\circ\sigma_n$, then s_n is a stable section of the projection $F_n(L(X)) \longrightarrow F_n(L(X))/F_{n-1}(L(X))$. Thus we have a stable splitting

$$\Sigma^\infty F_n(L(X)) \simeq \Sigma^\infty F_{n-1}(L(X)) \vee \Sigma^\infty(F_0(n)_+\wedge_{\mathbb{Z}_n} X^{(n)}) \simeq$$

$$\Sigma^\infty F_{n-1}(L(X)) \vee \Sigma^\infty(S^1_+\wedge_{\mathbb{Z}_n} X^{(n)}).$$

Piecing these splittings together yields the splitting of theorem 1.6.

We end this section by describing a calculation of $H_*(L(X))$, which is implied by the splitting in theorem 1.6.

Let G be any graded abelian group and let G^n be the n-fold tensor product $G\otimes\ldots\otimes G$. Let $H_*(S^1;\{G^n\})$ denote the homology of S^1 with twisted coefficients defined by the action of $\pi_1 S^1 = \mathbb{Z}$ on G^n induced by the projection of $\mathbb{Z}$ on $\mathbb{Z}_n$ and letting $\mathbb{Z}_n$ act on G^n by the rule

$$t_n(g_1\otimes\ldots\otimes g_n) = (-1)^{n-1}(g_n\otimes g_1\otimes\ldots\otimes g_{n-1}).$$

Here $t_n \in \mathbb{Z}_n$ is the image of $1 \in \mathbb{Z}$ under the projection $\mathbb{Z} \longrightarrow \mathbb{Z}_n$. The following is an easy consequence of the analysis of the Serre spectral sequence for the fibration

$$X^n \hookrightarrow S^1 \times_{\mathbb{Z}_n} X^n \longrightarrow S^1.$$

We refer the reader to [2] for the details of this analysis.

<u>Proposition 18</u>. (i) $H_*(S^1\times_{\mathbb{Z}_n} X^n) \simeq H_*(S^1;\{H_*(X)\}^n)$, and

$$H_q(S^1;\{H_*(X)^n\}) = \begin{cases} H_*(X)^n/(1-t_n) & \text{if } q = 0 \\ (H_*(X)^n)^{t_n} & \text{if } q = 1 \\ 0 & \text{if } q > 1 \end{cases}$$

(ii) $H_*(S^1_+\wedge_{\mathbb{Z}_n} X^{(n)}) = H_*(S^1;\{\tilde{H}_*(X)^n\})$, and

$$H_q(S^1;\{\tilde{H}_*(X)^n\}) = \begin{cases} \tilde{H}_*(X)^n/(1-t_n) & \text{if } q = 0 \\ (\tilde{H}_*(X)^n)^{t_n} & \text{if } q = 1 \\ 0 & \text{if } q > 1 \end{cases}$$

Here $(G^n)/(1-t_n)$ and $(G^n)^{t_n}$ denote the co-invariants and the invariants of the $\mathbb{Z}_n$-action, respectively.

The following calculation is now a corollary of (1.8(ii)) and theorem 1.6.

Corollary 1.9. $H_*(L(X)) = \bigoplus_{n\geq 0} \tilde{H}_*(X)^n/(1-t_n) \oplus \bigoplus_{n\geq 1} (\tilde{H}_*(X)^n)^{t_n}$.

§ 2. Proof of Theorem 1.5

Our goal in this section is to prove theorem 1.5; i.e. that $L(X)$ is equivalent to $\Lambda\Sigma X$ when X is connected. We begin by defining a map

$$h: L(X) \longrightarrow \Lambda\Sigma X$$

that will induce the equivalence.

Let $F_0(S^1,n)$ be as in section 1, and consider the map

$$h_n: F_0(S^1,n)\times_{\mathbb{Z}_n} X^n \longrightarrow \Lambda\Sigma X$$

defined as follows. Let $\alpha: X \hookrightarrow \Omega\Sigma X$ be the adjoint of the identity. So for $x \in X$, $\alpha(x): S^1 \longrightarrow S^1\wedge X$ is given by the formula $\alpha(x)(t) = t\wedge x$. Given $(a_1,\dots,a_n) \in F_0(S^1,n)$, define a pinch map

$$\pi(a_1,\dots,a_n): S^1 \longrightarrow \bigvee_n S^1$$

defined as the composition

$$S^1 \xrightarrow[p(a_1,\dots,a_n)]{} S^1_{r_1}\vee\dots\vee S^1_{r_n} \xrightarrow[\vee i_{r_j}]{} \bigvee_n S^1$$

where r_i is the length of the arc a_i, $S^1_{r_i}$ is the circle of circumference r_i, $p(a_1,\dots,a_n)$ is the pinch map that identifies all points in S^1 that do not lie in the interior of one of the a_i's to a point, and where $i_r: S^1_r \longrightarrow S^1$ is the linear stretching map, induced by the map of intervals $[0,r] \longrightarrow [0,2\pi]$ given by multiplication by $2\pi/r$. (Note: Since all of the arcs a_i are nondegenerate, each r_i is positive.)

Said another, but equivalent way, $\pi(a_1,\dots,a_n): S^1 \longrightarrow \bigvee_n S^1$ maps the arc a_i to the ith circle in the wedge in a degree one, affine, orientation preserving manner, and it maps all points not lying in any of the a_i's to the basepoint in $\bigvee_n S^1$.

We define the map $h_n: F_0(S^1,n)\times_{\mathbb{Z}_n} X^n \longrightarrow \Lambda\Sigma X$ by letting

$h_n((a_1,\dots,a_n)\times_{\mathbb{Z}_n}(x_1,\dots,x_n))\colon S^1 \longrightarrow S^1\wedge X$ be given by the composition

$$(2.1)\quad h_n((a_1,\dots,a_n)\times_{\mathbb{Z}_n}(x_1,\dots,x_n))\colon S^1 \xrightarrow[\pi(a_1,\dots,a_n)]{} \bigvee_n S^1 \xrightarrow[\alpha(x_1)\vee\dots\vee\alpha(x_n)]{} S^1\wedge X.$$

We leave it to the reader to verify that this formula defines a well defined, continuous map $h_n\colon F_0(S^1,n)\times_{\mathbb{Z}_n} X^n \longrightarrow \Lambda\Sigma X$, and that the disjoint union of the h_n's respect the equivalence relation in the definition of $L(X) = \coprod_n F_0(S^1,n)\times_{\mathbb{Z}_n} X^n/\sim$, and so defines a map

$$h\colon L(X) \longrightarrow \Lambda\Sigma X.$$

We will prove theorem 1.5 by proving the following

Claim 2.2. For X connected, $h\colon L(X) \longrightarrow \Lambda\Sigma X$ is a homotopy equivalence.

The main step in proving this claim is the calculation of the homomorphism h induces in homology. To compute this, notice that the identification map

$$\rho\colon \coprod_n F_0(S^1,n)\times_{\mathbb{Z}_n} X^n \longrightarrow L(X)$$

induces a well understood surjection in homology (prop. 1.8). Thus it is sufficient to compute the homology homomorphism induced by each map

$$h_n\colon F_0(S^1,n)\times_{\mathbb{Z}_n} X^n \longrightarrow \Lambda\Sigma X,$$

defined above. Let $\bar{h}_n\colon S^1\times_{\mathbb{Z}_n} X^n \longrightarrow \Lambda\Sigma X$ be the composition of h_n with the equivalence $S^1\times_{\mathbb{Z}_n} X^n \xrightarrow{\simeq} F_0(X^1,n)\times_{\mathbb{Z}_n} X^n$ given by lemma 1.4. Notice that $\bar{h}_n$ is given up to homotopy by the composition

$$\bar{h}_n\colon S^1\times_{\mathbb{Z}_n} X^n \xrightarrow[1\times\alpha_n]{} S^1\times_{\mathbb{Z}_n}\Lambda\Sigma X \xrightarrow[e]{} \Lambda\Sigma X$$

where e is the evaluation map and where $\alpha_n\colon X^n \longrightarrow \Omega\Sigma X \hookrightarrow \Lambda\Sigma X$ is the

James map. In our terminology α_n is defined on an n-tuple $(x_1,\dots,x_n)$ to be the composition

$$\alpha_n(x_1,\dots,x_n)\colon S^1 \xrightarrow[\pi_n]{} \bigvee_n S^1 \xrightarrow[\bigvee_j \alpha(x_j)]{} S^1 \wedge X$$

where π_n is the pinch map determined by identifying the nth roots of unity to the basepoint.

The homomorphism

$$\bar{h}_{n*}\colon H_*(S^1 \times_{\mathbb{Z}_n} X^n; k) \longrightarrow H_*(\Sigma X; k)$$

where k is a field, was computed in [2]. We now recall that calculation.

Let Y be any simply connected space. In [5] Goodwillie proved that there is, an isomorphism from $H_*(\Lambda Y; k)$ to the Hochschild homology of the differential graded algebra of singular chains on the Moore loop space $M(X)$.

$$\Phi\colon H_*(\Lambda Y; k) \xrightarrow{\cong} HH_*(\mathcal{S}_*(M(Y); k))$$

(Recall that the Moore-loops is a strictly associated H-space of the homotopy type of ΩY. The singular chains with coefficients in the field k, $\mathcal{S}_*(M(Y);k)$ is therefore an associative differential graded algebra.)

Now when $Y = \Sigma X$, the James construction $J(X)$ induces a chain homotopy equivalence of differential graded algebras

$$J\colon T(\tilde{H}_*(X; k)) \longrightarrow \mathcal{S}_*(M(\Sigma X); k)$$

where if V is a graded vector space over k, then $T(V)$ is the tensor algebra on V:

$$T(V) = \bigoplus_{n \geq 0} V^n$$

where V^n is the n-fold tensor product on V with V^0 denoting the ground field k.

Finally using calculations of Loday and Quillen [7], the Hochschild homology of a tensor algebra is known. Namely,

(2.3)
$$HH_q(T(V)) = \begin{cases} \bigoplus_{n\geq 0} V^n/(1-t_n) & \text{if } q = 0 \\ \bigoplus_n (V^n)^{t_n} & \text{if } q = 1 \\ 0 & \text{if } q > 1 . \end{cases}$$

where $t_n \in \mathbb{Z}_n$ is a generator, and where, as above, $V^n/(1-t_n)$ and $(V^n)^{t_n}$ denote the co-invariants and the invariants of V^n under the $\mathbb{Z}_n$-action given by

$$t_n(v_1\otimes\ldots\otimes v_n) = (-1)^{n-1}(v_n\otimes v_1\otimes\ldots\otimes v_{n-1}).$$

Putting these isomorphisms together we get the following calculation of $H_*(\Lambda\Sigma X; k)$.

Lemma 2.4. There is an isomorphism

$$\psi: H_*(\Lambda\Sigma X; k) \xrightarrow{\cong} \bigoplus_{m\geq 0} (\tilde{H}_*X)^m/(1-t_m) \oplus \bigoplus_{m\geq 1} ((\tilde{H}_*X)^m)^{t_m}.$$

Notice that by comparison with corollary 1.9 we have the following.

Corollary 2.5. For any field k, $H_*(L(X); k)$ and $H_*(\Lambda\Sigma X; k)$ are isomorphic.

We need to show that $h: L(X) \longrightarrow \Lambda\Sigma X$ induces such an isomorphism in homology. This will be an easy consequence of the following calculation which was done in [2, lemma 3.15].

Lemma 2.5. The following diagram commutes:

$$\begin{array}{ccc} H_*(S^1\times_{\mathbb{Z}_n} X^n) & \xrightarrow{\tilde{h}_{n*}} & H_*(\Lambda\Sigma X) \\ \cong\big\downarrow\gamma & & \cong\big\downarrow\psi \\ H_*(X)^n/(1-t_n) \oplus (H_*(X)^n)^{t_n} & \xrightarrow[r_n]{} & \bigoplus_{m\geq 0} \tilde{H}_*(X)^m/(1-t_m) \oplus \bigoplus_{m\geq 1} (\tilde{H}_*(X)^m)^{t_m} \end{array}$$

where γ is the isomorphism of (1.8), ψ is the isomorphism of 2.4, and r_n is induced by the reduction map $H_*X \longrightarrow \tilde{H}_*(X)$ and the inclusion of the appropriate direct summand.

We refer the reader to [2] for the details of all of the above calculations.

Now since $r_{n*}: H_*(X)^n/(1-t_n) \oplus (H_*(X)^n)^{t_n} \longrightarrow \underset{m}{\oplus}\tilde{H}_*(X)^m/(1-t_m) \oplus \underset{m}{\oplus}(\tilde{H}_*(X)^m)^{t_m}$ is a surjection onto the nth summand, the following is a corollary of lemma 2.5.

Corollary 2.6. The direct sum homomorphism

$$\underset{n\geq 0}{\oplus} \bar{h}_{n*}: \underset{n\geq 0}{\oplus} H_*(S^1 \times_{\mathbb{Z}_n} X^n; k) \longrightarrow H_*(\Lambda\Sigma X; k)$$

is a surjection for any field k.

We are now ready to prove the following.

Proposition 2.7. $h: L(X) \longrightarrow \Lambda\Sigma X$ is a homology equivalence.

Proof. Recall that the sum of the homomorphisms $\bar{h}_{n*}$ factors as follows:

$$\oplus h_{n*}: \underset{n\geq 0}{\oplus} H_*(S^1 \times_{\mathbb{Z}_n} X^n; k) \xrightarrow{\cong} \underset{n\geq 0}{\oplus} H_*(F_0(S^1,n) \times_{\mathbb{Z}_n} X^n; k)$$

$$\xrightarrow[\rho_*]{} H_*(L(X); k) \xrightarrow[h_*]{} H_*(\Lambda\Sigma X; k),$$

where $\rho: \amalg F_0(S^1,n) \times_{\mathbb{Z}_n} X^n \longrightarrow L(X)$ is the identification map as above.

By 2.5, 2.6 and this factorization we see that $h_*: H_*(L(X)); k) \longrightarrow H_*(\Lambda\Sigma X; k)$ is a surjection between isomorphic graded vector spaces. Now if X is a finite C.W. complex these graded vector spaces are of finite type and hence h_* is an isomorphism. In general, X is the direct limit of its finite subcomplexes, and since homology commutes with direct limits, we may still conclude that h_* is an isomorphism. Since the above argument holds for any field k, h must be an integral homology equivalence. h is in fact a homotopy equivalence because as is easily seen, both $L(X)$ and $\Lambda\Sigma X$ are simple spaces when X is connected.

Bibliography

1. M.G. Barratt, A free group functor for stable homotopy, Proc. Symp. Pure. Math. Vol. XXII (1971), 31-36.

2. G. Carlsson and R.L. Cohen, The cyclic groups and the free loop space, to appear.

3. G. Carlsson, R. Cohen, T. Goodwillie, and W.c. Hsiang, The free loop space and the algebraic K-theory of spaces, to appear.

4. R.L. Cohen, Stable proofs of stable splittings, Math. Proc. Camb. Phil. Soc. 88 (1980), 149-151.

5. T. Goodwillie, Cyclic homology, derivations, and the free loop space, Topology 24 (1985), 187-215.

6. I.M. James, Reduced product spaces, Annals of Math (2) 62 (1955), 170-197.

7. J.L. Loday and D.G. Quillen, Cyclic homology and the Lie algebra homology of matrices, Commentarii Math. Helvetici 59 (1984), 565-591.

8. J.P. May, The Geometry of Iterated Loop Spaces, Springer Lecture Notes Vol. 271, (1972).

9. J.P. May, Equivariant constructions of nonequivariant spectra, to appear in Proc. Conf. in honor of J. Moore, Annals of Math. studies.

10. R.J. Milgram, Iterated Loop spaces, Annals of Math. 84 (1966), 386-403.

11. J. Milnor, On the construction FK, London Math. Soc. Lecture Notes 4 (1972), 119-36.

12. V.P. Snaith, A stable decomposition for $\Omega^n S^n X$, J. London Math. Soc. 7 (194), 577-583.

Department of Mathematics
Stanford University
Stanford, California 94305

It was recently pointed out to the author that a construction similar to L(X) was known to G. Segal, and has been studied by C.F. Bödigheimer.

Calculations of Unstable Adams E_2 Terms for Spheres

Edward B. Curtis
Paul Goerss
Mark Mahowald
R. James Milgram

1. Introduction

We have made computer calculations of the E_2 terms of the unstable Adams spectral sequences which converge to the homotopy groups of the spheres. We use the Λ algebra and an algorithm based on EHP sequences to calculate the unstable Adams E_2 term through stem dimension 51 (added in proof: stem 58) for each sphere S^n. After some preliminaries in §2 and §3, we describe the EHP algorithm for Λ algebra calculations in §4. In §5, we show how periodicity can be used to shorten the calculations. The computer programs are sketched in §6. Appendix A has an example of assembly code for Λ algebra manipulations. The results of the calculations of the $E_2(S^n)$ through stem 51 are given in the tables in Appendix B.

In this paper, all spaces and groups are to be localized at the prime 2. There are analogous methods for odd primes. The end of a proof is indicated by ♦.

EHP Sequences

The EHP sequences in homotopy groups of spheres are the following. For each $n \geq 1$, there is a long exact sequence (abbr: LES)

$$(1.1)\quad \ldots \to \pi_{n+q+2}(S^{2n+1}) \xrightarrow{P} \pi_{n+q}(S^n) \xrightarrow{E} \pi_{n+q+1}(S^{n+1}) \xrightarrow{H} \pi_{n+q+1}(S^{2n+1}) \xrightarrow{P} \ldots$$

The calculations of $\pi_*(S^n)$, as carried out by Barratt [B] and Toda [Tod], proceed by a double induction. To calculate the groups in the q-stem $\pi_{n+q}(S^n)$, assume (inductively on q) that the groups of the p-stem $\pi_{n+p}(S^n)$ are known for all $p < q$, and all spheres S^n. The q-stem $\pi_{n+q}(S^n)$ is then calculated (inductively on n) starting from the fact that $\pi_{1+q}(S^1) = 0$ for all $q > 0$. The suspension homomorphism E is affected by $\pi_{n+q+2}(S^{2n+1})$ and $\pi_{n+q+1}(S^{2n+1})$ as follows. The exactness of the EHP sequence implies that the elements in image(P) vanish under suspension, and that the elements in kernel(P) must be adjoined to $\pi_{n+q}(S^n)$ to obtain $\pi_{n+q+1}(S^{n+1})$. It has long been recognized that the difficulty in this approach is that of determining the homomorphism P.

The Unstable Adams Spectral Sequence

For each space X, the (stable or unstable) Adams spectral sequence is a sequence of groups $E_r^{*,*}(X)$, which approximate and with increasing r converge to the (stable or ordinary) homotopy groups of X. For X the stable sphere, Adams himself made calculations of the first few groups of the E_2 term using elementary homological algebra. These calculations were extended by Whitehead, May, Mahowald, Tangora and others; [May], [M,T], [T1]. In [C1], was given a recursive algorithm, based on EHP sequences of the UASS, for calculating the E_2 terms of the finite spheres S^n, and calculations were made in low (≤ 17) stem dimensions. This algorithm was used by Whitehead [GWW], who made pencil and paper calculations which were complete through stem 34. In [T2], [T3] and [T4], Tangora showed how to program the algorithm on a computer. His calculations agreed with, and extended those of Whitehead. More important, Tangora proved the validity of the algorithm, and introduced some simplifications into the computations, which he called shortcuts. We are indebted to Tangora for these shortcuts, as well as his insights into the difficulties that arise in these programs.

The present work uses the same algorithm, and builds on [GWW], [T2], [T3], and [T4]; much of what we have done overlaps those. We will concentrate on the changes we have made. The main difference is that we do not attempt to incorporate the entire algorithm into one program. Instead, there is one main file, some subsidiary files, and several executable programs which operate on these files. The information as it is calculated is stored in the main file (called WFILE, for working file). This file can grow to be quite large. One program (*difftag*) extracts information from WFILE in order to manufacture files G(q), one for each positive integer q, which contain the computed image(P) in the q-stem. Other programs use the G(q) to modify WFILE according to the EHP process. Another difference between our programs and those of Tangora is that our programs are semi-interactive in that we are able to intervene in the construction of the files G(q). In this way, we sometimes save the computer a lot of time.

The tables at the end contain our computations of the unstable Adams E_2 terms through stem 51 for each sphere. The stable Adams E_2 term may also be read from the tables (the untagged terms), and they agree with the tables of Mahowald and Tangora ([M,T] and [T1]).

Some programs to do Λ algebra and Steenrod algebra calculations were initially written in BASIC and in assembly language for an IBM PC; they were used to explore the interactive programs and to do calculations through stem 32; this version is available from the last named author (at Stanford University) if you send him a blank $5^1/2$" floppy disc. We have included a printout of the assembly language Λ-algebra manipulator.

The programs as they are presented here were written in C and run through stem 51 on a VAX 11-750 operating under UNIX. We thank the University of Washington, the NSF, the Digital Equipment Company, for making this computer (uw-entropy) available to us; also the systems programmer, Steve Hubert, for showing us how to use it. The programs have been also been run on a SUN Model 2, as well as an ATT microcomputer operating under UNIX.

2. The Unstable Adams E_2 Term

The E_2 term of the UASS for the sphere S^n may be calculated by homological algebra, as follows. In [6A], it was shown that $E_2^{s,t}(S^n)$ is isomorphic to the homology of a differential module $\Lambda(n)$, obtained as a submodule of the Lambda algebra Λ. Λ is (defined as) the algebra (over $Z/2$) with a generator λ_i for each integer $i \geq 0$, and

(2.1) relations: whenever $2i < j$,

$$\lambda_i \lambda_j = \sum_{k \geq 0} C(j-2i-2-k, k)\, \lambda_{j-i-k-1} \lambda_{2i+k+1}$$

Then Λ becomes a differential algebra, with

(2.2) $$d\lambda_i = \sum_{k \geq 1} C(i-k, k)\, \lambda_{i-k} \lambda_{k-1}$$

Here $C(n,q)$ stands for the binomial coefficient reduced mod 2.

For each sequence $I = (i, j, \ldots, m)$ of non-negative integers, λ_I denotes the product $\lambda_i \lambda_j \cdots \lambda_m$. A sequence $I = (i_1, i_2, \ldots i_s)$ is called admissable if for each j, $2i_j \geq i_{j+1}$. It follows immediately from the relations that Λ has for basis (over $\mathbf{Z}/2$) the set of all λ_I, where I is admissable. Λ is bigraded by length and dimension, where

$$\text{length}(\lambda_I) = s$$

$$\dim(\lambda_I) = i_1 + i_2 + \ldots + i_s$$

For each positive integer n, $\Lambda(n)$ is defined to be the submodule of Λ spanned by those λ_I which are admissable and for which $i_1 < n$. One of the main results of [6A] is that $\Lambda(n)$ with its differential serves as an E_1 term for the UASS for spheres. That is,

$$E_2^{*,*}(S^n) = H_*(\Lambda(n))$$

$\Lambda^{s,t}(n)$ will denote the submodule of $\Lambda(n)$ spanned by admissable λ_I of length s, and dimension $t - s$.

The EHP sequence methods and the UASS methods are related by the fact that the $E_2^{*,*}(S^n)$ satisfy EHP sequences similar to those of (1.1). That is, for each n, there is a LES:

(2.3) $$\ldots \to E_2^{*,*}(S^{2n+1}) \xrightarrow{E} E_2^{*,*}(S^n) \xrightarrow{H} E_2^{*,*}(S^{n+1}) \xrightarrow{P} E_2^{*,*}(S^{2n+1}) \to \ldots$$

These LES's come about as follows. For each n, there is a short exact sequence:

(2.4) $$0 \to \Lambda(n) \xrightarrow{i} \Lambda(n+1) \xrightarrow{h} \Lambda(2n+1) \to 0$$

where i is the inclusion and h is defined on the admissable basis by

$$h(\lambda_i\lambda_j\cdots\lambda_m) = \begin{cases} \lambda_j\cdots\lambda_m, & \text{if } i=n \\ 0, & \text{if } i<n \end{cases}$$

The EHP sequence (2.3) for the unstable E_2 terms is the LES in homology of this short exact sequence (2.4) of differential modules. Thus the homomorphism P arises from the differential d in Λ. The EHP process for calculating the unstable E_2 terms is more tractable than the EHP process for the homotopy groups of spheres because the homorphism P for the unstable Adams E_2 terms, while still difficult, is more easily computable than the homomorphism P in homotopy. In what follows, we shall show how this can be done.

3. Notation and Conventions

Before presenting the algorithm for calculating $E_2^{*,*}(S^n)$, we will describe some techniques from [C1] and [T4] that have been useful.

Ordering

The monomials λ_I of each fixed bidegree (s, t) in Λ are ordered, lexicographically from the left. This induces a total order on each of the vector spaces $\Lambda^{s,t}$, by first expressing each polynomial as a sum of admissable monomials in decreasing order, and then comparing two such polynomials lexicographically. For a sum of admissable monomials, the term which is largest in the lexicographic order will be called the leading term. In a given homology class, the polynomial in the class which is least in the total order will be called the minimal representative.

If λ_I is the leading term of a minimal representative of some (nonzero) homology class, let c(I) stand for the minimal polynomial which is a cycle and which has λ_I for leading term. We seek a basis of each $E_2^{s,t}(S^n)$ consisting of such basis elements c(I), represented by their leading terms.

Odd Endings

Let Λ_o be the submodule of Λ generated by all admissable monomials λ_I which end with an odd index. Then Λ_o is closed under the differential, and the inclusion $\Lambda_o \to \Lambda$ induces an isomorphism in homology except in stem(0); the tower $\{\lambda_o{}^k\}$ is not present in $H_*(\Lambda_o)$. The inclusions $\Lambda_o(n) \to \Lambda(n)$ also induce isomorphisms except on the towers (which occur in stem(0) and, if n is even, in stem(n-1)). Furthermore, the $\Lambda_o(n)$ satisfy a modified EHP property as follows. For each n, there is a short exact sequence:

$$(3.1) \qquad 0 \to \Lambda_0(n) \xrightarrow{i} \Lambda_0(n+1) \xrightarrow{h} \Lambda_0{}^{\#}(2n+1) \to 0$$

where the unit is included in $\Lambda_0{}^{\#}(2n+1)$ when n is odd, but not included in $\Lambda_0{}^{\#}(2n+1)$ when n is even . Thus all the λ_{2k+1} are present in filtration one of Λ_0, but none of the λ_{2k}. The result of this is that if we may restrict attention to the λ_I with odd endings. With this convention, the spectral sequences converge to the finite summands of the 2-primary components of $\pi_*(S^n)$.

Notation

Henceforth for convenience of notation, Λ will stand for Λ_0, that is, the submodule of (what was previously called) Λ spanned by admissable λ_I with odd endings. The initial of a sequence I is its first index. $\Lambda(n)$ will denote the submodule of Λ spanned by admissable λ_I with odd endings and initial $i_1 < n$.

4.The EHP Process for Λ

We next describe the EHP process for finding a basis of each $E_2^{s,t}(S^n)$. Assume inductively that such a basis has been found in all dimensions $< q$ for all spheres, and also in dimension q for spheres S^m, where m is less than n. To obtain the basis for $E_2^{s,t}(S^n)$, for dimension q, for each filtration $s > 0$, take $t = s + q$, and we must

(i) delete a basis for the image of $P : E_2^{s-2,t-n+1}(S^{2n+1}) \to E_2^{s,t}(S^n)$

(ii) adjoin a basis for the kernel of $P : E_2^{s,t}(S^{2n+1}) \to E_2^{s+1,t}(S^n)$

In our situation, we obtain a first quadrant table $\{T^{*,*}\}$. At each integer lattice point $(t - s, s)$, $T^{s,t}$ will consists of a list of elements called rows. Each row is either a sequence:

$$(i_1, i_2, \ldots, i_s)$$

or a pair of sequences:

$$(i_1, i_2, \ldots, i_s) \leftarrow (j_1, j_2, \ldots, j_{s-1})$$

Here, $\dim(I) = t - s$, and if J is present, $\dim(J) = \dim(I) + 1$. If I appears in the latter form as $I \leftarrow J$, then I is said to be tagged by J. Each I that appears is the leading term of a cycle. As above, let c(I) be the minimal cycle in $\Lambda(i_1+1)$ which has λ_I for leading term, and without ambiguity, let c(I) also stand for the homology class in $H_*(\Lambda(i_1+1))$. The notation $I \leftarrow J$ means that

$$(i_1, i_2, \ldots, i_s) + \text{lower terms} = d\,(j_1, j_2, \ldots, j_{s-1} + \text{lower terms})$$

(The complications arising from the lower terms that may occur on the right hand side is discussed in [T4].)

Constructing the Table

The table $T^{*,*}$ starts out empty, changes at each stage, and when completed, gives a basis for the $E_2^{*,*}(S^n)$ as described by the theorem above. First, each of the odd-indexed lambdas is placed in the table. That is, (2k+1) is placed in $T^{1,2k+2}$; the rest of the table starts out empty. Assume inductively that the table has been made correct, i. e., gives a basis for $E_2^{*,*}(S^n)$, for all spheres S^n in all stems less than q, and for all filtrations s. For each $K = (k_1, \ldots)$ which is in the table in stem (q-1), adjoin to the table all (m, K), where subject to the restrictions that:

(4.1a) If K is untagged, then $2m \geq k_1$

(4.1b) If K is tagged by $N = (n_1, \ldots)$, then $2m \geq k_1$ and $2m < n_1$

If K is in $T^{s,t}$ (where t - s = q -1), the term (m, K) will be placed in $T^{s+1,\, t+m+1}$. Next consider, in increasing order, each J which is in the table in stem(q+1); suppose J is in $T^{s,t}$, where t - s = q+1. We must compute P(J) in $T^{s+1,t}$.

The LTO algorithm.

This algorithm computes P(J). This was implicit in [C1], and was described and proven correct in [T4]. In order to make the present treatment as self-contained as possible, we sketch here the algorithm in the form that we need it. Beause it depends on keeping track of only the leading terms of polynomials in Λ, we follow Tangora in calling it the Leading Term Only (LTO) algorithm.

(1) Calculate d(J) as a sum of terms, each in admissable form; find the leading term of the sum (call it I).

(2a) If I is present in $T^{s+1,t}$ and is not yet untagged, then J tags I; replace I by $I \leftarrow J$ and delete J from $T^{s,t}$

(2b) If I is present in $T^{s+1,t}$ and is already tagged, say $I \leftarrow K$, add d(K) to d(J); reduce mod 2; return to step(1) and continue.

(2c) If I is not present in the table, then the LTO algorithm asserts that some tail of I must be in the table and is tagged; find the shortest tail of I that is tagged, say

$$(i_p, \ldots, i_{s+1}) \leftarrow (m_p, \ldots, m_s)$$

Then take $K = (i_1, \ldots, i_{p-1}, m_p, \ldots, m_s)$; add d(K) to d(J); reduce mod 2; return to step (1) and continue. The LTO algorithm assertion is that eventually, either

$$J + \text{lower terms} \qquad \text{is a cycle}$$

or

$$d(J + \text{lower terms}) = I + \text{lower terms}$$

where I is present in $T^{s+1,t}$. In the first case, we say that J completes to a cycle. In the latter case, J tags I; replace I by $I \leftarrow J$ in $T^{s+1,t}$, and delete J from $T^{s,t}$. When this has been done for all J in $T^{s,t}$, the tags in box $T^{s+1,t}$ are correct, and go on to the next higher filtration. When this has been done for all filtrations s (there are only a finite number, by the vanishing theorm), the table $T^{*,*}$ has been made correct in stem(q), and we go on to the next stem. This completes the inductive step of the LTO algorithm.

Theorem A basis for $E_2^{s,t}(S^n)$ consists of all $c(I)$, where I is in $T^{s,t}$ with initial $i_1 < n$, which are either untagged or which have a tag J with initial $j_1 \geq n$.

The proof of this is in [T4]. ♦

5. Periodicity

There are two types of periodicity that shorten the calculations. The first is horizontal periodicity of bidegree $(2^k, 0)$ in the $(t - s, s)$ plane, analogous to James periodicity for truncated projective spaces. The other is of bidegree $(8, 4)$ along the upper edge, and is a version of Adams periodicity for the unstable Adams E_2 terms.

Horizontal Periodicity.

Suppose that $I = (i_1, i_2, \ldots, i_s)$ is tagged by $J = (j_1\, j_2, \ldots, j_{s-1})$. Let n be the least power of 2 which is greater than the difference of initials $j_1 - i_1$. Let $I^* = (i_1 + n, i_2, \ldots, i_s)$ and $J^* = (j_1 + n, j_2, \ldots, j_{s-1})$. The assertion is that if I^* is a cycle that is not tagged by some term less than J^*, then I^* will be tagged by J^*.

At present we cannot prove the full strength of this periodicity assertion, but we want to use it anyway. For this purpose, we define two integers index and flag as follows. For each admissable sequence $K = (k_1, k_2, \ldots k_s)$, let

$$\begin{aligned} \text{index}(K) &= k_2 - k_1 - 1, && \text{if } 2k_1 < k_2 \\ &= 0 && \text{otherwise} \end{aligned}$$

Suppose that I is tagged by J, with $\Sigma_\alpha I_\alpha = d(\Sigma_\beta J_\beta)$. Then let flag($I \leftarrow J$) be the maximum of index(K), where K appears in any relation that is used to express d($\Sigma_\beta J_\beta$) as a linear combination of admisssable monomials. That is flag(I ←J) is the largest initial that is affected by the relations in the first position.

Lemma Suppose that $I = (i_1, i_2, \ldots, i_s)$ is tagged by $J = (j_1, j_2, \ldots, j_{s-1})$, and suppose that flag(I ←J) < i_1. Let $n = 2^k$ be the least power of 2 for which $2^k > j_1 - i_1$ and take $I^* = (i_1 + n, i_2, \ldots, i_s)$ and $J^* = (j_1 + n, j_2, \ldots, j_{s-1})$. Suppose also that I^* and J^* both appear in the table; and that I^* is not tagged by some term earlier than J^*. Then J^* will tag I^*.

Proof. Let $M = H_*(RP^\infty)$, as a module with the Steenrod algebra acting on the right. As a vector space, M has a generator e_n in each positive dimension n. Consider the chain complex $M \otimes \Lambda$. with differential

$$d(x \otimes \lambda_I) = \Sigma_j (xSq^j) \otimes \lambda_{j-1}\lambda_I + x \otimes d(\lambda_I)$$

For any sequence $I = (i_1, i_2, \ldots i_s)$, let PI stand for $e_{i1} \otimes (i_2, \ldots i_s)$. The map $M \otimes \Lambda \to \Lambda$ which sends PI to I is a map of chain complexes because of the formulas for $d(e_j)$ and $d(\lambda_j)$. For each positive integer m, let $M(m, \infty) = H_*(RP_m{}^\infty)$. The assumption that flag(I ← J) < n implies that $\Sigma_\alpha PI_\alpha = d(\Sigma_\beta PJ_\beta)$ in $M(i_1,\infty) \otimes \Lambda$. James Periodicity for truncated projective spaces implies that $\Sigma_\alpha PI_\alpha{}^* = d(\Sigma_\beta PJ_\beta{}^*)$ in $M(i_1 + n,\infty) \otimes \Lambda$. This shows that $\Sigma_\alpha I_\alpha{}^* = d(\Sigma_\beta J_\beta{}^*)$ in $\Lambda/\Lambda(i_1)$. Hence I^* will be tagged by J^* in Λ. ♦

We use this flag(I ← J) to check validity of the periodicity assertion each time we want to use it. While the program *difftag* is calculating that I is tagged by J, we have *difftag* keep track of this integer flag (I ← J). If flag(I ←J) is smaller than the initial of I, the term with tag $I^* \leftarrow J^*$ is placed in a file called STORE, for use at a later time.

In the simplest version of the program (below), the program *kill* does not make use of this horizontal periodicity. A faster version (also below) takes account of and stores the valid cases of horizontal periodicity, as checked by *difftag*. For this we use two more programs *postpone*, and *perkill* which take account of the (validated) horizontal periodicities of period 2, 4, 8, 16, in increasing order. We have observed that in most cases, *difftag* calculates that flag(I ←J) is less than the initial of I, so that the periodicity is valid; furthermore, we have found no instances (through stem 51), where the strong form of the horizontal periodicity assertion is not valid.

Adams Periodicity.

Along the vanishing line of slope 1/2, there is a recurrent pattern of period (8, 4). The elements

(4 1 1 2 4 1 1 . . .)
(5 1 2 4 1 1 1)
(6 2 4 1 1 . . .)
(8 4 1 1 . . .)
(12, 1 1 . . .)
(13, 1 . . .)

together with similar previous elements, generate by the EHP process, an almost closed portion of the table. That is, each of the elements has for Hopf Invariant another member of the pattern. The stable survivors in this pattern are given in the following table.

							241124...
					41124...		1234 ...
			1124...		5124...	234...	
		124...	224...	344...	624...		
	24 ...	44...			8...		
4...							
5...							
6...							
8...							

Recurrent Pattern (6.1)

Adams periodicity for the unstable E_2 terms would assert that this pattern, including also the classes that are tagged (suppressed from the table for lack of space), repeats with period (8, 4) along the upper edge. We do not make use of Adams periodicity directly. Rather, in making the calculations near the edge we find that there is no interference from below, and that this pattern results. In making these calculations, we make repeated use of the observations of [T4; §3.9], in particular,

(i) If no untagged cycle at the target is smaller than the leading term of d(I), then I must complete to a cycle.

(ii) If K is a cycle, and i = 1, 3, 7, or 15, then (i, K) is also a cycle.

(iii) (2i-1, j-1, 1, K) is tagged by (2i, j, K), provided that both monomials are admissable, and present in the table.

<u>Use of Relations</u>

The differentials emanating from elements in the verticals in dimensions 4k - 1 can be difficult to compute directly. In these cases, we use the shortcut advocated by [T4; §3.9.7]. For example,

$$d(14, 2\,4\,1\,1\,2\,4\,1\,1\,2\,4\,1\,1\,1) = 2........2\,4\,5\,3\,3\,3$$

where the dots represent 2's. This differential is obtained from:

$$d(39) = 31, 7$$

and repeated multiplications on the left by 0 (i.e., by λ_0). Information of this sort is calculated by a program called *relation* (which is very similar to *difftag*), and is then put into the files by hand.

6. The Programs

The programs manufacture a file called WFILE. This file consists of a list of rows. Each row is either a sequence of integers (called a term):

$$(i_1, i_2, \ldots, i_s)$$

or a pair of such sequences (a term with a tag):

$$(i_1, i_2, \ldots, i_s) \leftarrow (j_1, j_2, \ldots, j_{s-1})$$

WFILE will be built up in stages as follows. A maximum stem dimension N is chosen, and all terms constructed are to have dimension $\leq$ N. We start by placing in WFILE all odd positive integers $i \leq N$; each is a term (consisting of a single integer) on a separate row. Then having completed q-1 stages, the qth stage takes four steps, as follows.

(1) Find all the terms I of dimension q that are to be tagged by some J of dimension q+1, and place these $I \leftarrow J$ in a file called G(q). For this step a program *difftag* calculates for each J, the leading term of d(J).

(2) A program *kill* uses G(q) to tag terms in WFILE. For each occurrence of $I \leftarrow J$ in G(q), it looks for I and J in WFILE, and if it finds them both untagged, it tags I by J and deletes J from WFILE.

(3) A program *showstem* searches WFILE for all terms that have dimension q, and places them in a file called S(q) .

(4) A program *loadstem* uses the terms of S(q) to make new rows according to (4.1), and appends these rows to WFILE.

Once the files G(q) are known through some dimension N, the file WFILE, which will contain the computed E_2 terms to stem N, is made by a shell script. This is a sort of master program which calls on the executable programs *kill, showstem,* and *loadstem* to perform steps (2), (3), (4) for each integer q from 1 to N. This is very straightforward, and no more will be said.

The program *difftag* is more complicated, so we give a sketch of it. Using the differential (2.2), *difftag* first computes d(J) as a linear combination of I's (possibly inadmissable), and places them in a list of rows called LIST; each row is assigned coefficient 1. This LIST is traversed sequentially from the beginning, and each I is tested for inadmissability from the left. If I is inadmissable at position j, then the relations (2.1) are used to express I as a linear combination of K's which are admissable at position j, and these K's are appended to LIST, each with coefficient 1; the coefficient of I is set to 0. Continue (just once) through the list until all rows are admissable (which must occur after a finite number of steps). LIST is next searched sequentially, keeping track of (by a pointer) the largest row I with non-zero coefficient. Initially, I is taken as the first row; if a larger row is encountered, the pointer is changed to point to this one. Each later occurrence of the same I is assigned coefficient 0 and the coefficient of the first occurrence is incremented. This coefficient is now reduced mod 2; if the coefficient becomes 1, I is the leading term, and the search terminates. If the coefficient is 0, the list is searched again for the largest I. If all coefficients become 0, J is a cycle, and the program exits. Otherwise, WFILE is searched for I. If I is found untagged, then J tags I, and the program terminates. If $I \leftarrow K$ is found, then d(K) is appended to LIST; again LIST is searched for the largest term, and the process continues as before. If I is not present, then WFILE is searched for shorter and shorter tails of I until finally some tail of I is found that is tagged, say

$$(i_p, \ldots, i_{s+1}) \leftarrow (m_p, \ldots, m_s)$$

Then K is taken to be $(i_1, \ldots, i_{p-1}, m_p, \ldots, m_s)$ and d(K) is appended to LIST, which is again searched for the largest term, and the process continues as before. ♦

<u>Modifications due to Periodicity</u>

If the programs were run as above, *difftag* would spend a lot of time on terms whose outcome is (correctly) predicted by periodicity. Therefore, we modify *difftag* as follows. The program keeps track (by an integer flag), of the largest initial of a term which is affected by a relation in the first position. If this flag is less than i_1, then $I^* \leftarrow J^*$ is placed in a file called STORE for later use. The valid cases of periodicity are incorporated into the programs as follows. Step (2) is replaced by two steps:

(2a)The program *kill* uses G(q) to tag terms in WFILE as before.

(2b)A program *postpone* uses G(q), to create a file STORE. For each occurrence of I←J in G(q), flagged as valid under horizontal periodicity, *postpone* places in STORE all $I^* \leftarrow J^*$ obtained by increasing the initials by the appropriate power of 2, and multiples thereof.

(2c) A program *perkill* uses G(q) and STORE to tag terms of dimension q in WFILE. For each occurrence of I ← J (in increasing order from either G(q) or STORE), *perkill* looks for I and J in WFILE, and if it finds them both untagged, it tags I by J and deletes J from WFILE.

Steps (3) and (4) proceed as before. ♦

Appendix A: 8086 Assembly code

We also include a printout of the 8086 assembly code for a Λ algebra manipulator, which can be used to put each monomial into admissable form. (Note: in moving this code to other CPU's, beware that they may handle flags differently).

```
;the two terms being fixed are stored in lambda2 and lambda1
; the routine expands these two into admissable form.

lamex0a:   mov    al,lambda2                 ;the two terms to put into
           mov    ah,lambda1                 ;admissable form
           shl    ah,1
           inc    ah
           sub    al,ah                      ;n = lambda2 - 2(lambda1) -1
           mov    variable_j,al
           jc     lamex1                     ;exit
           jz     lamex1
           mov    first_coefficient,ah       ;store it
           mov    dl,lambda1
           add    dl,al                      ;n+i in dl
           mov    second_coefficient,dl
           mov    constant_n,al

lamex2a:   test   constant_n,0ffh            ;Beginning of the loop
           jz     lamex1
           dec    constant_n
```

```
lamex2:     mov     ah,constant_n                    ;put the next terms in the
            mov     al,variable_j                    ;expansion into al,ah
            sub     ah,al                            ;do we continue?
            jc      lamex3
            call    binomial                         ;check  binomial coefficient

lamex3a:    call    load_buffer_with_coefficient     ;if non-zero, store in buffer

lamex3:     dec     variable_j                       ;continue processing
            test    variablej,80h                    ;check the sign
            jz      lamex2                           ;if non_negative continue

lamex1:     ret                                      ;otherwise exit

;the next routine gets the binomial coefficient. Assume that ah=n, al= m
;computes (n,m) mod(2). Returns z set if result is zero, zn set otherwise

binomial:   push    ax
            and     ah,al
            sub     ah,al
            jz      bin1
            sub     ah,ah
            pop     ax
            ret
bin1:       cmp     ah,1
            pop
            ret
```

Appendix B The Tables

Using these programs, we have calculated the unstable Adams E_2 terms for all spheres S^n, and all stems through stem 51. To save space, we have used a compressed notation as follows. The symbol * stands for the sequence 2 4 1 1; a sequence of dots with 2's at each end stands for repeated 2's, where each dot substitutes for a missing 2. Certain subsequences occur so often that it is convenient to write them in compressed form: 6653 for 6 6 5 3; 24333 for 2 4 3 3 3; 45333 for 4 5 3 3 3; 35733 for 3 5 7 3 3; and 59777 for 5 9 7 7 7.

The E_2 term for the UASS for each sphere may be read from the tables as follows. A basis for $E_2^{s,t}(S^n)$ is in one-one correspondence with the terms I which are in $T^{s,t}$ with initial less than n, and which are either untagged or are tagged by J with initial greater than or equal to n.

TABLE of $E_2{}^{s,t}(S^n)$

5						12111←2311
4				1111←221 211←41	2111←411 1211←231	3111←421
3			111	121←23	311←51	321←43
2		11	21	31←5		33
1	1		3			

Stems 1 - 6

6				124111
5			24111	34111←4511 11233←8111
4	4111	5111←621	6111←811 1233	7111←821 3511←461 2233←911
3	511	611←81 521←63 233	711←91 333	721←83 433←10 1 361←47
2	61	71←9 53		73←11
1	7			

Stems 7 - 10

8		111*1←22*1	211*1←41*1 121*1←23*1
7	11*1	21*1←4*1 12*1←244111	31*1←5*1 1211233←231233
6	2*1 111233←22233	3*1←64111 211233←41233 121233←23233	311233←42233
5	44111 21233←4233 12233←2433	54111←6511 36111←4711 31233←5233 12333←2353	32233←4433
4	3611←471 3233←633 2333←453	9111←10 21 5511←661 3333←553	10 1 1 1←12 1 1 4333←833 3433←473
3	533←93 353←57	10 1 1←12 1 921←10 3 561←67	11 1 1←13 1 733←11 3
2		11 1←13	
1			
	Stem 11	Stem 12	Stem 13

9	1 2 1 1 * 1←2 3 1 * 1	
8	3 1 1 * 1←4 2 * 1	4 1 1 * 1
7	3 2 * 1←4 4 4 1 1 1	5 1 * 1
6	3 4 4 1 1 1	6 * 1 1 2 3 3 3 3←2 3 5 3 3
5	7 4 1 1 1←8 5 1 1 5 1 2 3 3 2 3 3 3 3←4 5 3 3	8 4 1 1 1 6 1 2 3 3←8 2 3 3 24333
4	11 1 1 1←12 2 1 7 5 1 1←8 6 1 6 2 3 3 5 3 3 3←9 3 3 3 5 3 3←5 7 3	12 1 1 1 7 2 3 3←10 3 3 6 3 3 3←8 5 3
3	11 2 1←12 3 7 6 1←8 7 6 5 3	13 1 1 7 5 3←9 7
2	7 7	14 1
1		15

Stem 14 Stem 15

9		2 4 1 1 * 1
8	5 1 1 * 1←6 2 * 1	6 1 1 * 1←8 1 * 1 1 2 3 4 4 1 1 1
7	6 1 * 1←8 * 1 5 2 * 1←6 4 4 1 1 1 2 3 4 4 1 1 1	7 1 * 1←9 * 1 3 3 4 4 1 1 1←4 5 1 2 3 3 1 1 24333
6	7 * 1←10 4 1 1 1 5 4 4 1 1 1←8 1 2 3 3 1 24333	3 5 1 2 3 3←4 6 2 3 3 2 24333
5	9 4 1 1 1←10 5 1 1 7 1 2 3 3←9 2 3 3 3 4 3 3 3←4 7 3 3	3 6 2 3 3 3 5 3 3 3←4 6 5 3
4	13 1 1 1←14 2 1 9 5 1 1←10 6 1 7 3 3 3←9 5 3	14 1 1 1←16 1 1 8 3 3 3 5 10 1 1←6 11 1 3 6 5 3←4 7 7
3	14 1 1←16 1 13 2 1←14 3 9 6 1←10 7	15 1 1←17 1 11 3 3
2	15 1←17 13 3	
1		

Stem 16 Stem 17

11		1 1 * * 1
10	1 * * 1	2 * * 1 1 1 1 2 3 4 4 1 1 1←2 2 2 3 4 4 1 1 1
9	3 4 1 1 * 1←4 5 1 * 1 1 1 2 3 4 4 1 1 1←8 1 1 * 1	4 4 1 1 * 1 2 1 2 3 4 4 1 1 1←4 2 3 4 4 1 1 1 1 2 2 3 4 4 1 1 1←2 4 3 4 4 1 1 1
8	7 1 1 * 1←8 2 * 1 3 5 1 * 1←4 6 * 1 2 2 3 4 4 1 1 1←9 1 * 1 1 1 1 24333←2 2 24333	3 6 1 * 1←4 7 * 1 3 2 3 4 4 1 1 1←6 3 4 4 1 1 1 2 1 1 24333←4 1 24333 1 2 1 24333←2 3 24333
7	7 2 * 1←8 4 4 1 1 1 4 3 4 4 1 1 1←10 * 1 3 6 * 1←4 8 4 1 1 1 2 1 24333←4 24333 1 2 24333←2 3 6 2 3 3	5 3 4 4 1 1 1←6 5 1 2 3 3 3 5 4 4 1 1 1←4 7 1 2 3 3 3 1 24333←5 24333
6	7 4 4 1 1 1←12 4 1 1 1 3 6 1 2 3 3←4 7 2 3 3 3 24333←5 6 2 3 3	5 5 1 2 3 3←6 6 2 3 3
5	11 4 1 1 1←12 5 1 1 9 1 2 3 3←16 1 1 1 5 10 1 1 1←6 11 1 1 4 5 3 3 3 3 6 3 3 3←4 7 5 3	10 1 2 3 3←12 2 3 3
4	15 1 1 1←16 2 1 11 5 1 1←12 6 1 10 2 3 3←17 1 1 9 3 3 3 5 7 3 3	11 2 3 3←14 3 3 10 3 3 3←12 5 3 5 6 5 3←6 7 7
3	15 2 1←16 3 12 3 3←18 1 11 6 1←12 7 10 5 3	13 3 3←17 3 11 5 3←13 7 5 7 7
2	15 3←19 11 7	
1		
	Stem 18	Stem 19

12	1 1 1 * * 1 ← 2 2 * * 1	2 1 1 * * 1 ← 4 1 * * 1 1 2 1 * * 1 ← 2 3 * * 1
11	2 1 * * 1 ← 4 * * 1 1 2 * * 1 ← 2 4 4 1 1 * 1	3 1 * * 1 ← 5 * * 1 1 2 1 1 2 3 4 4 1 1 1 ← 2 3 1 2 3 4 4 1 1 1
10	3 * * 1 ← 6 4 1 1 * 1 2 1 1 2 3 4 4 1 1 1 ← 4 1 2 3 4 4 1 1 1 1 2 1 2 3 4 4 1 1 1 ← 2 3 2 3 4 4 1 1 1	3 1 1 2 3 4 4 1 1 1 ← 4 2 2 3 4 4 1 1 1
9	5 4 1 1 * 1 ← 6 5 1 * 1 3 6 1 1 * 1 ← 4 7 1 * 1 3 1 2 3 4 4 1 1 1 ← 5 2 3 4 4 1 1 1 1 2 1 1 24333 ← 2 3 1 24333	3 2 2 3 4 4 1 1 1 ← 4 4 3 4 4 1 1 1
8	9 1 1 * 1 ← 10 2 * 1 5 5 1 * 1 ← 6 6 * 1 3 1 1 24333 ← 4 2 24333	10 1 1 * 1 ← 12 1 * 1 4 1 1 24333 ← 8 3 4 4 1 1 1 3 4 3 4 4 1 1 1 ← 4 7 4 4 1 1 1
7	10 1 * 1 ← 12 * 1 9 2 * 1 ← 10 4 4 1 1 1 5 6 * 1 ← 6 8 4 1 1 1 3 2 24333 ← 4 3 6 2 3 3	11 1 * 1 ← 13 * 1 7 3 4 4 1 1 1 ← 8 5 1 2 3 3 5 1 24333 ← 11 4 4 1 1 1 1 2 45333 ← 2 4 7 3 3 3
6	11 * 1 ← 14 4 1 1 1 9 4 4 1 1 1 ← 12 1 2 3 3 5 8 4 1 1 1 ← 6 12 1 1 1 3 3 6 2 3 3 ← 4 8 3 3 3 2 45333	7 5 1 2 3 3 ← 8 6 2 3 3 6 24333 ← 13 1 2 3 3 3 45333 ← 4 5 7 3 3
5	13 4 1 1 1 ← 14 5 1 1 11 1 2 3 3 ← 13 2 3 3 4 7 3 3 3	7 6 2 3 3 ← 14 2 3 3 3 5 7 3 3
4	17 1 1 1 ← 18 2 1 13 5 1 1 ← 14 6 1 11 3 3 3 ← 13 5 3 6 6 5 3	18 1 1 1 ← 20 1 1 12 3 3 3 ← 16 3 3 7 6 5 3 ← 8 7 7
3	18 1 1 ← 20 1 17 2 1 ← 18 3 13 6 1 ← 14 7	19 1 1 ← 21 1 15 3 3 ← 19 3 7 7 7
2	19 1 ← 21	
1		
	Stem 20	Stem 21

13	1 2 1 1 * * 1 ← 2 3 1 * * 1	
12	3 1 1 * * 1 ← 4 2 * * 1	4 1 1 * * 1
11	3 2 * * 1 ← 4 4 4 1 1 * 1	5 1 * * 1
10	3 4 4 1 1 * 1	6 * * 1
9	7 4 1 1 * 1 ← 8 5 1 * 1 5 1 2 3 4 4 1 1 1	8 4 1 1 * 1 6 1 2 3 4 4 1 1 1 ← 8 2 3 4 4 1 1 1 2 4 1 1 24333
8	11 1 1 * 1 ← 12 2 * 1 7 5 1 * 1 ← 8 6 * 1 6 2 3 4 4 1 1 1 5 1 1 24333 ← 6 2 24333	12 1 1 * 1 7 2 3 4 4 1 1 1 ← 10 3 4 4 1 1 1 6 1 1 24333 ← 8 1 24333 1 2 2 45333 ← 2 4 45333
7	11 2 * 1 ← 12 4 4 1 1 1 7 6 * 1 ← 8 8 4 1 1 1 6 1 24333 ← 8 24333 5 2 24333 ← 6 3 6 2 3 3 2 2 45333 ← 14 * 1	13 1 * 1 9 3 4 4 1 1 1 ← 10 5 1 2 3 3 7 1 24333 ← 9 24333 5 7 4 4 1 1 1 ← 6 9 1 2 3 3 3 2 45333 ← 4 4 7 3 3 3
6	7 8 4 1 1 1 ← 8 12 1 1 1 7 24333 ← 9 6 2 3 3 5 3 6 2 3 3 ← 6 8 3 3 3 4 45333 ← 16 4 1 1 1	9 5 1 2 3 3 ← 10 6 2 3 3 5 9 1 2 3 3 ← 6 10 2 3 3 5 45333 ← 6 5 7 3 3 3 4 7 3 3 3 ← 4 6653 2 35733
5	15 4 1 1 1 ← 16 5 1 1 7 12 1 1 1 ← 8 13 1 1 5 8 3 3 3 ← 20 1 1 1	14 1 2 3 3 ← 16 2 3 3 5 10 2 3 3 ← 6 12 3 3 5 9 3 3 3 ← 6 10 5 3 5 5 7 3 3 ← 10 6 5 3 3 6653
4	19 1 1 1 ← 20 2 1 15 5 1 1 ← 16 6 1 13 3 3 3 ← 17 3 3 8 6 5 3 ← 21 1 1 7 13 1 1 ← 8 14 1 3 5 7 7	15 2 3 3 ← 18 3 3 14 3 3 3 ← 16 5 3 9 6 5 3 ← 10 7 7 7 14 1 1 ← 8 15 1 6 11 3 3 ← 8 13 3 5 10 5 3 ← 6 11 7 4 5 7 7
3	19 2 1 ← 20 3 15 6 1 ← 16 7 14 5 3 ← 22 1 7 14 1 ← 8 15	15 5 3 ← 17 7 9 7 7 ← 21 3 7 13 3 ← 9 15
2	15 7 ← 23	
1		
	Stem 22	Stem 23

13		2 4 1 1 * * * 1
12	5 1 1 * * 1 ← 6 2 * * 1	6 1 1 * * 1 ← 8 1 * * 1 1 2 3 4 4 1 1 * 1
11	6 1 * * 1 ← 8 * * 1 5 2 * * 1 ← 6 4 4 1 1 * 1 2 3 4 4 1 1 * 1	7 1 * * 1 ← 9 * * 1 3 3 4 4 1 1 * 1 ← 4 5 1 2 3 4 4 1 1 1 1 1 * 24333
10	7 * * 1 ← 10 4 1 1 * 1 5 4 4 1 1 * 1 ← 8 1 2 3 4 4 1 1 1 1 * 24333	3 5 1 2 3 4 4 1 1 1 ← 4 6 2 3 4 4 1 1 1 2 * 24333
9	9 4 1 1 * 1 ← 10 5 1 * 1 7 1 2 3 4 4 1 1 1 ← 9 2 3 4 4 1 1 1 3 4 1 1 24333 ← 4 5 1 24333	3 6 2 3 4 4 1 1 1 1 2 2 2 45333 ← 2 4 2 45333
8	13 1 1 * 1 ← 14 2 * 1 9 5 1 * 1 ← 10 6 * 1 7 1 1 24333 ← 8 2 24333 3 5 1 24333 ← 4 6 24333 2 2 2 45333 ← 9 1 24333	14 1 1 * 1 ← 16 1 * 1 8 1 1 24333 5 10 1 * 1 ← 6 11 * 1 3 6 1 24333 ← 4 7 24333 3 2 2 45333 ← 4 4 45333
7	14 1 * 1 ← 16 * 1 13 2 * 1 ← 14 4 4 1 1 1 9 6 * 1 ← 10 8 4 1 1 1 7 2 24333 ← 8 3 6 2 3 3 4 2 45333 ← 10 24333 3 6 24333 ← 4 7 6 2 3 3 1 2 35733 ← 2 3 6653	15 1 * 1 ← 17 * 1 11 3 4 4 1 1 1 ← 12 5 1 2 3 3 5 9 4 4 1 1 1 ← 6 11 1 2 3 3 5 2 45333 ← 6 4 7 3 3 3 3 4 45333 ← 4 5 8 3 3 3 2 2 35733 ← 8 45333
6	15 * 1 ← 18 4 1 1 1 13 4 4 1 1 1 ← 16 1 2 3 3 9 8 4 1 1 1 ← 10 12 1 1 1 7 3 6 2 3 3 ← 8 8 3 3 3 6 45333 ← 11 6 2 3 3 5 10 1 2 3 3 ← 6 11 2 3 3 3 35733 ← 5 6653	11 5 1 2 3 3 ← 12 6 2 3 3 7 9 1 2 3 3 ← 8 10 2 3 3 7 45333 ← 8 5 7 3 3 5 4 7 3 3 3 ← 6 6653 4 35733 ← 9 8 3 3 3 3 5 8 3 3 3 ← 4 8 6 5 3 1 2 3 5 7 7 ← 8 9 3 3 3
5	17 4 1 1 1 ← 18 5 1 1 15 1 2 3 3 ← 17 2 3 3 9 12 1 1 1 ← 10 13 1 1 7 14 1 1 1 ← 8 15 1 1 7 8 3 3 3 ← 16 3 3 3 6 9 3 3 3 ← 8 11 3 3 5 10 3 3 3 ← 6 11 5 3 2 3 5 7 7	7 10 2 3 3 ← 8 12 3 3 7 9 3 3 3 ← 8 10 5 3 7 5 7 3 3 ← 12 6 5 3 3 3 5 7 7 ← 17 3 3 3 2 4 5 7 7 ← 9 11 3 3
4	21 1 1 1 ← 22 2 1 17 5 1 1 ← 18 6 1 15 3 3 3 ← 17 5 3 9 13 1 1 ← 10 14 1 7 11 3 3 ← 9 13 3 5 5 7 7 ← 19 3 3	22 1 1 1 ← 24 1 1 11 6 5 3 ← 12 7 7 7 12 3 3 ← 8 15 3 7 10 5 3 ← 8 11 7 6 5 7 7 ← 18 5 3 4 7 7 7 ← 10 13 3
3	22 1 1 ← 24 1 21 2 1 ← 22 3 17 6 1 ← 18 7 9 14 1 ← 10 15	23 1 1 ← 25 1 11 7 7 ← 19 7 7 11 7 ← 11 15
2	23 1 ← 25	
1		
	Stem 24	Stem 25

8	15 1 1 * 1←16 2 * 1 11 5 1 * 1←12 6 * 1 10 2 3 4 4 1 1 1←17 1 * 1 9 1 1 24333←10 2 24333 5 5 1 24333←6 6 24333 4 2 2 45333 1 2 2 35733←2 4 35733	11 2 3 4 4 1 1 1←14 3 4 4 1 1 1 10 1 1 24333←12 1 24333 5 2 2 45333←6 4 45333 3 4 2 45333←4 6 45333 2 2 2 35733←8 2 45333 1 2 3 35733←2 3 5 5 7 3 3
7	15 2 * 1←16 4 4 1 1 1 12 3 4 4 1 1 1←18 * 1 11 6 * 1←12 8 4 1 1 1 10 1 24333←12 24333 9 2 24333←10 3 6 2 3 3 6 2 45333 5 6 24333←6 7 6 2 3 3 3 2 35733←4 3 6653 2 3 35733←4 5 5 7 3 3 1 1 2 3 5 7 7←2 2 4 5 7 7	13 3 4 4 1 1 1←14 5 1 2 3 3 11 1 24333←13 24333 7 2 45333←8 4 7 3 3 3 5 4 45333←6 5 8 3 3 3 4 2 35733←10 45333 3 6 45333←4 7 8 3 3 3 2 1 2 3 5 7 7←4 2 3 5 7 7 1 2 2 3 5 7 7←2 3 4 5 7 7
6	15 4 4 1 1 1←20 4 1 1 1 11 8 4 1 1 1←12 12 1 1 1 11 24333←13 6 2 3 3 9 3 6 2 3 3←10 8 3 3 3 5 7 6 2 3 3←6 12 3 3 3 5 35733←9 5 7 3 3 3 5 5 7 3 3←5 8 6 5 3 3 3 6653 2 2 3 5 7 7←4 4 5 7 7 1 2 4 5 7 7←2 4 7 7 7	13 5 1 2 3 3←14 6 2 3 3 9 45333←10 5 7 3 3 7 4 7 3 3 3←8 6653 6 35733←11 8 3 3 3 5 5 8 3 3 3←6 8 6 5 3 3 6 9 3 3 3←4 7 11 3 3 3 2 3 5 7 7←5 4 5 7 7 2 3 3 5 7 7←4 5 5 7 7
5	19 4 1 1 1←20 5 1 1 17 1 2 3 3←24 1 1 1 11 12 1 1 1←12 13 1 1 6 11 3 3 3←8 13 3 3 4 3 5 7 7←8 5 7 7 3 6 11 3 3←4 7 13 3 3 4 5 7 7←6 7 7 7	18 1 2 3 3←20 2 3 3 9 9 3 3 3←10 10 5 3 7 6653←14 6 5 3 5 3 5 7 7←9 5 7 7 3 5 5 7 7←5 9 7 7
4	23 1 1 1←24 2 1 19 5 1 1←20 6 1 18 2 3 3←25 1 1 11 13 1 1←12 14 1 7 13 3 3←9 15 3 7 5 7 7←13 7 7 5 7 7 7←9 11 7	19 2 3 3←22 3 3 18 3 3 3←20 5 3 13 6 5 3←14 7 7 10 11 3 3←12 13 3 9 10 5 3←10 11 7
3	23 2 1←24 3 20 3 3←26 1 19 6 1←20 7 11 14 1←12 15	21 3 3←25 3 19 5 3←21 7 11 13 3←13 15
2	23 3←27	
1		
	Stem 26 (filtrations ≦ 8)	Stem 27 (filtrations ≦ 8)

15		1 1 * * * 1
14	1 * * * 1	2 * * * 1 1 1 1 2 3 4 4 1 1 * 1←2 2 2 3 4 4 1 1 * 1
13	3 4 1 1 * * 1←4 5 1 * * 1 1 1 2 3 4 4 1 1 * 1←8 1 1 * * 1	4 4 1 1 * * 1 2 1 2 3 4 4 1 1 * 1←4 2 3 4 4 1 1 * 1 1 2 2 3 4 4 1 1 * 1←2 4 3 4 4 1 1 * 1
12	7 1 1 * * 1←8 2 * * 1 3 5 1 * * 1←4 6 * * 1 2 2 3 4 4 1 1 * 1←9 1 * * 1 1 1 1 * 24333←2 2 * 24333	3 6 1 * * 1←4 7 * * 1 3 2 3 4 4 1 1 * 1←6 3 4 4 1 1 * 1 2 1 1 * 24333←4 1 * 24333 1 2 1 * 24333←2 3 * 24333
11	7 2 * * 1←8 4 4 1 1 * 1 4 3 4 4 1 1 * 1←10 * * 1 3 6 * * 1←4 8 4 1 1 * 1 2 1 * 24333←4 * 24333 1 2 * 24333←2 3 6 2 3 4 4 1 1 1	5 3 4 4 1 1 * 1←6 5 1 2 3 4 4 1 1 1 3 5 4 4 1 1 * 1←4 7 1 2 3 4 4 1 1 1 3 1 * 24333←5 * 24333
10	7 4 4 1 1 * 1←12 4 1 1 * 1 3 6 1 2 3 4 4 1 1 1←4 7 2 3 4 4 1 1 1 3 * 24333←5 6 2 3 4 4 1 1 1	5 5 1 2 3 4 4 1 1 1←6 6 2 3 4 4 1 1 1 1 2 2 2 45333←2 4 2 2 45333
9	11 4 1 1 * 1←12 5 1 * 1 9 1 2 3 4 4 1 1 1←16 1 1 * 1 5 10 1 1 * 1←6 11 1 * 1 3 6 1 1 24333←4 7 1 24333 2 2 2 2 45333	10 1 2 3 4 4 1 1 1←12 2 3 4 4 1 1 1 3 2 2 2 45333←4 4 2 45333

Stem 26 (filtrations ≥ 9) Stem 27 (filtrations ≥ 9)

7	18 1 * 1←20 * 1 17 2 * 1←18 4 4 1 1 1 13 6 * 1←14 8 4 1 1 1 11 2 24333←12 3 6 2 3 3 5 2 35733←6 3 6653 3 4 35733←4 7 5 7 3 3 3 1 2 3 5 7 7←4 2 4 5 7 7 2 3 3 6653←8 35733 1 2 3 3 5 7 7←2 3 5 5 7 7	19 1 * 1←21 * 1 15 3 4 4 1 1 1←16 5 1 2 3 3 13 1 24333←19 4 4 1 1 1 9 2 45333←10 4 7 3 3 3 7 4 45333←8 5 8 3 3 3 6 2 35733 5 6 45333←6 7 8 3 3 3 3 5 35733←5 7 5 7 3 3 3 3 3 6653←9 35733 1 2 4 3 5 7 7←2 3 6 5 7 7
6	19 * 1←22 4 1 1 1 17 4 4 1 1 1←20 1 2 3 3 13 8 4 1 1 1←14 12 1 1 1 11 3 6 2 3 3←12 8 3 3 3 7 35733←9 6653 5 3 6653←11 5 7 3 3 3 3 3 5 7 7←5 5 5 7 7 3 2 4 5 7 7←4 4 7 7 7 2 4 3 5 7 7←4 6 5 7 7	15 5 1 2 3 3←16 6 2 3 3 14 24333←21 1 2 3 3 11 45333←12 5 7 3 3 9 4 7 3 3 3←10 6653 7 5 8 3 3 3←8 8 6 5 3 5 7 8 3 3 3←8 13 3 3 3 5 2 3 5 7 7←13 8 3 3 3 4 3 3 5 7 7←8 3 5 7 7 3 6 11 3 3 3←4 7 13 3 3 3 4 3 5 7 7←4 7 5 7 7
5	21 4 1 1 1←22 5 1 1 19 1 2 3 3←21 2 3 3 13 12 1 1 1←14 13 1 1 10 9 3 3 3←12 11 3 3 7 12 3 3 3←8 15 3 3 6 3 5 7 7←10 5 7 7 3 6 5 7 7←6 9 7 7 3 4 7 7 7←4 7 11 7	15 6 2 3 3←22 2 3 3 11 9 3 3 3←12 10 5 3 7 13 3 3 3←8 14 5 3 7 8 6 5 3←9 15 3 3 7 3 5 7 7←11 5 7 7 6 4 5 7 7←16 6 5 3 3 5 7 7 7←5 7 11 7
4	25 1 1 1←26 2 1 21 5 1 1←22 6 1 19 3 3 3←21 5 3 13 13 1 1←14 14 1 11 11 3 3←13 13 3 7 7 7 7←11 11 7	26 1 1 1←28 1 1 20 3 3 3←24 3 3 15 6 5 3←16 7 7 11 10 5 3←12 11 7 9 18 1 1←10 19 1 8 7 7 7←22 5 3 7 14 5 3←8 15 7
3	26 1 1←28 1 25 2 1←26 3 21 6 1←22 7 13 14 1←14 15	27 1 1←29 1 23 3 3←27 3 15 7 7←23 7
2	27 1←29	
1		

Stem 28 (filtrations ≦ 8) Stem 29 (filtrations ≦ 8)

16	1 1 1 * * * 1←2 2 * * * 1	2 1 1 * * * 1←4 1 * * * 1 1 2 1 * * * 1←2 3 * * * 1
15	2 1 * * * 1←4 * * * 1 1 2 * * * 1←2 4 4 1 1 * * 1	3 1 * * * 1←5 * * * 1 1 2 1 1 2 3 4 4 1 1 * 1←2 3 1 2 3 4 4 1 1 * 1
14	3 * * * 1←6 4 1 1 * * 1 2 1 1 2 3 4 4 1 1 * 1←4 1 2 3 4 4 1 1 * 1 1 2 1 2 3 4 4 1 1 * 1←2 3 2 3 4 4 1 1 * 1	3 1 1 2 3 4 4 1 1 * 1←4 2 2 3 4 4 1 1 * 1
13	5 4 1 1 * * 1←6 5 1 * * 1 3 6 1 1 * * 1←4 7 1 * * 1 3 1 2 3 4 4 1 1 * 1←5 2 3 4 4 1 1 * 1 1 2 1 1 * 24333←2 3 1 * 24333	3 2 2 3 4 4 1 1 * 1←4 4 3 4 4 1 1 * 1
12	9 1 1 * * 1←10 2 * * 1 5 5 1 * * 1←6 6 * * 1 3 1 1 * 24333←4 2 * 24333	10 1 1 * * 1←12 1 * * 1 4 1 1 * 24333←8 3 4 4 1 1 * 1 3 4 3 4 4 1 1 * 1←4 7 4 4 1 1 * 1
11	10 1 * * 1←12 * * 1 9 2 * * 1←10 4 4 1 1 * 1 5 6 * * 1←6 8 4 1 1 * 1 3 2 * 24333←4 3 6 2 3 4 4 1 1 1	11 1 * * 1←13 * * 1 7 3 4 4 1 1 * 1←8 5 1 2 3 4 4 1 1 1 5 1 * 24333←11 4 4 1 1 * 1 1 2...2 45333←2 4 2 2 2 45333
10	11 * * 1←14 4 1 1 * 1 9 4 4 1 1 * 1←12 1 2 3 4 4 1 1 1 5 8 4 1 1 * 1←6 12 1 1 * 1 3 3 6 2 3 4 4 1 1 1←4 8 1 1 24333 2...2 45333	7 5 1 2 3 4 4 1 1 1←8 6 2 3 4 4 1 1 1 6 * 24333←13 1 2 3 4 4 1 1 1 3 2..2 45333←4 4 2 2 45333
9	13 4 1 1 * 1←14 5 1 * 1 11 1 2 3 4 4 1 1 1←13 2 3 4 4 1 1 1 4 2 2 2 45333 1 2 2 2 35733←2 4 2 35733	7 6 2 3 4 4 1 1 1←14 2 3 4 4 1 1 1 3 4 2 2 45333←4 6 2 45333 2..2 35733
8	17 1 1 * 1←18 2 * 1 13 5 1 * 1←14 6 * 1 11 1 1 24333←12 2 24333 6 2 2 45333 3 2 2 35733←4 4 35733 1 2 1 2 3 5 7 7←2 3 2 3 5 7 7	18 1 1 * 1←20 1 * 1 12 1 1 24333←16 3 4 4 1 1 1 3 6 2 45333 1 2 3 3 6653←4 5 35733

Stem 28 (filtrations ≥ 9) Stem 29 (filtrations ≥ 9)

8	19 1 1 * 1 ← 20 2 * 1 15 5 1 * 1 ← 16 6 * 1 13 1 1 24333 ← 14 2 24333 7 13 1 * 1 3 4 2 35733 ← 4 6 35733 2 2 3 3 6653 ← 8 2 35733	20 1 1 * 15 2 3 4 4 1 1 1 ← 18 3 4 4 1 1 14 1 1 24333 ← 16 1 24333 7 14 1 * 1 ← 8 15 * 1 5 10 1 24333 ← 6 11 24333 5 6 2 45333 3 2 3 3 6653 ← 4 7 35733 2 3 3 3 6653 ← 4 5 3 6653 1 2 3 3 3 5 7 7 ← 2 3 5 3 5 7 7
7	19 2 * 1 ← 20 4 4 1 1 1 15 6 * 1 ← 16 8 4 1 1 1 14 1 24333 ← 16 24333 13 2 24333 ← 14 3 6 2 3 3 10 2 45333 7 2 35733 ← 8 3 6653 4 3 3 6653 ← 10 35733 3 6 35733 ← 4 7 6653 2 3 3 3 5 7 7 ← 4 5 3 5 7 7	21 1 * 1 17 3 4 4 1 1 1 ← 18 5 1 2 3 3 15 1 24333 ← 17 24333 11 2 45333 ← 12 4 7 3 3 3 7 13 4 4 1 1 1 ← 8 15 1 2 3 3 5 3 3 6653 ← 8 2 3 5 7 7 3 5 3 6653 ← 5 7 6653 2 4 3 3 5 7 7 ← 4 6 3 5 7 7
6	15 8 4 1 1 1 ← 16 12 1 1 1 15 24333 ← 17 6 2 3 3 13 3 6 2 3 3 ← 14 8 3 3 3 12 45333 7 14 1 2 3 3 ← 8 15 2 3 3 7 3 6653 ← 11 6653 6 2 3 5 7 7 ← 8 4 5 7 7 5 3 3 5 7 7 ← 6 6 5 7 7 3 5 3 5 7 7 ← 5 7 5 7 7	22 * 1 17 5 1 2 3 3 ← 18 6 2 3 3 13 45333 ← 14 5 7 3 3 11 4 7 3 3 3 ← 12 6653 9 5 8 3 3 3 ← 10 8 6 5 3 7 7 8 3 3 3 ← 8 15 3 3 3 7 2 3 5 7 7 ← 9 4 5 7 7 6 3 3 5 7 7 ← 8 5 5 7 7 3 6 3 5 7 7 ← 6 7 5 7 7 2 3 5 7 7 7 ← 10 3 5 7 7
5	23 4 1 1 1 ← 24 5 1 1 15 12 1 1 1 ← 16 13 1 1 12 9 3 3 3 9 18 1 1 1 ← 10 19 1 1 7 14 3 3 3 ← 8 15 5 3 7 4 5 7 7 ← 10 7 7 7 6 5 5 7 7 ← 8 9 7 7 5 6 5 7 7 ← 6 11 7 7 4 5 7 7 7 ← 12 5 7 7	24 4 1 1 1 22 1 2 3 3 ← 24 2 3 3 13 9 3 3 3 ← 14 10 5 3 13 5 7 3 3 ← 18 6 5 3 9 8 6 5 3 ← 10 14 5 3 9 3 5 7 7 7 5 5 7 7 ← 9 9 7
4	27 1 1 1 ← 28 2 1 23 5 1 1 ← 24 6 1 21 3 3 3 ← 25 3 3 15 13 1 1 ← 16 14 1 13 11 3 3 9 7 7 7 ← 17 7 7 7 9 7 7 ← 9 15 7	28 1 1 1 23 2 3 3 ← 26 3 3 22 3 3 3 ← 24 5 3 17 6 5 3 ← 18 7 7 14 11 3 3 ← 16 13 3 13 10 5 3 ← 14 11 7 9 14 5 3 ← 10 15 7
3	27 2 1 ← 28 3 23 6 1 ← 24 7 15 14 1 ← 16 15 14 13 3	29 1 1 23 5 3 ← 25 7 15 13 3 ← 17 15 13 11 7
2	15 15	30 1
1		31
	Stem 30 (filtrations ≤ 8)	Stem 31 (filtrations ≤ 8)

	Stem 30 (filtrations ≧ 9)	Stem 31 (filtrations ≧ 9)
17	1 2 1 1 * * * 1 ← 2 3 1 * * * 1	
16	3 1 1 * * * 1 ← 4 2 * * * 1	4 1 1 * * * 1
15	3 2 * * * 1 ← 4 4 4 1 1 * * 1	5 1 * * * 1
14	3 4 4 1 1 * * 1	6 * * * 1
13	7 4 1 1 * * 1 ← 8 5 1 * * 1 5 1 2 3 4 4 1 1 * 1	8 4 1 1 * * 1 6 1 2 3 4 4 1 1 * 1 ← 8 2 3 4 4 1 1 * 1 * * 24333
12	11 1 1 * * 1 ← 12 2 * * 1 7 5 1 * * 1 ← 8 6 * * 1 6 2 3 4 4 1 1 * 1 5 1 1 * 24333 ← 6 2 * 24333	12 1 1 * * 1 7 2 3 4 4 1 1 * 1 ← 10 3 4 4 1 1 * 1 6 1 1 * 24333 ← 8 1 * 24333 1 2....2 45333 ← 2 4 2..2 45333
11	11 2 * * 1 ← 12 4 4 1 1 * 1 7 6 * * 1 ← 8 8 4 1 1 * 1 6 1 * 24333 ← 8 * 24333 5 2 * 24333 ← 6 3 6 2 3 4 4 1 1 1 2....2 45333	13 1 * * 1 9 3 4 4 1 1 * 1 ← 10 5 1 2 3 4 4 1 1 1 7 1 * 24333 ← 9 * 24333 5 7 4 4 1 1 * 1 ← 6 9 1 2 3 4 4 1 1 1 3 2...2 45333 ← 4 4 2 2 2 45333
10	7 8 4 1 1 * 1 ← 8 12 1 1 * 1 7 * 24333 ← 9 6 2 3 4 4 1 1 1 5 3 6 2 3 4 4 1 1 1 ← 6 8 1 1 24333 4 2..2 45333 1 2..2 35733 ← 2 3 6 2 45333	14 * * 1 9 5 1 2 3 4 4 1 1 1 ← 10 6 2 3 4 4 1 1 1 5 9 1 2 3 4 4 1 1 1 ← 6 10 2 3 4 4 1 1 1 3 4 2 2 2 45333 ← 4 6 2 2 45333 5 2..2 45333 ← 6 4 2 2 45333 2...2 35733
9	15 4 1 1 * 1 ← 16 5 1 * 1 7 12 1 1 * 1 ← 8 13 1 * 1 5 8 1 1 24333 3 2 2 2 35733 ← 4 4 2 35733 1 1 2 3 3 6653 ← 2 3 5 35733	16 4 1 1 * 1 14 1 2 3 4 4 1 1 1 ← 16 2 3 4 4 1 1 1 5 10 2 3 4 4 1 1 1 ← 6 12 3 4 4 1 1 1 5 4 2 2 45333 ← 6 6 2 45333 3 6 2 2 45333 1 2 2 3 3 6653 ← 2 4 3 3 6653

Stem 30 (filtrations ≧ 9) Stem 31 (filtrations ≧ 9)

8	21 1 1 * 1 ← 22 2 * 1 17 5 1 * 1 ← 18 6 * 1 15 1 1 24333 ← 16 2 24333 9 13 1 * 1 ← 17 1 24333 3 6 2 35733 3 3 3 3 6653 ← 4 5 2 3 5 7 7 1 2 4 3 3 5 7 7 ← 2 3 6 3 5 7 7	22 1 1 * 1 ← 24 1 * 1 7 12 3 4 4 1 1 1 ← 8 15 4 4 1 1 1 7 6 2 45333 ← 14 2 45333 4 3 3 3 6653 ← 8 3 3 6653 3 4 3 3 6653 ← 4 7 3 6653 1 1 2 3 5 7 7 7 ← 2 2 4 5 7 7 7
7	22 1 * 1 ← 24 * 1 21 2 * 1 ← 22 4 4 1 1 1 17 6 * 1 ← 18 8 4 1 1 1 15 2 24333 ← 16 3 6 2 3 3 12 2 45333 ← 18 24333 9 2 35733 ← 10 3 6653 6 3 3 6653 5 6 35733 ← 6 7 6653 3 5 7 8 3 3 3 ← 4 7 8 6 5 3 3 5 2 3 5 7 7 ← 4 6 4 5 7 7 3 4 3 3 5 7 7 ← 4 7 3 5 7 7 1 2 3 5 7 7 7 ← 8 3 3 5 7 7	23 1 * 1 ← 25 * 1 19 3 4 4 1 1 1 ← 20 5 1 2 3 3 13 2 45333 ← 14 4 7 3 3 3 10 2 35733 ← 16 45333 7 3 3 6653 ← 11 3 6653 5 5 3 6653 3 6 2 3 5 7 7 ← 4 7 4 5 7 7 2 2 3 5 7 7 7 ← 4 4 5 7 7 7 1 2 4 5 7 7 7 ← 2 4 7 7 7 7
6	23 * 1 ← 26 4 1 1 1 21 4 4 1 1 1 ← 24 1 2 3 3 17 8 4 1 1 1 ← 18 12 1 1 1 15 3 6 2 3 3 ← 16 8 3 3 3 14 45333 ← 19 6 2 3 3 11 35733 ← 13 6653 9 3 6653 7 3 3 5 7 7 ← 8 6 5 7 7 3 6 4 5 7 7 ← 4 8 7 7 7 2 4 5 7 7 7 ← 9 5 5 7 7	19 5 1 2 3 3 ← 20 6 2 3 3 15 45333 ← 16 5 7 3 3 13 4 7 3 3 3 ← 14 6653 12 35733 ← 17 8 3 3 3 9 2 3 5 7 7 ← 16 9 3 3 3 5 10 9 3 3 3 ← 6 11 11 3 3 3 6 5 5 7 7 ← 4 7 9 7 7 3 4 5 7 7 7 ← 5 8 7 7 7
5	25 4 1 1 1 ← 26 5 1 1 23 1 2 3 3 ← 25 2 3 3 17 12 1 1 1 ← 18 13 1 1 15 8 3 3 3 ← 24 3 3 3 14 9 3 3 3 ← 16 11 3 3 7 6 5 7 7 ← 8 11 7 7 5 10 11 3 3 ← 6 11 13 3 4 7 7 7 7 ← 10 9 7 7	15 9 3 3 3 ← 16 10 5 3 15 5 7 3 3 ← 20 6 5 3 11 3 5 7 7 10 4 5 7 7 ← 17 11 3 3 5 7 7 7 7 ← 9 11 7 7
4	29 1 1 1 ← 30 2 1 25 5 1 1 ← 26 6 1 23 3 3 3 ← 25 5 3 17 13 1 1 ← 18 14 1 15 11 3 3 ← 17 13 3 13 5 7 7 11 7 7 7 ← 19 7 7 7 11 7 7 ← 11 15 7	30 1 1 1 ← 32 1 1 19 6 5 3 ← 20 7 7 15 10 5 3 ← 16 11 7 14 5 7 7 12 7 7 7 ← 18 13 3
3	30 1 1 ← 32 1 29 2 1 ← 30 3 25 6 1 ← 26 7 17 14 1 ← 18 15	31 1 1 ← 33 1 27 3 3 15 11 7 ← 19 15
2	31 1 ← 33 29 3	
1		
	Stem 32 (filtrations ≦ 8)	Stem 33 (filtrations ≦ 8)

	Stem 32 (filtrations ≥ 9)	Stem 33 (filtrations ≥ 9)
17		2 4 1 1 * * * 1
16	5 1 1 * * * 1 ← 6 2 * * * 1	6 1 1 * * * 1 ← 8 1 * * * 1 1 2 3 4 4 1 1 * * 1
15	6 1 * * * 1 ← 8 * * * 1 5 2 * * * 1 ← 6 4 4 1 1 * * 1 2 3 4 4 1 1 * * 1	7 1 * * * 1 ← 9 * * * 1 3 3 4 4 1 1 * * 1 ← 4 5 1 2 3 4 4 1 1 * 1 1 1 * * 24333
14	7 * * * 1 ← 10 4 1 1 * * 1 5 4 4 1 1 * * 1 ← 8 1 2 3 4 4 1 1 * 1 1 * * 24333	3 5 1 2 3 4 4 1 1 * 1 ← 4 6 2 3 4 4 1 1 * 1 2 * * 24333
13	9 4 1 1 * * 1 ← 10 5 1 * * 1 7 1 2 3 4 4 1 1 * 1 ← 9 2 3 4 4 1 1 * 1 3 4 1 1 * 24333 ← 4 5 1 * 24333	3 6 2 3 4 4 1 1 * 1 1 2.....2 45333 ← 2 4 2...2 45333
12	13 1 1 * * 1 ← 14 2 * * 1 9 5 1 * * 1 ← 10 6 * * 1 7 1 1 * 24333 ← 8 2 * 24333 3 5 1 * 24333 ← 4 6 * 24333 2.....2 45333 ← 9 1 * 24333	14 1 1 * * 1 ← 16 1 * * 1 8 1 1 * 24333 5 10 1 * * 1 ← 6 11 * * 1 3 6 1 * 24333 ← 4 7 * 24333 3 2....2 45333 ← 4 4 2..2 45333
11	14 1 * * 1 ← 16 * * 1 13 2 * * 1 ← 14 4 4 1 1 * 1 9 6 * * 1 ← 10 8 4 1 1 * 1 7 2 * 24333 ← 8 3 6 2 3 4 4 1 1 1 4 2...2 45333 ← 10 * 24333 3 6 * 24333 ← 4 7 6 2 3 4 4 1 1 1 1 2...2 35733 ← 2 3 6 2 2 45333	15 1 * * 1 ← 17 * * 1 11 3 4 4 1 1 * 1 ← 12 5 1 2 3 4 4 1 1 1 5 9 4 4 1 1 * 1 ← 6 11 1 2 3 4 4 1 1 1 5 2...2 45333 ← 6 4 2 2 2 45333 3 4 2..2 45333 ← 4 5 8 1 1 24333 2....2 35733 ← 8 2..2 45333
10	15 * * 1 ← 18 4 1 1 * 1 13 4 4 1 1 * 1 ← 16 1 2 3 4 4 1 1 1 9 8 4 1 1 * 1 ← 10 12 1 1 * 1 7 3 6 2 3 4 4 1 1 1 ← 8 8 1 1 24333 6 2..2 45333 ← 11 6 2 3 4 4 1 1 1 5 10 1 2 3 4 4 1 1 1 ← 6 11 2 3 4 4 1 1 1 3 2..2 35733 ← 4 3 6 2 45333	11 5 1 2 3 4 4 1 1 1 ← 12 6 2 3 4 4 1 1 1 7 9 1 2 3 4 4 1 1 1 ← 8 10 2 3 4 4 1 1 1 7 2..2 45333 ← 8 4 2 2 45333 5 4 2 2 2 45333 ← 6 6 2 2 45333 4 2..2 35733 ← 9 8 1 1 24333 3 5 8 1 1 24333 ← 4 7 13 1 * 1 1 2 2 2 3 3 6653 ← 2 3 6 2 35733
9	17 4 1 1 * 1 ← 18 5 1 * 1 15 1 2 3 4 4 1 1 1 ← 17 2 3 4 4 1 1 1 9 12 1 1 * 1 ← 10 13 1 * 1 7 14 1 1 * 1 ← 8 15 1 * 1 7 8 1 1 24333 ← 16 1 1 24333 5 10 1 1 24333 ← 6 11 1 24333 3 3 6 2 45333 ← 4 6 2 35733 2 2 2 3 3 6653 1 2 3 3 3 6653 ← 2 3 5 3 6653	7 10 2 3 4 4 1 1 1 ← 8 12 3 4 4 1 1 1 7 4 2 2 45333 ← 8 6 2 45333 5 6 2 2 45333 ← 11 13 1 * 1 3 2 2 3 3 6653 ← 4 4 3 3 6653

Stem 32 (filtrations ≥ 9) Stem 33 (filtrations ≥ 9)

8	23 1 1 * 1←24 2 * 1 19 5 1 * 1←20 6 * 1 18 2 3 4 4 1 1 1←25 1 * 1 17 1 1 24333←18 2 24333 5 6 2 35733 3 5 3 3 6653←4 7 2 3 5 7 7 2 1 2 3 5 7 7 7←4 2 3 5 7 7 7 1 2 2 3 5 7 7 7←2 3 4 5 7 7 7	19 2 3 4 4 1 1 1←22 3 4 4 1 1 1 18 1 1 24333←20 1 24333 9 6 2 45333←16 2 45333 5 10 2 45333←6 12 45333 3 6 3 3 6653 3 1 2 3 5 7 7 7←4 2 4 5 7 7 7
7	23 2 * 1←24 4 4 1 1 1 20 3 4 4 1 1 1←26 * 1 19 6 * 1←20 8 4 1 1 1 18 1 24333←20 24333 17 2 24333←18 3 6 2 3 3 11 2 35733←12 3 6653 5 5 2 3 5 7 7←6 6 4 5 7 7 3 6 3 3 5 7 7←4 7 5 5 7 7 3 2 3 5 7 7 7←5 4 5 7 7 7	21 3 4 4 1 1 1←22 5 1 2 3 3 19 1 24333←21 24333 15 2 45333←16 4 7 3 3 3 12 2 35733←18 45333 9 15 4 4 1 1 1←10 17 1 2 3 3 9 3 3 6653←12 2 3 5 7 7 6 5 2 3 5 7 7 5 7 3 6653←6 12 9 3 3 3 3 2 4 5 7 7 7←4 4 7 7 7 7
6	23 4 4 1 1 1←28 4 1 1 1 19 8 4 1 1 1←20 12 1 1 1 19 24333←21 6 2 3 3 17 3 6 2 3 3←18 8 3 3 3 13 35733←17 5 7 3 3 10 2 3 5 7 7←12 4 5 7 7 5 7 3 5 7 7 5 6 4 5 7 7←6 8 7 7 7	21 5 1 2 3 3←22 6 2 3 3 17 45333←18 5 7 3 3 15 4 7 3 3 3←16 6653 14 35733←19 8 3 3 3 11 2 3 5 7 7←13 4 5 7 7 9 17 1 2 3 3←10 18 2 3 3 3 4 7 7 7 7←4 7 11 7 7
5	27 4 1 1 1←28 5 1 1 25 1 2 3 3←32 1 1 1 19 12 1 1 1←20 13 1 1 12 3 5 7 7←16 5 7 7 11 4 5 7 7←14 7 7 7	26 1 2 3 3←28 2 3 3 17 9 3 3 3←18 10 5 3 15 6653←22 6 5 3 13 3 5 7 7←17 5 7 7 9 18 2 3 3←10 20 3 3 59777
4	31 1 1 1←32 2 1 27 5 1 1←28 6 1 26 2 3 3←33 1 1 25 3 3 3 19 13 1 1←20 14 1 15 5 7 7←21 7 7 13 7 7 7←17 11 7	27 2 3 3←30 3 3 26 3 3 3←28 5 3 21 6 5 3←22 7 7 18 11 3 3←20 13 3 17 10 5 3←18 11 7 11 22 1 1←12 23 1
3	31 2 1←32 3 28 3 3←34 1 27 6 1←28 7 26 5 3 19 14 1←20 15	29 3 3←33 3 27 5 3←29 7 19 13 3←21 15
2	31 3←35 27 7	
1		

Stem 34 (filtrations ≤ 8)   Stem 35 (filtrations ≤ 8)

19		1 1 * * * * 1
18	1 * * * * 1	2 * * * * 1 1 1 1 2 3 4 4 1 1 * * 1←2 2 2 3 4 4 1 1 * * 1
17	3 4 1 1 * * * 1←4 5 1 * * * 1 1 1 2 3 4 4 1 1 * * 1←8 1 1 * * * 1	4 4 1 1 * * * 1 2 1 2 3 4 4 1 1 * * 1←4 2 3 4 4 1 1 * * 1 1 2 2 3 4 4 1 1 * * 1←2 4 3 4 4 1 1 * * 1
16	7 1 1 * * * 1←8 2 * * * 1 3 5 1 * * * 1←4 6 * * * 1 2 2 3 4 4 1 1 * * 1←9 1 * * * 1 1 1 1 * * 24333←2 2 * * 24333	3 6 1 * * * 1←4 7 * * * 1 3 2 3 4 4 1 1 * * 1←6 3 4 4 1 1 * * 1 2 1 1 * * 24333←4 1 * * 24333 1 2 1 * * 24333←2 3 * * 24333
15	7 2 * * * 1←8 4 4 1 1 * * 1 4 3 4 4 1 1 * * 1←10 * * * 1 3 6 * * * 1←4 8 4 1 1 * * 1 2 1 * * 24333←4 * * 24333 1 2 * * 24333←2 3 6 2 3 4 4 1 1 * 1	5 3 4 4 1 1 * * 1←6 5 1 2 3 4 4 1 1 * 1 3 5 4 4 1 1 * * 1←4 7 1 2 3 4 4 1 1 * 1 3 1 * * 24333←5 * * 24333
14	7 4 4 1 1 * * 1←12 4 1 1 * * 1 3 6 1 2 3 4 4 1 1 * 1←4 7 2 3 4 4 1 1 * 1 3 * * 24333←5 6 2 3 4 4 1 1 * 1	5 5 1 2 3 4 4 1 1 * 1←6 6 2 3 4 4 1 1 * 1 1 2……2 45333←2 4 2….2 45333
13	11 4 1 1 * * 1←12 5 1 * * 1 9 1 2 3 4 4 1 1 * 1←16 1 1 * * 1 5 10 1 1 * * 1←6 11 1 * * 1 3 6 1 1 * 24333←4 7 1 * 24333 2……2 45333	10 1 2 3 4 4 1 1 * 1←12 2 3 4 4 1 1 * 1 3 2…..2 45333←4 4 2…2 45333
12	15 1 1 * * 1←16 2 * * 1 11 5 1 * * 1←12 6 * * 1 10 2 3 4 4 1 1 * 1←17 1 * * 1 9 1 1 * 24333←10 2 * 24333 5 5 1 * 24333←6 6 * 24333 4 2….2 45333 1 2….2 35733←2 4 2..2 35733	11 2 3 4 4 1 1 * 1←14 3 4 4 1 1 * 1 10 1 1 * 24333←12 1 * 24333 5 2….2 45333←6 4 2..2 45333 3 4 2…2 45333←4 6 2..2 45333 2…..2 35733←8 2…2 45333
11	15 2 * * 1←16 4 4 1 1 * 1 12 3 4 4 1 1 * 1←18 * * 1 11 6 * * 1←12 8 4 1 1 * 1 10 1 * 24333←12 * 24333 9 2 * 24333←10 3 6 2 3 4 4 1 1 1 6 2…2 45333 5 6 * 24333←6 7 6 2 3 4 4 1 1 1 3 2…2 35733←4 3 6 2 2 45333	13 3 4 4 1 1 * 1←14 5 1 2 3 4 4 1 1 1 11 1 * 24333←13 * 24333 7 2…2 45333←8 4 2 2 2 45333 5 4 2..2 45333←6 5 8 1 1 24333 4 2…2 35733←10 2..2 45333 3 6 2..2 45333←4 7 8 1 1 24333 1 2..2 3 3 6653←2 3 5 6 2 45333
10	15 4 4 1 1 * 1←20 4 1 1 * 1 11 8 4 1 1 * 1←12 12 1 1 * 1 11 * 24333←13 6 2 3 4 4 1 1 1 9 3 6 2 3 4 4 1 1 1←10 8 1 1 24333 5 7 6 2 3 4 4 1 1 1←6 12 1 1 24333 5 2..2 35733←6 3 6 2 45333 3 3 6 2 2 45333←4 5 6 2 45333 2..2 3 3 6653	13 5 1 2 3 4 4 1 1 1←14 6 2 3 4 4 1 1 1 9 2..2 45333←10 4 2 2 45333 7 4 2 2 2 45333←8 6 2 2 45333 6 2..2 35733←11 8 1 1 24333 5 5 8 1 1 24333←6 7 13 1 * 1 3 2 2 2 3 3 6653←4 3 6 2 35733
9	19 4 1 1 * 1←20 5 1 * 1 17 1 2 3 4 4 1 1 1←24 1 1 * 1 11 12 1 1 * 1←12 13 1 * 1 5 3 6 2 45333←6 6 2 35733 3 5 6 2 45333	18 1 2 3 4 4 1 1 1←20 2 3 4 4 1 1 1 9 4 2 2 45333←10 6 2 45333 7 6 2 2 45333←13 13 1 * 1 5 7 13 1 * 1←6 10 2 45333 3 3 6 2 35733←4 6 3 3 6653 2 4 3 3 3 6653 1 2 1 2 3 5 7 7 7←2 3 2 3 5 7 7 7
	Stem 34 (filtrations ≥ 9)	Stem 35 (filtrations ≥ 9)

9	21 4 1 1 * 1←22 5 1 * 1 19 1 2 3 4 4 1 1 1←21 2 3 4 4 1 1 1 13 12 1 1 * 1←14 13 1 * 1 7 12 1 1 24333←8 13 1 24333 7 3 6 2 45333←8 6 2 35733 5 5 6 2 45333←11 6 2 45333 3 4 3 3 3 6653←4 5 5 3 6653	15 6 2 3 4 4 1 1 1←22 2 3 4 4 1 1 1 11 4 2 2 45333←12 6 2 45333 7 7 13 1 * 1←8 10 2 45333 5 9 13 1 * 1←6 12 2 45333 5 3 6 2 35733←6 6 3 3 6653 3 5 6 2 35733
8	25 1 1 * 1←26 2 * 1 21 5 1 * 1←22 6 * 1 19 1 1 24333←20 2 24333 7 13 1 24333←8 14 24333 7 6 2 35733←14 2 35733 3 5 5 3 6653←5 9 3 6653	26 1 1 * 1←28 1 * 1 20 1 1 24333←24 3 4 4 1 1 1 9 18 1 * 1←10 19 * 1 7 14 1 24333←8 15 24333 7 10 2 45333←8 12 45333 5 6 3 3 6653 4 7 3 3 6653
7	26 1 * 1←28 * 1 25 2 * 1←26 4 4 1 1 1 21 6 * 1←22 8 4 1 1 1 19 2 24333←20 3 6 2 3 3 13 2 35733←14 3 6653 10 3 3 6653←16 35733 7 14 24333←8 15 6 2 3 3	27 1 * 1←29 * 1 23 3 4 4 1 1 1←24 5 1 2 3 3 21 1 24333←27 4 4 1 1 1 17 2 45333←18 4 7 3 3 3 11 3 3 6653←17 35733 9 17 4 4 1 1 1←10 19 1 2 3 3 7 12 45333 3 5 7 3 5 7 7
6	27 * 1←30 4 1 1 1 25 4 4 1 1 1←28 1 2 3 3 21 8 4 1 1 1←22 12 1 1 1 19 3 6 2 3 3←20 8 3 3 3 15 35733←17 6653 13 3 6653←19 5 7 3 3 9 18 1 2 3 3←10 19 2 3 3 7 6 4 5 7 7←8 8 7 7 7 5 9 3 5 7 7 3 5 7 7 7 7←5 7 11 7 7	23 5 1 2 3 3←24 6 2 3 3 22 24333←29 1 2 3 3 19 45333←20 5 7 3 3 17 4 7 3 3 3←18 6653 13 2 3 5 7 7 11 17 1 2 3 3←12 18 2 3 3 7 12 9 3 3 3←8 13 11 3 3 6 9 3 5 7 7←16 3 5 7 7 4 5 7 7 7 7←8 9 7 7 7
5	29 4 1 1 1←30 5 1 1 27 1 2 3 3←29 2 3 3 21 12 1 1 1←22 13 1 1 18 9 3 3 3←20 11 3 3 14 3 5 7 7←18 5 7 7 11 22 1 1 1←12 23 1 1 9 18 3 3 3←10 19 5 3 7 8 7 7 7←8 15 7 7 6 9 7 7 7←16 7 7 7	23 6 2 3 3←30 2 3 3 19 9 3 3 3←20 10 5 3 15 3 5 7 7←19 5 7 7 14 4 5 7 7 11 18 2 3 3←12 20 3 3 7 13 11 3 3←8 14 13 3 7 9 7 7 7←9 15 7 7
4	33 1 1 1←34 2 1 29 5 1 1←30 6 1 27 3 3 3←29 5 3 21 13 1 1←22 14 1 19 11 3 3←21 13 3 15 7 7 7←19 11 7	34 1 1 1←36 1 1 28 3 3 3←32 3 3 23 6 5 3←24 7 7 19 10 5 3←20 11 7 11 20 3 3←12 23 3 7 14 13 3←8 15 15
3	34 1 1←36 1 33 2 1←34 3 29 6 1←30 7 21 14 1←22 15	35 1 1←37 1 31 3 3←35 3 23 7 7
2	35 1←37	
1		

Stem 36 (filtrations ≤ 9) Stem 37 (filtrations ≤ 9)

20	1 1 1 * * * * 1←2 2 * * * * 1	2 1 1 * * * * 1←4 1 * * * * 1 1 2 1 * * * * 1←2 3 * * * * 1
19	2 1 * * * * 1←4 * * * * 1 1 2 * * * * 1←2 4 4 1 1 * * * 1	3 1 * * * * 1←5 * * * * 1 1 2 1 1 2 3 4 4 1 1 * * 1←2 3 1 2 3 4 4 1 1 * * 1
18	3 * * * * 1←6 4 1 1 * * * 1 2 1 1 2 3 4 4 1 1 * * 1←4 1 2 3 4 4 1 1 * * 1 1 2 1 2 3 4 4 1 1 * * 1←2 3 2 3 4 4 1 1 * * 1	3 1 1 2 3 4 4 1 1 * * 1←4 2 2 3 4 4 1 1 * * 1
17	5 4 1 1 * * * 1←6 5 1 * * * 1 3 6 1 1 * * * 1←4 7 1 * * * 1 3 1 2 3 4 4 1 1 * * 1←5 2 3 4 4 1 1 * * 1 1 2 1 1 * * 24333←2 3 1 * * 24333	3 2 2 3 4 4 1 1 * * 1←4 4 3 4 4 1 1 * * 1
16	9 1 1 * * * 1←10 2 * * * 1 5 5 1 * * * 1←6 6 * * * 1 3 1 1 * * 24333←4 2 * * 24333	10 1 1 * * * 1←12 1 * * * 1 4 1 1 * * 24333←8 3 4 4 1 1 * * 1 3 4 3 4 4 1 1 * * 1←4 7 4 4 1 1 * * 1
15	10 1 * * * 1←12 * * * 1 9 2 * * * 1←10 4 4 1 1 * * 1 5 6 * * * 1←6 8 4 1 1 * * 1 3 2 * * 24333←4 3 6 2 3 4 4 1 1 * 1	11 1 * * * 1←13 * * * 1 7 3 4 4 1 1 * * 1←8 5 1 2 3 4 4 1 1 * 1 5 1 * * 24333←11 4 4 1 1 * * 1 1 2.......2 45333←2 4 2.....2 45333
14	11 * * * 1←14 4 1 1 * * 1 9 4 4 1 1 * * 1←12 1 2 3 4 4 1 1 * 1 5 8 4 1 1 * * 1←6 12 1 1 * * 1 3 3 6 2 3 4 4 1 1 * 1←4 8 1 1 * 24333 2.......2 45333	7 5 1 2 3 4 4 1 1 * 1←8 6 2 3 4 4 1 1 * 1 6 * * 24333←13 1 2 3 4 4 1 1 * 1 3 2......2 45333←4 4 2....2 45333
13	13 4 1 1 * * 1←14 5 1 * * 1 11 1 2 3 4 4 1 1 * 1←13 2 3 4 4 1 1 * 1 4 2.....2 45333 1 2.....2 35733←2 4 2...2 35733	7 6 2 3 4 4 1 1 * 1←14 2 3 4 4 1 1 * 1 3 4 2....2 45333←4 6 2...2 45333 2......2 35733
12	17 1 1 * * 1←18 2 * * 1 13 5 1 * * 1←14 6 * * 1 11 1 1 * 24333←12 2 * 24333 6 2....2 45333 3 2....2 35733←4 4 2..2 35733	18 1 1 * * 1←20 1 * * 1 12 1 1 * 24333←16 3 4 4 1 1 * 1 7 2....2 45333←8 4 2..2 45333 3 6 2...2 45333 1 2...2 3 3 6653←2 4 2 2 2 3 3 6653
11	18 1 * * 1←20 * * 1 17 2 * * 1←18 4 4 1 1 * 1 13 6 * * 1←14 8 4 1 1 * 1 11 2 * 24333←12 3 6 2 3 4 4 1 1 1 5 2...2 35733←6 3 6 2 2 45333 3 4 2..2 35733←4 5 6 2 2 45333 2...2 3 3 6653←8 2..2 35733	19 1 * * 1←21 * * 1 15 3 4 4 1 1 * 1←16 5 1 2 3 4 4 1 1 1 13 1 * 24333←19 4 4 1 1 * 1 9 2...2 45333←10 4 2 2 2 45333 7 4 2..2 45333←8 5 8 1 1 24333 6 2...2 35733 5 6 2..2 45333←6 7 8 1 1 24333 3 2..2 3 3 6653←4 3 5 6 2 45333
10	19 * * 1←22 4 1 1 * 1 17 4 4 1 1 * 1←20 1 2 3 4 4 1 1 1 13 8 4 1 1 * 1←14 12 1 1 * 1 11 3 6 2 3 4 4 1 1 1←12 8 1 1 24333 7 2..2 35733←8 3 6 2 45333 5 3 6 2 2 45333←6 5 6 2 45333 4 2 2 2 3 3 6653←9 6 2 2 45333 3 5 6 2 2 45333←4 7 6 2 45333 1 24333 6653←2 3 6 3 3 6653	15 5 1 2 3 4 4 1 1 1←16 6 2 3 4 4 1 1 1 14 * 24333←21 1 2 3 4 4 1 1 1 11 2..2 45333←12 4 2 2 45333 9 4 2 2 2 45333←10 6 2 2 45333 7 5 8 1 1 24333←8 7 13 1 * 1 5 7 8 1 1 24333←6 9 13 1 * 1 5 2 2 2 3 3 6653←6 3 6 2 35733 3 3 5 6 2 45333←4 5 6 2 35733 2 24333 6653

Stem 36 (filtrations ≥ 10)   Stem 37 (filtrations ≥ 10)

7	27 2 * 1←28 4 4 1 1 1 23 6 * 1←24 8 4 1 1 1 22 1 24333←24 24333 21 2 24333←22 3 6 2 3 3 18 2 45333←30 * 1 15 2 35733←16 3 6653 12 3 3 6653←18 35733 9 14 24333←10 15 6 2 3 3 6 11 35733←8 13 5 7 3 3 6 9 3 6653 5 9 2 3 5 7 7←6 10 4 5 7 7 4 5 7 3 5 7 7←8 9 3 5 7 7 2 3 5 7 7 7 7←4 59777	25 3 4 4 1 1 1←26 5 1 2 3 3 23 1 24333←25 24333 19 2 45333←20 4 7 3 3 3 13 3 3 6653←16 2 3 5 7 7 7 14 45333←8 15 8 3 3 3 7 9 3 6653←10 12 9 3 3 3 5 10 2 3 5 7 7←6 11 4 5 7 7 5 5 7 3 5 7 7←9 9 3 5 7 7 3 5 9 3 5 7 7 2 4 5 7 7 7 7←4 6 9 7 7 7
6	23 8 4 1 1 1←24 12 1 1 1 23 24333←25 6 2 3 3 21 3 6 2 3 3←22 8 3 3 3 20 45333←32 4 1 1 1 15 3 6653←19 6653 14 2 3 5 7 7←16 4 5 7 7 9 15 6 2 3 3←10 20 3 3 3 8 12 9 3 3 3 7 13 5 7 3 3←10 8 7 7 7 7 9 3 5 7 7←17 3 5 7 7 5 10 4 5 7 7←6 12 7 7 7 3 59777	25 5 1 2 3 3←26 6 2 3 3 21 45333←22 5 7 3 3 19 4 7 3 3 3←20 6653 15 2 3 5 7 7←17 4 5 7 7 9 12 9 3 3 3←10 13 11 3 3 7 14 9 3 3 3←8 15 11 3 3 6 11 3 5 7 7←8 13 5 7 7 3 6 9 7 7 7←8 11 7 7 7
5	31 4 1 1 1←32 5 1 1 23 12 1 1 1←24 13 1 1 21 8 3 3 3←36 1 1 1 20 9 3 3 3 15 4 5 7 7←18 7 7 7 10 19 3 3 3←12 21 3 3 9 8 7 7 7←10 15 7 7 7 14 11 3 3←8 15 13 3 6 11 7 7 7←8 13 11 7	30 1 2 3 3←32 2 3 3 21 9 3 3 3←22 10 5 3 21 5 7 3 3←26 6 5 3 9 13 11 3 3←10 14 13 3 7 13 5 7 7 7 11 7 7 7←9 13 11 7
4	35 1 1 1←36 2 1 31 5 1 1←32 6 1 29 3 3 3←33 3 3 24 6 5 3←37 1 1 23 13 1 1←24 14 1 21 11 3 3 17 7 7 7 11 21 3 3←13 23 3 7 13 11 7←9 15 15	31 2 3 3←34 3 3 30 3 3 3←32 5 3 25 6 5 3←26 7 7 22 11 3 3←24 13 3 21 10 5 3←22 11 7 20 5 7 7 9 14 13 3←10 15 15
3	35 2 1←36 3 31 6 1←32 7 30 5 3←38 1 23 14 1←24 15 22 13 3	31 5 3←33 7 25 7 7←37 3 23 13 3←25 15 21 11 7
2	31 7←39 23 15	
1		

Stem 38 (filtrations ≤ 7) Stem 39 (filtrations ≤ 7)

12	19 1 1 * * 1←20 2 * * 1 15 5 1 * * 1←16 6 * * 1 13 1 1 * 24333←14 2 * 24333 8 2....2 45333←21 1 * * 1 7 13 1 * * 1←8 14 * * 1 3 4 2...2 35733←4 6 2..2 3 5 7 3 3 2....2 3 3 6653←8 2...2 35733	15 2 3 4 4 1 1 * 1←18 3 4 4 1 1 * 1 14 1 1 * 24333←16 1 * 24333 9 2....2 45333←10 4 2..2 45333 7 14 1 * * 1←8 15 * * 1 5 10 1 * 24333←6 11 * 24333 5 6 2...2 45333 3 2...2 3 3 6653←4 4 2 2 2 3 3 6653
11	19 2 * * 1←20 4 4 1 1 * 1 15 6 * * 1←16 8 4 1 1 * 1 14 1 * 24333←16 * 24333 13 2 * 24333←14 3 6 2 3 4 4 1 1 1 10 2...2 45333←22 * * 1 7 14 * * 1←8 16 4 1 1 * 1 7 2...2 35733←8 3 6 2 2 45333 4 2..2 3 3 6653←10 2..2 35733 3 6 2..2 35733←4 7 6 2 2 45333 1 2 24333 6653←2 3 5 6 2 35733	17 3 4 4 1 1 * 1←18 5 1 2 3 4 4 1 1 1 15 1 * 24333←17 * 24333 11 2...2 45333←12 4 2 2 2 45333 9 4 2..2 45333←10 5 8 1 1 24333 7 13 4 4 1 1 * 1←8 15 1 2 3 4 4 1 1 1 5 2..2 3 3 6653←6 3 5 6 2 45333 3 4 2 2 2 3 3 6653←4 5 5 6 2 45333 2 2 24333 6653←8 2 2 2 3 3 6653
10	15 8 4 1 1 * 1←16 12 1 1 * 1 15 * 24333←17 6 2 3 4 4 1 1 1 13 3 6 2 3 4 4 1 1 1←14 8 1 1 24333 12 2..2 45333←24 4 1 1 * 1 9 2..2 35733←10 3 6 2 45333 7 14 1 2 3 4 4 1 1 1←8 15 2 3 4 4 1 1 1 7 3 6 2 2 45333←8 5 6 2 45333 6 2 2 2 3 3 6653←11 6 2 2 45333 5 5 6 2 2 45333←6 7 6 2 45333 3 24333 6653←4 3 6 3 3 6653	17 5 1 2 3 4 4 1 1 1←18 6 2 3 4 4 1 1 1 13 2..2 45333←14 4 2 2 45333 11 4 2 2 2 45333←12 6 2 2 45333 9 5 8 1 1 24333←10 7 13 1 * 1 7 7 8 1 1 24333←8 9 13 1 * 1 7 2 2 2 3 3 6653←8 3 6 2 35733 5 3 5 6 2 45333←6 5 6 2 35733 4 24333 6653←9 5 6 2 45333 3 5 5 6 2 45333←4 7 6 2 35733 1 2 3 5 5 3 6653←2 3 6 5 2 3 5 7 7
9	23 4 1 1 * 1←24 5 1 * 1 15 12 1 1 * 1←16 13 1 * 1 13 8 1 1 24333←28 1 1 * 1 9 18 1 1 * 1←10 19 1 * 1 9 3 6 2 45333←10 6 2 35733 7 14 1 1 24333←8 15 1 24333 7 5 6 2 45333←13 6 2 45333 5 7 6 2 45333←6 10 2 35733 3 3 6 3 3 6653←4 6 5 2 3 5 7 7 2 3 5 5 3 6653	22 1 2 3 4 4 1 1 1←24 2 3 4 4 1 1 1 13 4 2 2 45333←14 6 2 45333 9 7 13 1 * 1←10 10 2 45333 7 9 13 1 * 1←8 12 2 45333 7 3 6 2 3 5 7 3 3←8 6 3 3 6653 5 5 6 2 35733←11 6 2 3 5 7 3 3 2 4 7 3 3 6653
8	27 1 1 * 1←28 2 * 1 23 5 1 * 1←24 6 * 1 21 1 1 24333←22 2 24333 15 13 1 * 1←29 1 * 1 9 13 1 24333←10 14 24333 9 6 2 35733←16 2 3 5 7 3 3 5 10 2 35733←6 12 3 5 7 3 3 5 7 3 3 6653←6 9 2 3 5 7 7 5 5 5 3 6653←9 12 45333 3 6 5 2 3 5 7 7	23 2 3 4 4 1 1 1←26 3 4 4 1 1 1 22 1 1 24333←24 1 24333 9 10 2 45333←10 12 45333 7 12 2 45333←8 14 45333 7 6 3 3 6653←14 3 3 6653 6 5 5 3 6653←8 9 3 6653 2 3 5 7 3 5 7 7←4 5 9 3 5 7 7 1 2 3 5 7 7 7 7←2 3 59777

Stem 38 (filtrations 8 to 12)   Stem 39 (filtrations 8 to 12)

21	1 2 1 1 * * * * 1←2 3 1 * * * * 1	
20	3 1 1 * * * * 1←4 2 * * * * 1	4 1 1 * * * * 1
19	3 2 * * * * 1←4 4 4 1 1 * * * 1	5 1 * * * * 1
18	3 4 4 1 1 * * * 1	6 * * * * 1
17	7 4 1 1 * * * 1←8 5 1 * * * 1 5 1 2 3 4 4 1 1 * * 1	8 4 1 1 * * * 1 6 1 2 3 4 4 1 1 * * 1←8 2 3 4 4 1 1 * * 1 2 4 1 1 * * 24333
16	11 1 1 * * * 1←12 2 * * * 1 7 5 1 * * * 1←8 6 * * * 1 6 2 3 4 4 1 1 * * 1 5 1 1 * * 24333←6 2 * * 24333	12 1 1 * * * 1 7 2 3 4 4 1 1 * * 1←10 3 4 4 1 1 * * 1 6 1 1 * * 24333←8 1 * * 24333 1 2........2 45333←2 4 2......2 45333
15	11 2 * * * 1←12 4 4 1 1 * * 1 7 6 * * * 1←8 8 4 1 1 * * 1 6 1 * * 24333←8 * * 24333 5 2 * * 24333←6 3 6 2 3 4 4 1 1 * 1 2........2 45333←14 * * * 1	13 1 * * * 1 9 3 4 4 1 1 * * 1←10 5 1 2 3 4 4 1 1 * 1 7 1 * * 24333←9 * * 24333 5 7 4 4 1 1 * * 1←6 9 1 2 3 4 4 1 1 * 1 3 2.......2 45333←4 4 2.....2 45333
14	7 8 4 1 1 * * 1←8 12 1 1 * * 1 7 * * 24333←9 6 2 3 4 4 1 1 * 1 5 3 6 2 3 4 4 1 1 * 1←6 8 1 1 * 24333 4 2......2 45333←16 4 1 1 * * 1 1 2......2 35733←2 3 6 2...2 45333	9 5 1 2 3 4 4 1 1 * 1←10 6 2 3 4 4 1 1 * 1 5 9 1 2 3 4 4 1 1 * 1←6 10 2 3 4 4 1 1 * 1 5 2......2 45333←6 4 2....2 45333 3 4 2.....2 45333←4 6 2....2 45333 2.......2 3 5 7 3 3
13	15 4 1 1 * * 1←16 5 1 * * 1 7 12 1 1 * * 1←8 13 1 * * 1 5 8 1 1 * 24333←20 1 1 * * 1 3 2.....2 35733←4 4 2...2 3 5 7 3 3	14 1 2 3 4 4 1 1 * 1←16 2 3 4 4 1 1 * 1 5 10 2 3 4 4 1 1 * 1←6 12 3 4 4 1 1 * 1 5 4 2....2 45333←6 6 2...2 45333 3 6 2....2 45333 1 2....2 3 3 6653←2 4 2..2 3 3 6653

Stem 38 (filtrations ≧ 13) Stem 39 (filtrations ≧ 13)

7	30 1 * 1←32 * 1 29 2 * 1←30 4 4 1 1 1 25 6 * 1←26 8 4 1 1 1 23 2 24333←24 3 6 2 3 3 20 2 45333←26 24333 17 2 35733←18 3 6653 7 12 35733←8 15 5 7 3 3 7 9 2 3 5 7 7←8 10 4 5 7 7 6 5 7 3 5 7 7←10 9 3 5 7 7 3 6 9 3 5 7 7←8 11 3 5 7 7 3 4 5 7 7 7 7←4 7 9 7 7 7	31 1 * 1←33 * 1 27 3 4 4 1 1 1←28 5 1 2 3 3 21 2 45333←22 4 7 3 3 3 18 2 35733←24 45333 15 3 3 6653←19 3 6653 11 12 45333 9 14 45333←10 15 8 3 3 3 7 13 35733←9 15 5 7 3 3 7 5 7 3 5 7 7←11 9 3 5 7 7 3 3 59777←6 6 9 7 7 7
6	31 * 1←34 4 1 1 1 29 4 4 1 1 1←32 1 2 3 3 25 8 4 1 1 1←26 12 1 1 1 23 3 6 2 3 3←24 8 3 3 3 22 45333←27 6 2 3 3 19 35733←21 6653 17 3 6653 7 11 3 5 7 7←8 14 5 7 7 7 10 4 5 7 7←8 12 7 7 7 6 12 3 5 7 7←9 13 5 7 7 5 59777	27 5 1 2 3 3←28 6 2 3 3 23 45333←24 5 7 3 3 21 4 7 3 3 3←22 6653 20 35733←25 8 3 3 3 17 2 3 5 7 7←24 9 3 3 3 11 12 9 3 3 3←12 13 11 3 3 9 15 8 3 3 3←12 21 3 3 3 7 12 3 5 7 7←8 15 5 7 7 6 59777←10 13 5 7 7 5 6 9 7 7 7←8 13 7 7 7 3 6 11 7 7 7←4 7 13 11 7
5	33 4 1 1 1←34 5 1 1 31 1 2 3 3←33 2 3 3 25 12 1 1 1←26 13 1 1 23 8 3 3 3←32 3 3 3 22 9 3 3 3←24 11 3 3 18 3 5 7 7 11 20 3 3 3←12 23 3 3 7 14 5 7 7 7 12 7 7 7←8 15 11 7	23 9 3 3 3←24 10 5 3 23 5 7 3 3←28 6 5 3 19 3 5 7 7←33 3 3 3 18 4 5 7 7←25 11 3 3 11 21 3 3 3←13 23 3 3 11 13 11 3 3←12 14 13 3 9 11 7 7 7 7 13 7 7 7←9 15 11 7
4	37 1 1 1←38 2 1 33 5 1 1←34 6 1 31 3 3 3←33 5 3 25 13 1 1←26 14 1 23 11 3 3←25 13 3 21 5 7 7←35 3 3 19 7 7 7 11 15 7 7	38 1 1 1←40 1 1 27 6 5 3←28 7 7 23 10 5 3←24 11 7 22 5 7 7←34 5 3 20 7 7 7←26 13 3 13 26 1 1←14 27 1 11 14 13 3←12 15 15 10 13 11 7
3	38 1 1←40 1 37 2 1←38 3 33 6 1←34 7 25 14 1←26 15	39 1 1←41 1 27 7 7←35 7 23 11 7←27 15 11 15 15
2	39 1←41	
1		
	Stem 40 (filtrations ≦ 7)	Stem 41 (filtrations ≦ 7)

	Stem 40	Stem 41
12	21 1 1 * * 1←22 2 * * 1 17 5 1 * * 1←18 6 * * 1 15 1 1 * 24333←16 2 * 24333 10 2....2 45333←17 1 * 24333 9 13 1 * * 1←10 14 * * 1 3 6 2...2 35733 1 2 2 24333 6653←2 4 24333 6653	22 1 1 * * 1←24 1 * * 1 11 2....2 45333←12 4 2..2 45333 7 12 3 4 4 1 1 * 1←8 15 4 4 1 1 * 1 7 6 2...2 45333←14 2...2 45333 3 4 2..2 3 3 6653←4 6 2 2 2 3 3 6653 2..2 4 3 3 3 6653←8 2..2 3 3 6653
11	22 1 * * 1←24 * * 1 21 2 * * 1←22 4 4 1 1 * 1 17 6 * * 1←18 8 4 1 1 * 1 15 2 * 24333←16 3 6 2 3 4 4 1 1 1 12 2....4 5 3 3 3←18 * 24333 9 14 * * 1←10 16 4 1 1 * 1 9 2...2 35733←10 3 6 2 2 45333 6 2..2 3 3 6653 5 6 2..2 35733←6 7 6 2 2 45333 3 2 24333 6653←4 3 5 6 2 35733	23 1 * * 1←25 * * 1 19 3 4 4 1 1 * 1←20 5 1 2 3 4 4 1 1 1 13 2...2 45333←14 4 2 2 2 45333 11 4 2..2 45333←12 5 8 1 1 24333 10 2...2 35733←16 2..2 45333 7 2..2 3 3 6653←8 3 5 6 2 45333 4 2 24333 6653←10 2 2 2 3 3 6653 3 6 2 2 2 3 3 6653←4 7 5 6 2 45333 1 2 2 3 5 5 3 6653←2 3 4 7 3 3 6653 1 1 2 4 7 3 3 6653
10	23 * * 1←26 4 1 1 * 1 21 4 4 1 1 * 1←24 1 2 3 4 4 1 1 1 17 8 4 1 1 * 1←18 12 1 1 * 1 15 3 6 2 3 4 4 1 1 1←16 8 1 1 24333 14 2..2 45333←19 6 2 3 4 4 1 1 1 11 2..2 35733←12 3 6 2 45333 9 16 4 1 1 * 1←10 20 1 1 * 1 9 3 6 2 2 45333←10 5 6 2 45333 5 7 6 2 2 45333←6 9 6 2 45333 5 24333 6653←6 3 6 3 3 6653 3 3 5 6 2 35733←4 5 6 3 3 6653 2 2 3 5 5 3 6653←4 4 7 3 3 6653 1 2 4 7 3 3 6653	19 5 1 2 3 4 4 1 1 1←20 6 2 3 4 4 1 1 1 15 2..2 45333←16 4 2 2 45333 13 4 2 2 2 45333←14 6 2 2 45333 12 2..2 35733←17 8 1 1 24333 11 5 8 1 1 24333←12 7 13 1 * 1 9 2 2 2 3 3 6653←10 3 6 2 35733 7 3 5 6 2 45333←8 5 6 2 35733 6 24333 6653←11 5 6 2 45333 5 5 5 6 2 45333←6 7 6 2 35733 3 2 3 5 5 3 6653←4 3 6 5 2 3 5 7 7 2 2 4 7 3 3 6653
9	25 4 1 1 * 1←26 5 1 * 1 23 1 2 3 4 4 1 1 1←25 2 3 4 4 1 1 1 17 12 1 1 * 1←18 13 1 * 1 15 8 1 1 24333←24 1 1 24333 11 3 6 2 45333←12 6 2 35733 5 9 6 2 45333←6 12 2 35733 5 3 6 3 3 6653←6 6 5 2 3 5 7 7 3 5 6 3 3 6653←4 7 12 45333 3 4 7 3 3 6653←8 5 5 3 6653 1 2 3 5 7 3 5 7 7←2 3 5 9 3 5 7 7	15 4 2 2 45333←16 6 2 45333 13 6 2 2 45333←19 13 1 * 1 11 7 13 1 * 1←12 10 2 45333 9 3 6 2 35733←10 6 3 3 6653 7 5 6 2 35733←13 6 2 35733 5 7 6 2 35733←6 10 3 3 6653 3 5 5 5 3 6653←4 6 9 3 6653 3 3 6 5 2 3 5 7 7←5 7 12 45333 1 2 4 5 7 3 5 7 7←2 3 6 9 3 5 7 7
8	29 1 1 * 1←30 2 * 1 25 5 1 * 1←26 6 * 1 23 1 1 24333←24 2 24333 17 13 1 * 1←25 1 24333 7 10 2 35733←8 12 35733 7 5 5 3 6653←9 9 3 6653 5 9 3 3 6653←6 11 2 3 5 7 7 5 6 5 2 3 5 7 7 3 3 5 7 3 5 7 7←5 5 9 3 5 7 7 2 4 5 7 3 5 7 7←4 6 9 3 5 7 7 1 2 4 5 7 7 7 7←2 3 6 9 7 7 7	30 1 1 * 1←32 1 * 1 15 6 2 45333←22 2 45333 11 10 2 45333←12 12 45333 9 6 3 3 6653←16 3 3 6653 5 10 3 3 6653←8 13 35733 4 3 5 7 3 5 7 7←8 5 7 3 5 7 7 3 6 11 35733←4 7 13 5 7 3 3 3 6 9 3 6653←4 8 12 9 3 3 3 3 4 5 7 3 5 7 7←4 7 9 3 5 7 7

Stem 40 (filtrations 8 to 12) Stem 41 (filtrations 8 to 12)

21		2 4 1 1 * * * * 1
20	5 1 1 * * * * 1←6 2 * * * * 1	6 1 1 * * * * 1←8 1 * * * * 1 1 2 3 4 4 1 1 * * * 1
19	6 1 * * * * 1←8 * * * * 1 5 2 * * * * 1←6 4 4 1 1 * * * 1 2 3 4 4 1 1 * * * 1	7 1 * * * * 1←9 * * * * 1 3 3 4 4 1 1 * * * 1←4 5 1 2 3 4 4 1 1 * * 1 1 1 * * * 24333
18	7 * * * * 1←10 4 1 1 * * * 1 5 4 4 1 1 * * * 1←8 1 2 3 4 4 1 1 * * 1 1 * * * 24333	3 5 1 2 3 4 4 1 1 * * 1←4 6 2 3 4 4 1 1 * * 1 2 * * * 24333
17	9 4 1 1 * * * 1←10 5 1 * * * 1 7 1 2 3 4 4 1 1 * * 1←9 2 3 4 4 1 1 * * 1 3 4 1 1 * * 24333←4 5 1 * * 24333	3 6 2 3 4 4 1 1 * * 1 1 2.........2 45333←2 4 2.......2 45333
16	13 1 1 * * * 1←14 2 * * * 1 9 5 1 * * * 1←10 6 * * * 1 7 1 1 * * 24333←8 2 * * 24333 3 5 1 * * 24333←4 6 * * 24333 2.........2 45333←9 1 * * 24333	14 1 1 * * * 1←16 1 * * * 1 8 1 1 * * 24333 5 10 1 * * * 1←6 11 * * * 1 3 6 1 * * 24333←4 7 * * 24333 3 2........2 45333←4 4 2......2 45333
15	14 1 * * * 1←16 * * * 1 13 2 * * * 1←14 4 4 1 1 * * 1 9 6 * * * 1←10 8 4 1 1 * * 1 7 2 * * 24333←8 3 6 2 3 4 4 1 1 * 1 4 2.......2 45333←10 * * 24333 3 6 * * 24333←4 7 6 2 3 4 4 1 1 * 1 1 2.......2 35733←2 3 6 2....2 45333	15 1 * * * 1←17 * * * 1 11 3 4 4 1 1 * * 1←12 5 1 2 3 4 4 1 1 * 1 5 9 4 4 1 1 * * 1←6 11 1 2 3 4 4 1 1 * 1 5 2.......2 45333←6 4 2.....2 45333 3 4 2......2 45333←4 5 8 1 1 * 24333 2........2 35733←8 2......2 45333
14	15 * * * 1←18 4 1 1 * * 1 13 4 4 1 1 * * 1←16 1 2 3 4 4 1 1 * 1 9 8 4 1 1 * * 1←10 12 1 1 * * 1 7 3 6 2 3 4 4 1 1 * 1←8 8 1 1 * 24333 6 2......2 45333←11 6 2 3 4 4 1 1 * 1 5 10 1 2 3 4 4 1 1 * 1←6 11 2 3 4 4 1 1 * 1 3 2......2 35733←4 3 6 2...2 45333	11 5 1 2 3 4 4 1 1 * 1←12 6 2 3 4 4 1 1 * 1 7 9 1 2 3 4 4 1 1 * 1←8 10 2 3 4 4 1 1 * 1 7 2......2 45333←8 4 2....2 45333 5 4 2.....2 45333←6 6 2....2 45333 4 2......2 35733←9 8 1 1 * 24333 3 5 8 1 1 * 24333←4 8 2....2 45333 1 2.....2 3 3 6653←2 3 6 2...2 35733
13	17 4 1 1 * * 1←18 5 1 * * 1 15 1 2 3 4 4 1 1 * 1←17 2 3 4 4 1 1 * 1 9 12 1 1 * * 1←10 13 1 * * 1 7 14 1 1 * * 1←8 15 1 * * 1 7 8 1 1 * 24333←16 1 1 * 24333 5 10 1 1 * 24333←6 11 1 * 24333 3 3 6 2...2 45333←4 6 2...2 35733 2.....2 3 3 6653	7 10 2 3 4 4 1 1 * 1←8 12 3 4 4 1 1 * 1 7 4 2....2 45333←8 6 2...2 45333 5 6 2....2 45333←12 2....2 45333 3 2....2 3 3 6653←4 4 2..2 3 3 6653

Stem 40 (filtrations ≥ 13) Stem 41 (filtrations ≥ 13)

7	31 2 * 1←32 4 4 1 1 1 28 34411 1←34 * 1 27 6 * 1←28 8 4 1 1 1 26 1 24333←28 24333 25 2 24333←26 3 6 2 3 3 19 2 35733←20 3 6653 11 22 * 1←12 24 4 1 1 1 10 9 3 6653 7 14 35733←8 15 6653 6 5 9 3577←12 9 3577 5 6 9 3577←8 13 3577 4 3 59777←8 59777 3 6 11 3577←4 7 13 5 7 7 3 3 6 9 7 7 7←4 7 11 7 7 7	29 34411 1←30 5 1 2 3 3 27 1 24333←29 24333 23 2 45333←24 4 7 3 3 3 20 2 35733←26 45333 17 3 3 6653←20 23577 11 21 4 4 1 1 1←12 23 1 2 3 3 11 9 3 6653←14 12 9 3 3 3 7 13 3 6653←9 15 6653 7 5 9 3577←13 9 3577 6 6 9 3577←8 14 3577 5 3 59777←9 59777 3 6 12 3577←4 7 14 5 7 7 3 5 59777←5 7 13 5 7 7
6	31 4 4 1 1 1←36 4 1 1 1 27 8 4 1 1 1←28 12 1 1 1 27 24333←29 6 2 3 3 25 3 6 2 3 3←26 8 3 3 3 21 35733←25 5 7 3 3 18 23577←20 4 5 7 7 12 12 9 3 3 3 11 22 1 2 3 3←12 23 2 3 3 9 11 3577←10 14 5 7 7 7 13 3577←9 15 5 7 7 7 59777←11 13 5 7 7	29 5 1 2 3 3←30 6 2 3 3 25 45333←26 5 7 3 3 23 4 7 3 3 3←24 6653 22 35733←27 8 3 3 3 19 23577←21 4 5 7 7 13 12 9 3 3 3←14 13 11 3 3 11 15 8 3 3 3←12 23 3 3 3 10 11 3577←12 13 5 7 7 7 14 3577←10 15 5 7 7 7 6 9 7 7 7←8 15 7 7 7 5 10 19 3 3 3←6 11 21 3 3
5	35 4 1 1 1←36 5 1 1 33 1 2 3 3←40 1 1 1 27 12 1 1 1←28 13 1 1 20 3577←24 5 7 7 19 4 5 7 7←22 7 7 7 13 26 1 1 1←14 27 1 1 11 22 3 3 3←12 23 5 3 9 14 5 7 7←12 13 11 7	34 1 2 3 3←36 2 3 3 25 9 3 3 3←26 10 5 3 23 6653←30 6 5 3 21 3577←25 5 7 7 13 13 11 3 3←14 14 13 3
4	39 1 1 1←40 2 1 35 5 1 1←36 6 1 34 2 3 3←41 1 1 27 13 1 1←28 14 1 23 5 7 7←29 7 7 21 7 7 7←25 11 7 11 13 11 7←13 15 15	35 2 3 3←38 3 3 34 3 3 3←36 5 3 29 6 5 3←30 7 7 26 11 3 3←28 13 3 25 10 5 3←26 11 7 13 14 13 3←14 15 15
3	39 2 1←40 3 36 3 3←42 1 35 6 1←36 7 27 14 1←28 15	37 3 3←41 3 35 5 3←37 7 27 13 3←29 15
2	39 3←43	
1		

Stem 42 (filtrations ≤ 7) Stem 43 (filtrations ≤ 7)

11	23 2 * * 1←24 4 4 1 1 * 1 20 34411 * 1←26 * * 1 19 6 * * 1←20 8 4 1 1 * 1 18 1 * 24333←20 * 24333 17 2 * 24333←18 3 6 2 34411 1 11 14 * * 1←12 16 4 1 1 * 1 11 2...2 35733←12 3 6 2 2 45333 5 2 24333 6653←6 3 5 6 2 35733 3 4 24333 6653←4 5 5 6 2 35733 2 1 24733 6653←4 24733 6653 1 2 24733 6653←8 24333 6653	21 34411 * 1←22 5 1 2 34411 1 19 1 * 24333←21 * 24333 15 2...2 45333←16 4 2 2 2 45333 13 4 2..2 45333←14 5 8 1 1 24333 12 2...2 35733←18 2..2 45333 9 15 4 4 1 1 * 1←10 17 1 2 34411 1 9 2..2 3 3 6653←10 3 5 6 2 45333 6 2 24333 6653 5 6 2 2 2 3 3 6653←6 7 5 6 2 45333 3 1 24733 6653←5 24733 6653
10	23 4 4 1 1 * 1←28 4 1 1 * 1 19 8 4 1 1 * 1←20 12 1 1 * 1 19 * 24333←21 6 2 34411 1 17 3 6 2 34411 1←18 8 1 1 24333 13 2..2 35733←14 3 6 2 45333 11 16 4 1 1 * 1←12 20 1 1 * 1 11 3 6 2 2 45333←12 5 6 2 45333 7 24333 6653←8 3 6 3 3 6653 5 3 5 6 2 35733←6 5 6 3 3 6653 4 2 3 5 5 3 6653←9 5 6 2 35733 3 5 5 6 2 35733←4 7 6 3 3 6653 3 24733 6653←6 4 7 3 3 6653	21 5 1 2 34411 1←22 6 2 34411 1 17 2..2 45333←18 4 2 2 45333 15 4 2 2 2 45333←16 6 2 2 45333 14 2..2 35733←19 8 1 1 24333 13 5 8 1 1 24333←14 7 13 1 * 1 11 2 2 2 3 3 6653←12 3 6 2 35733 9 17 1 2 34411 1←10 18 2 34411 1 ·9 3 5 6 2 45333←10 5 6 2 35733 5 7 5 6 2 45333←6 9 6 2 35733 5 2 3 5 5 3 6653←6 3 6 5 23577 3 3 4 7 3 3 6653←4 5 6 5 23577 2 3 3 6 5 23577←4 7 5 5 3 6653 1 2 3 3 5 7 3577←2 3 5 5 7 3577
9	27 4 1 1 * 1←28 5 1 * 1 25 1 2 34411 1←32 1 1 * 1 19 12 1 1 * 1←20 13 1 * 1 13 3 6 2 45333←14 6 2 35733 11 20 1 1 * 1←12 21 1 * 1 7 9 6 2 45333←8 12 2 35733 7 3 6 3 3 6653←8 6 5 23577 5 5 6 3 3 6653←6 7 12 45333 5 4 7 3 3 6653←11 6 3 3 6653 4 5 5 5 3 6653 3 6 5 5 3 6653←4 7 9 3 6653 2 3 3 5 7 3577←4 5 5 7 3577	26 1 2 34411 1←28 2 34411 1 17 4 2 2 45333←18 6 2 45333 15 6 2 2 45333←21 13 1 * 1 13 7 13 1 * 1←14 10 2 45333 11 3 6 2 35733←12 6 3 3 6653 9 18 2 34411 1←10 20 34411 1 5 9 6 2 35733←6 12 3 3 6653 5 3 6 5 23577←6 6 9 3 6653 3 5 6 5 23577←9 6 5 23577 2 4 3 5 7 3577←4 6 5 7 3577
8	31 1 1 * 1←32 2 * 1 27 5 1 * 1←28 6 * 1 26 2 34411 1←33 1 * 1 25 1 1 24333←26 2 24333 11 21 1 * 1←12 22 * 1 9 5 5 3 6653 7 12 2 35733←8 14 35733 7 6 5 23577←13 12 45333 5 3 5 7 3577←9 5 7 3577 3 5 5 7 3577←5 7 9 3577 3 3 5 9 3577←5 8 12 9 3 3 3	27 2 34411 1←30 34411 1 26 1 1 24333←28 1 24333 17 6 2 45333←24 2 45333 13 10 2 45333←14 12 45333 11 22 1 * 1←12 23 * 1 10 5 5 3 6653←12 9 3 6653 9 18 1 24333←10 19 24333 7 10 3 3 6653←8 15 35733 6 11 3 3 6653←8 13 3 6653 6 3 5 7 3577←8 5 9 3577 5 6 9 3 6653←6 8 12 9 3 3 3 4 3 5 9 3577←10 5 7 3577 3 6 5 7 3577←6 7 9 3577 3 3 6 9 3577←4 6 12 3577 2 3 3 59777←4 5 59777
	Stem 42 (filtrations 8 to 11)	Stem 43 (filtrations 8 to 11)

	Stem 42 (filtrations 12 to 15)	Stem 43 (filtrations 12 to 15)
15	15 2 * * * 1←16 4 4 1 1 * * 1 12 34411 * * 1←18 * * * 1 11 6 * * * 1←12 8 4 1 1 * * 1 10 1 * * 24333←12 * * 24333 9 2 * * 24333←10 3 6 2 34411 * 1 6 2.......2 45333 5 6 * * 24333←6 7 6 2 34411 * 1 3 2.......2 35733←4 3 6 2....2 45333	13 34411 * * 1←14 5 1 2 34411 * 1 11 1 * * 24333←13 * * 24333 7 2.......2 45333←8 4 2.....2 45333 5 4 2......2 45333←6 5 8 1 1 * 24333 4 2.......2 35733←10 2......2 45333 3 6 2......2 45333←4 7 8 1 1 * 24333 1 2......2 3 3 6653←2 3 5 6 2...2 45333
14	15 4 4 1 1 * * 1←20 4 1 1 * * 1 11 8 4 1 1 * * 1←12 12 1 1 * * 1 11 * * 24333←13 6 2 34411 * 1 9 3 6 2 34411 * 1←10 8 1 1 * 24333 5 7 6 2 34411 * 1←6 12 1 1 * 24333 5 2......2 35733←6 3 6 2...2 45333 3 3 6 2....2 45333←4 5 6 2...2 45333 2......2 3 3 6653	13 5 1 2 34411 * 1←14 6 2 34411 * 1 9 2......2 45333←10 4 2....2 45333 7 4 2.....2 45333←8 6 2....2 45333 6 2......2 35733←11 8 1 1 * 24333 5 5 8 1 1 * 24333←6 8 2....2 45333 3 2.....2 3 3 6653←4 3 6 2...2 35733
13	19 4 1 1 * * 1←20 5 1 * * 1 17 1 2 34411 * 1←24 1 1 * * 1 11 12 1 1 * * 1←12 13 1 * * 1 5 3 6 2...2 45333←6 6 2...2 35733 3 5 6 2...2 45333 1 2..2 4 3 3 3 6653←2 4 2 24333 6653	18 1 2 34411 * 1←20 2 34411 * 1 9 4 2....2 45333←10 6 2...2 45333 7 6 2....2 45333←14 2....2 45333 5 8 2....2 45333←6 10 2...2 45333 3 3 6 2...2 35733←4 6 2..2 3 3 6653 2...2 4 3 3 3 6653
12	23 1 1 * * 1←24 2 * * 1 19 5 1 * * 1←20 6 * * 1 18 2 34411 * 1←25 1 * * 1 17 1 1 * 24333←18 2 * 24333 11 13 1 * * 1←12 14 * * 1 5 6 2...2 35733 3 2 2 24333 6653←4 4 24333 6653 1 1 1 24733 6653←2 2 24733 6653	19 2 34411 * 1←22 34411 * 1 18 1 1 * 24333←20 1 * 24333 13 2....2 45333←14 4 2..2 45333 9 6 2...2 45333←16 2...2 45333 5 10 2...2 45333←6 12 2..2 45333 3 6 2..2 3 3 6653 2 1 1 24733 6653←4 1 24733 6653 1 2 1 24733 6653←2 3 24733 6653 1 1 2 24733 6653←2 4 2 3 5 5 3 6653

Stem 42 (filtrations 12 to 15) Stem 43 (filtrations 12 to 15)

	Stem 42 (filtrations ≥ 16)	Stem 43 (filtrations ≥ 16)
23		1 1 * * * * * * 1
22	1 * * * * * * 1	2 * * * * * * 1 1 1 1 2 34411 * * * 1←2 2 2 34411 * * * 1
21	3 4 1 1 * * * * 1←4 5 1 * * * * 1 1 1 2 34411 * * * 1←8 1 1 * * * * 1	4 4 1 1 * * * * 1 2 1 2 34411 * * * 1←4 2 34411 * * * 1 1 2 2 34411 * * * 1←2 4 34411 * * * 1
20	7 1 1 * * * * 1←8 2 * * * * 1 3 5 1 * * * * 1←4 6 * * * * 1 2 2 34411 * * * 1←9 1 * * * * 1 1 1 1 * * * 24333←2 2 * * * 24333	3 6 1 * * * * 1←4 7 * * * * 1 3 2 34411 * * * 1←6 34411 * * * 1 2 1 1 * * * 24333←4 1 * * * 24333 1 2 1 * * * 24333←2 3 * * * 24333
19	7 2 * * * * 1←8 4 4 1 1 * * * 1 4 34411 * * * 1←10 * * * * 1 3 6 * * * * 1←4 8 4 1 1 * * * 1 2 1 * * * 24333←4 * * * 24333 1 2 * * * 24333←2 3 6 2 34411 * * 1	5 34411 * * * 1←6 5 1 2 34411 * * 1 3 5 4 4 1 1 * * * 1←4 7 1 2 34411 * * 1 3 1 * * * 24333←5 * * * 24333
18	7 4 4 1 1 * * * 1←12 4 1 1 * * * 1 3 6 1 2 34411 * * 1←4 7 2 34411 * * 1 3 * * * 24333←5 6 2 34411 * * 1	5 5 1 2 34411 * * 1←6 6 2 34411 * * 1 1 2.........2 45333←2 4 2........2 45333
17	11 4 1 1 * * * 1←12 5 1 * * * 1 9 1 2 34411 * * 1←16 1 1 * * * 1 5 10 1 1 * * * 1←6 11 1 * * * 1 3 6 1 1 * * 24333←4 7 1 * * 24333 2..........2 45333	10 1 2 34411 * * 1←12 2 34411 * * 1 3 2.........2 45333←4 4 2.......2 45333
16	15 1 1 * * * 1←16 2 * * * 1 11 5 1 * * * 1←12 6 * * * 1 10 2 34411 * * 1←17 1 * * * 1 9 1 1 * * 24333←10 2 * * 24333 5 5 1 * * 24333←6 6 * * 24333 4 2........2 45333 1 2........2 35733←2 4 2......2 35733	11 2 34411 * * 1←14 34411 * * 1 10 1 1 * * 24333←12 1 * * 24333 5 2........2 45333←6 4 2......2 45333 3 4 2.......2 45333←4 6 2......2 45333 2.........2 35733←8 2.......2 45333

Stem 42 (filtrations ≥ 16)　　Stem 43 (filtrations ≥ 16)

7	34 1 * 1←36 * 1 33 2 * 1←34 4 4 1 1 1 29 6 * 1←30 8 4 1 1 1 27 2 24333←28 3 6 2 3 3 21 2 35733←22 3 6653 18 3 3 6653←24 35733 13 22 * 1←14 24 4 1 1 1 9 14 35733←10 15 6653 7 13 2 3577←8 14 4 5 7 7 7 6 9 3577←8 15 3577 6 3 59777←10 59777 3 6 59777←6 7 13 5 7 7 3 5 6 9 7 7 7←4 7 13 7 7 7	35 1 * 1←37 * 1 31 34411 1←32 5 1 2 3 3 29 1 24333←35 4 4 1 1 1 25 2 45333←26 4 7 3 3 3 19 3 3 6653←25 35733 15 12 45333←28 45333 7 14 2 3577←8 15 4 5 7 7 7 8 12 9 3 3 3 7 7 9 3577←9 15 3577 7 3 59777←9 14 4 5 7 7 5 5 59777←6 7 14 5 7 7
6	35 * 1←38 4 1 1 1 33 4 4 1 1 1←36 1 2 3 3 29 8 4 1 1 1←30 12 1 1 1 27 3 6 2 3 3←28 8 3 3 3 23 35733←25 6653 21 3 6653←27 5 7 3 3 13 24 4 1 1 1←14 28 1 1 1 11 11 3577←12 14 5 7 7 7 14 4 5 7 7	31 5 1 2 3 3←32 6 2 3 3 30 24333←37 1 2 3 3 27 45333←28 5 7 3 3 25 4 7 3 3 3←26 6653 21 2 3577←29 8 3 3 3 15 12 9 3 3 3←16 13 11 3 3 14 9 3577←24 3577 12 11 3577 9 18 9 3 3 3←10 19 11 3 3 5 7 14 5 7 7←6 11 15 7 7
5	37 4 1 1 1←38 5 1 1 35 1 2 3 3←37 2 3 3 29 12 1 1 1←30 13 1 1 26 9 3 3 3←28 11 3 3 22 3577←26 5 7 7 11 14 5 7 7 9 18 11 3 3←10 19 13 3	31 6 2 3 3←38 2 3 3 27 9 3 3 3←28 10 5 3 23 3577←27 5 7 7 22 4 5 7 7←32 6 5 3 15 13 11 3 3←16 14 13 3 13 13 5 7 7 9 15 7 7 7
4	41 1 1 1←42 2 1 37 5 1 1←38 6 1 35 3 3 3←37 5 3 29 13 1 1←30 14 1 27 11 3 3←29 13 3 23 7 7 7←27 11 7 13 13 11 7	42 1 1 1←44 1 1 36 3 3 3←40 3 3 31 6 5 3←32 7 7 27 10 5 3←28 11 7 24 7 7 7←38 5 3 15 14 13 3←16 15 15 14 13 11 7
3	42 1 1←44 1 41 2 1←42 3 37 6 1←38 7 29 14 1←30 15	43 1 1←45 1 39 3 3←43 3 31 7 7←39 7 15 15 15
2	43 1←45	
1		

Stem 44 (filtrations ≦ 7) Stem 45 (filtrations ≦ 7)

	27 * * 1←30 4 1 1 * 1	
	25 4 4 1 1 * 1←28 1 2 34411 1	23 5 1 2 34411 1←24 6 2 34411 1
	21 8 4 1 1 * 1←22 12 1 1 * 1	22 * 24333←29 1 2 34411 1
	19 3 6 2 34411 1←20 8 1 1 24333	19 2..2 45333←20 4 2 2 45333
	15 2..2 35733←16 3 6 2 45333	17 4 2 2 2 45333←18 6 2 2 45333
	13 16 4 1 1 * 1←14 20 1 1 * 1	15 5 8 1 1 24333←16 7 13 1 * 1
	13 3 6 2 2 45333←14 5 6 2 45333	13 2 2 2 3 3 6653←14 3 6 2 35733
	12 2 2 2 3 3 6653←17 6 2 2 45333	11 17 1 2 34411 1←12 18 2 34411 1
	9 18 1 2 34411 1←10 19 2 34411 1	11 3 5 6 2 45333←12 5 6 2 35733
	9 24333 6653←10 3 6 3 3 6653	10 24333 6653←21 8 1 1 24333
	7 3 5 6 2 35733←8 5 6 3 3 6653	7 13 8 1 1 24333←8 15 13 1 * 1
	6 2 3 5 5 3 6653←8 4 7 3 3 6653	7 2 3 5 5 3 6653←8 3 6 5 2 3577
	5 5 5 6 2 35733←6 7 6 3 3 6653	3 5 4 7 3 3 6653←4 7 6 5 2 3577
	2 4 5 5 5 3 6653	3 4 5 5 5 3 6653←6 5 6 5 2 3577
10	1 2 4 3 5 7 3577←2 3 6 5 7 3577	1 2 3 3 5 9 3577←4 5 3 5 7 3577
	29 4 1 1 * 1←30 5 1 * 1	
	27 1 2 34411 1←29 2 34411 1	
	21 12 1 1 * 1←22 13 1 * 1	
	15 3 6 2 45333←16 6 2 35733	
	13 20 1 1 * 1←14 21 1 * 1	23 6 2 34411 1←30 2 34411 1
	13 5 6 2 45333←19 6 2 45333	19 4 2 2 45333←20 6 2 45333
	11 22 1 1 * 1←12 23 1 * 1	15 7 13 1 * 1←16 10 2 45333
	9 18 1 1 24333←10 19 1 24333	13 3 6 2 35733←14 6 3 3 6653
	9 3 6 3 3 6653←10 6 5 2 3577	11 18 2 34411 1←12 20 34411 1
	7 5 6 3 3 6653←8 7 12 45333	11 5 6 2 35733←23 13 1 * 1
	7 4 7 3 3 6653←12 5 5 3 6653	7 9 6 2 35733←8 12 3 3 6653
	5 7 6 3 3 6653←8 11 3 3 6653	7 3 6 5 2 3577←8 6 9 3 6653
	3 5 10 3 3 6653←4 7 13 35733	5 5 6 5 2 3577
	3 4 3 5 7 3577←4 7 5 7 3577	3 5 3 5 7 3577←4 6 5 9 3577
	2 3 3 5 9 3577←8 3 5 7 3577	2 4 3 5 9 3577←5 7 5 7 3577
9	1 2 3 3 59777←2 3 5 59777	1 2 4 3 59777←2 3 6 59777
	33 1 1 * 1←34 2 * 1	34 1 1 * 1←36 1 * 1
	29 5 1 * 1←30 6 * 1	28 1 1 24333←32 34411 1
	27 1 1 24333←28 2 24333	15 10 2 45333←16 12 45333
	15 6 2 35733←22 2 35733	13 6 3 3 6653←26 2 45333
	13 21 1 * 1←14 22 * 1	11 20 34411 1←12 23 4 4 1 1 1
	11 5 5 3 6653←13 9 3 6653	7 12 3 3 6653←8 15 3 6653
	7 11 3 3 6653←8 13 2 3577	7 6 9 3 6653←8 8 12 9 3 3 3
	7 7 12 45333←9 13 3 6653	6 3 5 9 3577←10 5 9 3577
	7 3 5 7 3577←9 5 9 3577	3 6 5 9 3577←8 7 9 3577
	5 3 5 9 3577←11 5 7 3577	3 5 6 9 3577←4 7 13 3577
8	2 4 3 59777←4 6 59777	3 4 3 59777←4 7 59777

Stem 44 (filtrations 8 to 10) Stem 45 (filtrations 8 to 10)

15	18 1 * * * 1←20 * * * 1 17 2 * * * 1←18 4 4 1 1 * * 1 13 6 * * * 1←14 8 4 1 1 * * 1 11 2 * * 24333←12 3 6 2 34411 * 1 5 2.......2 35733←6 3 6 2....2 45333 3 4 2......2 35733←4 5 6 2....2 45333 2.......2 3 3 6653←8 2......2 35733	19 1 * * * 1←21 * * * 1 15 34411 * * 1←16 5 1 2 34411 * 1 13 1 * * 24333←19 4 4 1 1 * * 1 9 2.......2 45333←10 4 2.....2 45333 7 4 2......2 45333←8 5 8 1 1 * 24333 6 2.......2 35733 5 6 2......2 45333←6 7 8 1 1 * 24333 3 2......2 3 3 6653←4 3 5 6 2...2 45333
14	19 * * * 1←22 4 1 1 * * 1 17 4 4 1 1 * * 1←20 1 2 34411 * 1 13 8 4 1 1 * * 1←14 12 1 1 * * 1 11 3 6 2 34411 * 1←12 8 1 1 * 24333 7 2......2 35733←8 3 6 2...2 45333 5 3 6 2....2 45333←6 5 6 2...2 45333 4 2.....2 3 3 6653←9 6 2....2 45333 3 5 6 2....2 45333←4 7 6 2...2 45333 1 2...2 4 3 3 3 6653←2 3 6 2..2 3 3 6653	15 5 1 2 34411 * 1←16 6 2 34411 * 1 14 * * 24333←21 1 2 34411 * 1 11 2......2 45333←12 4 2....2 45333 9 4 2.....2 45333←10 6 2....2 45333 7 5 8 1 1 * 24333←8 8 2....2 45333 5 7 8 1 1 * 24333←6 10 2....2 45333 5 2.....2 3 3 6653←6 3 6 2...2 35733 3 3 5 6 2...2 45333←4 5 6 2...2 35733 2....2 4 3 3 3 6653←13 8 1 1 * 24333
13	21 4 1 1 * * 1←22 5 1 * * 1 19 1 2 34411 * 1←21 2 34411 * 1 13 12 1 1 * * 1←14 13 1 * * 1 7 12 1 1 * 24333←8 13 1 * 24333 7 3 6 2...2 45333←8 6 2...2 35733 5 5 6 2...2 45333←11 6 2...2 45333 3 2..2 4 3 3 3 6653←4 4 2 24333 6653 1 2 1 1 24733 6653←2 3 1 24733 6653	15 6 2 34411 * 1←22 2 34411 * 1 11 4 2....2 45333←12 6 2...2 45333 7 8 2....2 45333←8 10 2...2 45333 5 10 2....2 45333←6 12 2...2 45333 5 3 6 2...2 35733←6 6 2..2 3 3 6653 3 5 6 2...2 35733←16 2....2 45333 1 2 1 2 24733 6653←2 3 2 24733 6653
12	25 1 1 * * 1←26 2 * * 1 21 5 1 * * 1←22 6 * * 1 19 1 1 * 24333←20 2 * 24333 13 13 1 * * 1←14 14 * * 1 7 13 1 * 24333←8 14 * 24333 7 6 2...2 35733←14 2...2 35733 3 4 2 24333 6653←4 6 24333 6653 3 1 1 24733 6653←4 2 24733 6653 2 1 2 24733 6653←5 1 24733 6653	26 1 1 * * 1←28 1 * * 1 20 1 1 * 24333←24 34411 * 1 15 2....2 45333←16 4 2..2 45333 9 18 1 * * 1←10 19 * * 1 7 14 1 * 24333←8 15 * 24333 7 10 2...2 45333←8 12 2..2 45333 5 6 2..2 3 3 6653←18 2...2 45333 4 1 1 24733 6653 3 1 2 24733 6653←4 4 2 3 5 5 3 6653
11	26 1 * * 1←28 * * 1 25 2 * * 1←26 4 4 1 1 * 1 21 6 * * 1←22 8 4 1 1 * 1 19 2 * 24333←20 3 6 2 34411 1 13 14 * * 1←14 16 4 1 1 * 1 13 2...2 35733←14 3 6 2 2 45333 10 2..2 3 3 6653←16 2..2 35733 7 14 * 24333←8 15 6 2 34411 1 7 2 24333 6653←8 3 5 6 2 35733 3 6 24333 6653←4 7 5 6 2 35733 3 2 24733 6653←6 24733 6653 1 2 3 3 6 5 2 3577←2 3 5 6 5 2 3577	27 1 * * 1←29 * * 1 23 34411 * 1←24 5 1 2 34411 1 21 1 * 24333←27 4 4 1 1 * 1 17 2...2 45333←18 4 2 2 2 45333 15 4 2..2 45333←16 5 8 1 1 24333 11 2..2 3 3 6653←12 3 5 6 2 45333 9 17 4 4 1 1 * 1←10 19 1 2 34411 1 8 2 24333 6653←20 2..2 45333 7 12 2..2 45333←8 13 8 1 1 24333 3 4 2 3 5 5 3 6653←4 5 4 7 3 3 6653 1 2 4 5 5 5 3 6653←4 4 5 5 5 3 6653
	Stem 44 (filtrations 11 to 15)	Stem 45 (filtrations 11 to 15)

24	1 1 1 * * * * * 1←2 2 * * * * * 1	2 1 1 * * * * * 1←4 1 * * * * * 1 1 2 1 * * * * * 1←2 3 * * * * * 1
23	2 1 * * * * * 1←4 * * * * * 1 1 2 * * * * * 1←2 4 4 1 1 * * * * 1	3 1 * * * * * 1←5 * * * * * 1 1 2 1 1 2 34411 * * * 1←2 3 1 2 34411 * * * 1
22	3 * * * * * 1←6 4 1 1 * * * * 1 2 1 1 2 34411 * * * 1←4 1 2 34411 * * * 1 1 2 1 2 34411 * * * 1←2 3 2 34411 * * * 1	3 1 1 2 34411 * * * 1←4 2 2 34411 * * * 1
21	5 4 1 1 * * * * 1←6 5 1 * * * * 1 3 6 1 1 * * * * 1←4 7 1 * * * * 1 3 1 2 34411 * * * 1←5 2 34411 * * * 1 1 2 1 1 * * * 24333←2 3 1 * * * 24333	3 2 2 34411 * * * 1←4 4 34411 * * * 1
20	9 1 1 * * * * 1←10 2 * * * * 1 5 5 1 * * * * 1←6 6 * * * * 1 3 1 1 * * * 24333←4 2 * * * 24333	10 1 1 * * * * 1←12 1 * * * * 1 4 1 1 * * * 24333←8 34411 * * * 1 3 4 34411 * * * 1←4 7 4 4 1 1 * * * 1
19	10 1 * * * * 1←12 * * * * 1 9 2 * * * * 1←10 4 4 1 1 * * * 1 5 6 * * * * 1←6 8 4 1 1 * * * 1 3 2 * * * 24333←4 3 6 2 34411 * * 1	11 1 * * * * 1←13 * * * * 1 7 34411 * * * 1←8 5 1 2 34411 * * 1 5 1 * * * 24333←11 4 4 1 1 * * * 1 1 2..........2 45333←2 4 2.........2 45333
18	11 * * * * 1←14 4 1 1 * * * 1 9 4 4 1 1 * * * 1←12 1 2 34411 * * 1 5 8 4 1 1 * * * 1←6 12 1 1 * * * 1 3 3 6 2 34411 * * 1←4 8 1 1 * * 24333 2..........2 45333	7 5 1 2 34411 * * 1←8 6 2 34411 * * 1 6 * * * 24333←13 1 2 34411 * * 1 3 2.........2 45333←4 4 2........2 45333
17	13 4 1 1 * * * 1←14 5 1 * * * 1 11 1 2 34411 * * 1←13 2 34411 * * 1 4 2.........2 45333 1 2.........2 35733←2 4 2.......2 35733	7 6 2 34411 * * 1←14 2 34411 * * 1 3 4 2........2 45333←4 6 2.......2 45333 2..........2 35733
16	17 1 1 * * * 1←18 2 * * * 1 13 5 1 * * * 1←14 6 * * * 1 11 1 1 * * 24333←12 2 * * 24333 6 2........2 45333 3 2........2 35733←4 4 2......2 35733	18 1 1 * * * 1←20 1 * * * 1 12 1 1 * * 24333←16 34411 * * 1 3 6 2.......2 45333 1 2.......2 3 3 6653←2 4 2.....2 3 3 6653

Stem 44 (filtrations ≧ 16)   Stem 45 (filtrations ≧ 16)

	Stem 46	Stem 47
7	35 2 * 1←36 4 4 1 1 1 31 6 * 1←32 8 4 1 1 1 30 1 24333←32 24333 29 2 24333←30 3 6 2 3 3 23 2 35733←24 3 6653 20 3 3 6653←26 35733 15 22 * 1←16 24 4 1 1 1 14 9 3 6653←38 * 1 12 5 7 3577←16 9 3577 9 13 2 3577←10 14 4 5 7 7 8 3 59777 6 5 59777←8 7 13 5 7 7 5 9 15 8 3 3 3←6 11 21 3 3 3	33 34411 1←34 5 1 2 3 3 31 1 24333←33 24333 27 2 45333←28 4 7 3 3 3 21 3 3 6653←24 2 3577 15 9 3 6653←18 12 9 3 3 3 13 23 4 4 1 1 1←14 25 1 2 3 3 13 5 7 3577←17 9 3577 11 5 9 3577←25 3 6653 9 15 3 6653←10 20 9 3 3 3 9 3 59777 7 5 59777←8 7 14 5 7 7
6	31 8 4 1 1 1←32 12 1 1 1 31 24333←33 6 2 3 3 29 3 6 2 3 3←30 8 3 3 3 23 3 6653←27 6653 22 2 3577←24 4 5 7 7 16 12 9 3 3 3←40 4 1 1 1 15 24 4 1 1 1←16 28 1 1 1 15 9 3577←25 3577 13 11 3577←14 14 5 7 7 11 59777 7 7 13 5 7 7←9 17 7 7 7 5 9 11 7 7 7←6 10 13 11 7	33 5 1 2 3 3←34 6 2 3 3 29 45333←30 5 7 3 3 27 4 7 3 3 3←28 6653 23 2 3577←25 4 5 7 7 17 12 9 3 3 3←18 13 11 3 3 14 11 3577←16 13 5 7 7 13 25 1 2 3 3←14 26 2 3 3 12 59777←26 3577 7 7 14 5 7 7←8 11 15 7 7 6 9 11 7 7 7 5 9 14 5 7 7←6 11 13 11 7
5	39 4 1 1 1←40 5 1 1 31 12 1 1 1←32 13 1 1 28 9 3 3 3←44 1 1 1 23 4 5 7 7←26 7 7 7 15 28 1 1 1←16 29 1 1 14 13 5 7 7←28 5 7 7 13 14 5 7 7←16 13 11 7 5 10 13 11 7←6 11 15 15	38 1 2 3 3←40 2 3 3 29 9 3 3 3←30 10 5 3 29 5 7 3 3←34 6 5 3 17 13 11 3 3←18 14 13 3 15 13 5 7 7←29 5 7 7 13 26 2 3 3←14 28 3 3 13 25 3 3 3←14 26 5 3 7 11 15 7 7
4	43 1 1 1←44 2 1 39 5 1 1←40 6 1 37 3 3 3←41 3 3 31 13 1 1←32 14 1 29 11 3 3←45 1 1 25 7 7 7←33 7 7 15 29 1 1←16 30 1 15 13 11 7←17 15 15	39 2 3 3←42 3 3 38 3 3 3←40 5 3 33 6 5 3←34 7 7 30 11 3 3←32 13 3 29 10 5 3←30 11 7 17 14 13 3←18 15 15 15 30 1 1←16 31 1 14 27 3 3←16 29 3 13 26 5 3←14 27 7
3	43 2 1←44 3 39 6 1←40 7 31 14 1←32 15 30 13 3←46 1 15 30 1←16 31	39 5 3←41 7 31 13 3←33 15 29 11 7←45 3 15 29 3←17 31
2	31 15←47	

Stem 46 (filtrations ≤ 7)   Stem 47 (filtrations ≤ 7)

11	27 2 * * 1←28 4 4 1 1 * 1 23 6 * * 1←24 8 4 1 1 * 1 22 1 * 24333←24 * 24333 21 2 * 24333←22 3 6 2 34411 1 15 14 * * 1←16 16 4 1 1 * 1 15 2...2 35733←16 3 6 2 2 45333 12 2..2 3 3 6653←18 2..2 35733 9 14 * 24333←10 15 6 2 34411 1 9 2 24333 6653←10 3 5 6 2 35733 6 1 24733 6653←8 24733 6653 5 6 24333 6653←6 7 5 6 2 35733 5 2 24733 6653←30 * * 1 2 2 4 5 5 5 3 6653 1 1 2 3 3 5 9 3577←2 2 4 3 5 9 3577	25 34411 * 1←26 5 1 2 34411 1 23 1 * 24333←25 * 24333 19 2...2 45333←20 4 2 2 2 45333 17 4 2..2 45333←18 5 8 1 1 24333 13 2..2 3 3 6653←14 3 5 6 2 45333 10 2 24333 6653←16 2 2 2 3 3 6653 9 12 2..2 45333←10 13 8 1 1 24333 7 14 2..2 45333←8 15 8 1 1 24333 7 1 24733 6653←9 24733 6653 3 6 2 3 5 5 3 6653←4 7 4 7 3 3 6653 3 2 4 5 5 5 3 6653←6 4 5 5 5 3 6653 1 2 2 3 3 5 9 3577←2 3 4 3 5 9 3577
10	23 8 4 1 1 * 1←24 12 1 1 * 1 23 * 24333←25 6 2 34411 1 21 3 6 2 34411 1←22 8 1 1 24333 17 2..2 35733←18 3 6 2 45333 15 16 4 1 1 * 1←16 20 1 1 * 1 15 3 6 2 2 45333←16 5 6 2 45333 14 2 2 2 3 3 6653←19 6 2 2 45333 11 24333 6653←12 3 6 3 3 6653 9 15 6 2 34411 1←10 20 1 1 24333 9 3 5 6 2 35733←10 5 6 3 3 6653 8 2 3 5 5 3 6653←32 4 1 1 * 1 7 24733 6653←10 4 7 3 3 6653 5 7 5 6 2 35733←6 9 6 3 3 6653 3 3 5 6 5 2 3577←5 7 6 5 2 3577 2 2 3 3 5 9 3577←4 4 3 5 9 3577 1 2 4 3 5 9 3577←2 3 6 5 9 3577	25 5 1 2 34411 1←26 6 2 34411 1 21 2..2 45333←22 4 2 2 45333 19 4 2 2 2 45333←20 6 2 2 45333 17 5 8 1 1 24333←18 7 13 1 * 1 15 2 2 2 3 3 6653←16 3 6 2 35733 13 3 5 6 2 45333←14 5 6 2 35733 12 24333 6653←17 5 6 2 45333 9 13 8 1 1 24333←10 15 13 1 * 1 9 2 3 5 5 3 6653←10 3 6 5 2 3577 5 5 4 7 3 3 6653←6 7 6 5 2 3577 5 4 5 5 5 3 6653←8 5 6 5 2 3577 4 3 5 6 5 2 3577 3 5 7 6 3 3 6653←4 7 7 12 45333 3 2 3 3 5 9 3577←4 7 3 5 7 3577
9	31 4 1 1 * 1←32 5 1 * 1 23 12 1 1 * 1←24 13 1 * 1 17 3 6 2 45333←18 6 2 35733 15 20 1 1 * 1←16 21 1 * 1 15 5 6 2 45333←21 6 2 45333 11 3 6 3 3 6653←12 6 5 2 3577 9 5 6 3 3 6653←10 7 12 45333 9 4 7 3 3 6653←36 1 1 * 1 5 9 6 3 3 6653←6 11 12 45333 3 6 11 3 3 6653←4 7 13 3 6653 3 6 3 5 7 3577←4 7 5 9 3577 3 4 3 5 9 3577←5 6 5 9 3577	30 1 2 34411 1←32 2 34411 1 21 4 2 2 45333←22 6 2 45333 17 7 13 1 * 1←18 10 2 45333 15 3 6 2 35733←16 6 3 3 6653 13 5 6 2 35733←19 6 2 35733 9 15 13 1 * 1←10 18 2 45333 9 3 6 5 2 3577←10 6 9 3 6653 7 5 6 5 2 3577←13 6 5 2 3577 5 9 5 5 3 6653←6 10 9 3 6653 3 5 3 5 9 3577←5 7 5 9 3577
8	35 1 1 * 1←36 2 * 1 31 5 1 * 1←32 6 * 1 29 1 1 24333←30 2 24333 17 6 2 35733←24 2 35733 15 21 1 * 1←16 22 * 1 13 5 5 3 6653←17 12 45333 11 6 5 2 3577←37 1 * 1 9 7 12 45333←10 13 2 3577 9 3 5 7 3577 7 13 3 3 6653←8 15 2 3577 7 3 5 9 3577←9 8 12 9 3 3 3 3 6 6 9 3577←4 7 14 3577 3 5 3 59777←5 7 59777	31 2 34411 1←34 34411 1 30 1 1 24333←32 1 24333 17 10 2 45333←18 12 45333 15 6 3 3 6653←22 3 3 6653 14 5 5 3 6653←16 9 3 6653 10 3 5 7 3577←12 5 9 3577 9 18 2 45333←10 20 45333 9 6 9 3 6653←10 8 12 9 3 3 3 8 3 5 9 3577 5 10 9 3 6653←6 12 12 9 3 3 3 4 5 3 59777 3 6 3 59777←4 7 14 4 5 7 7
	Stem 46 (filtrations 8 to 11)	Stem 47 (filtrations 8 to 11)

15	19 2 * * * 1←20 4 4 1 1 * * 1 15 6 * * * 1←16 8 4 1 1 * * 1 14 1 * * 24333←16 * * 24333 13 2 * * 24333←14 3 6 2 34411 * 1 10 2.......2 45333 7 2.......2 35733←8 3 6 2....2 45333 4 2......2 3 3 6653←10 2......2 35733 3 6 2......2 35733←4 7 6 2....2 45333 1 2....2 4 3 3 3 6653←2 3 5 6 2...2 35733	21 1 * * * 1 17 34411 * * 1←18 5 1 2 34411 * 1 15 1 * * 24333←17 * * 24333 11 2.......2 45333←12 4 2.....2 45333 7 13 4 4 1 1 * * 1←8 15 1 2 34411 * 1 5 2......2 3 3 6653←6 3 5 6 2...2 45333 3 4 2.....2 3 3 6653←4 5 5 6 2...2 45333 2.....2 4 3 3 3 6653←8 2.....2 3 3 6653
14	15 8 4 1 1 * * 1←16 12 1 1 * * 1 15 * * 24333←17 6 2 34411 * 1 13 3 6 2 34411 * 1←14 8 1 1 * 24333 12 2......2 45333 9 2......2 35733←10 3 6 2...2 45333 7 14 1 2 34411 * 1←8 15 2 34411 * 1 7 3 6 2....2 45333←8 5 6 2...2 45333 6 2.....2 3 3 6653←11 6 2....2 45333 5 5 6 2....2 45333←6 7 6 2...2 45333 3 2...2 4 3 3 3 6653←4 3 6 2..2 3 3 6653	22 * * * 1 17 5 1 2 34411 * 1←18 6 2 34411 * 1 13 2......2 45333←14 4 2....2 45333 11 4 2.....2 45333←12 6 2....2 45333 9 5 8 1 1 * 24333←10 8 2....2 45333 7 7 8 1 1 * 24333←8 10 2....2 45333 7 2.....2 3 3 6653←8 3 6 2...2 35733 5 3 5 6 2...2 45333←6 5 6 2...2 35733 4 2...2 4 3 3 3 6653←9 5 6 2...2 45333 3 5 5 6 2...2 45333←4 7 6 2...2 35733 1 2 2 1 2 24733 6653←2 3 6 2 24333 6653
13	23 4 1 1 * * 1←24 5 1 * * 1 15 12 1 1 * * 1←16 13 1 * * 1 9 18 1 1 * * 1←10 19 1 * * 1 9 3 6 2...2 45333←10 6 2...2 35733 7 14 1 1 * 24333←8 15 1 * 24333 7 5 6 2...2 45333←13 6 2...2 45333 5 7 6 2...2 45333←6 10 2...2 35733 3 3 6 2..2 3 3 6653←4 6 2 24333 6653 2 2 1 2 24733 6653←28 1 1 * * 1	24 4 1 1 * * 1 22 1 2 34411 * 1←24 2 34411 * 1 13 4 2....2 45333←14 6 2...2 45333 9 8 2....2 45333←10 10 2...2 45333 7 10 2....2 45333←8 12 2...2 45333 7 3 6 2...2 35733←8 6 2..2 3 3 6653 5 5 6 2...2 35733←11 6 2...2 35733 2 4 1 1 24733 6653
12	27 1 1 * * 1←28 2 * * 1 23 5 1 * * 1←24 6 * * 1 21 1 1 * 24333←22 2 * 24333 15 13 1 * * 1←16 14 * * 1 9 13 1 * 24333←10 14 * 24333 9 6 2...2 35733←16 2...2 35733 5 10 2...2 35733←6 12 2..2 35733 5 1 1 24733 6653←6 2 24733 6653 3 6 2 24333 6653←29 1 * * 1 1 1 2 4 5 5 5 3 6653←2 3 4 5 5 5 3 6653	23 2 34411 * 1←26 34411 * 1 22 1 1 * 24333←24 1 * 24333 17 2....2 45333←18 4 2..2 45333 9 10 2...2 45333←10 12 2..2 45333 7 12 2...2 45333←8 14 2..2 45333 7 6 2..2 3 3 6653←14 2..2 3 3 6653 6 1 1 24733 6653←8 1 24733 6653 1 2 2 4 5 5 5 3 6653←4 2 4 5 5 5 3 6653

Stem 46 (filtrations 12 to 15) Stem 47 (filtrations 12 to 15)

25	1 2 1 1 * * * * * 1←2 3 1 * * * * * 1	
24	3 1 1 * * * * * 1←4 2 * * * * * 1	4 1 1 * * * * * 1
23	3 2 * * * * * 1←4 4 4 1 1 * * * * 1	5 1 * * * * * 1
22	3 4 4 1 1 * * * * 1	6 * * * * * 1
21	7 4 1 1 * * * * 1←8 5 1 * * * * 1 5 1 2 34411 * * * 1	8 4 1 1 * * * * 1 6 1 2 34411 * * * 1←8 2 34411 * * * 1 2 4 1 1 * * * 24333
20	11 1 1 * * * * 1←12 2 * * * * 1 7 5 1 * * * * 1←8 6 * * * * 1 6 2 34411 * * * 1 5 1 1 * * * 24333←6 2 * * * 24333	12 1 1 * * * * 1 7 2 34411 * * * 1←10 34411 * * * 1 6 1 1 * * * 24333←8 1 * * * 24333 1 2...........2 45333←2 4 2.........2 45333
19	11 2 * * * * 1←12 4 4 1 1 * * * 1 7 6 * * * * 1←8 8 4 1 1 * * * 1 6 1 * * * 24333←8 * * * 24333 5 2 * * * 24333←6 3 6 2 34411 * * 1 2...........2 45333	13 1 * * * * 1 9 34411 * * * 1←10 5 1 2 34411 * * 1 7 1 * * * 24333←9 * * * 24333 5 7 4 4 1 1 * * * 1←6 9 1 2 34411 * * 1 3 2..........2 45333←4 4 2.........2 45333
18	7 8 4 1 1 * * * 1←8 12 1 1 * * * 1 7 * * * 24333←9 6 2 34411 * * 1 5 3 6 2 34411 * * 1←6 8 1 1 * * 24333 4 2..........2 45333 1 2.........2 35733←2 3 6 2.......2 45333	14 * * * * 1 9 5 1 2 34411 * * 1←10 6 2 34411 * * 1 5 9 1 2 34411 * * 1←6 10 2 34411 * * 1 5 2.........2 45333←6 4 2........2 45333 3 4 2........2 45333←4 6 2........2 45333 2..........2 35733
17	15 4 1 1 * * * 1←16 5 1 * * * 1 7 12 1 1 * * * 1←8 13 1 * * * 1 5 8 1 1 * * 24333 3 2.........2 35733←4 4 2.......2 35733	16 4 1 1 * * * 1 14 1 2 34411 * * 1←16 2 34411 * * 1 5 10 2 34411 * * 1←6 12 34411 * * 1 5 4 2........2 45333←6 6 2.......2 45333 3 6 2........2 45333 1 2........2 3 3 6653←2 4 2......2 3 3 6653
16	19 1 1 * * * 1←20 2 * * * 1 15 5 1 * * * 1←16 6 * * * 1 13 1 1 * * 24333←14 2 * * 24333 7 13 1 * * * 1 3 4 2.......2 35733←4 6 2......2 35733 2........2 3 3 6653←8 2.......2 35733	20 1 1 * * * 1 15 2 34411 * * 1←18 34411 * * 1 14 1 1 * * 24333←16 1 * * 24333 7 14 1 * * * 1←8 15 * * * 1 5 10 1 * * 24333←6 11 * * 24333 5 6 2.......2 45333 3 2.......2 3 3 6653←4 4 2.....2 3 3 6653

Stem 46 (filtrations ≥ 16) Stem 47 (filtrations ≥ 16)

7	38 1 * 1←40 * 1 37 2 * 1←38 4 4 1 1 1 33 6 * 1←34 8 4 1 1 1 31 2 24333←32 3 6 2 3 3 28 2 45333←34 24333 25 2 35733←26 3 6653 17 22 * 1←18 24 4 1 1 1 14 5 7 3577←18 9 3577 11 22 24333←12 23 6 2 3 3 11 13 23577←12 14 4 5 7 7 10 3 59777←16 11 3577 8 5 59777 5 10 11 3577	39 1 * 1←41 * 1 35 34411 1←36 5 1 2 3 3 29 2 45333←30 4 7 3 3 3 26 2 35733←32 45333 23 3 3 6653←27 3 6653 15 5 7 3577←19 9 3577 13 25 4 4 1 1 1←14 27 1 2 3 3 11 20 45333←12 21 8 3 3 3 11 8 12 9 3 3 3←20 12 9 3 3 3 11 3 59777←13 14 4 5 7 7 9 5 59777←10 7 14 5 7 7 7 12 12 9 3 3 3←12 20 9 3 3 3 5 7 14 4 5 7 7←6 11 14 5 7 7
6	39 * 1←42 4 1 1 1 37 4 4 1 1 1←40 1 2 3 3 33 8 4 1 1 1←34 12 1 1 1 31 3 6 2 3 3←32 8 3 3 3 30 45333←35 6 2 3 3 27 35733←29 6653 17 24 4 1 1 1←18 28 1 1 1 15 11 3577←16 14 5 7 7 13 26 1 2 3 3←14 27 2 3 3 13 59777←17 13 5 7 7 11 14 4 5 7 7←27 3577 9 7 13 5 7 7 7 9 11 7 7 7←8 10 13 11 7	35 5 1 2 3 3←36 6 2 3 3 31 45333←32 5 7 3 3 29 4 7 3 3 3←30 6653 28 35733←33 8 3 3 3 25 23577←32 9 3 3 3 19 12 9 3 3 3←20 13 11 3 3 15 25 1 2 3 3←16 26 2 3 3 14 59777←18 13 5 7 7 11 21 8 3 3 3←12 24 6 5 3 11 20 9 3 3 3←12 21 11 3 3 10 7 13 5 7 7←12 17 7 7 7 9 18 3577←16 25 3 3 3 9 7 14 5 7 7←10 11 15 7 7 8 9 11 7 7 7
5	41 4 1 1 1←42 5 1 1 39 1 2 3 3←41 2 3 3 33 12 1 1 1←34 13 1 1 31 8 3 3 3←40 3 3 3 30 9 3 3 3←32 11 3 3 17 28 1 1 1←18 29 1 1 15 30 1 1 1←16 31 1 1 15 14 5 7 7←30 5 7 7 14 25 3 3 3←16 27 3 3 13 26 3 3 3←14 27 5 3 10 17 7 7 7 7 10 13 11 7←8 11 15 15	31 9 3 3 3←32 10 5 3 31 5 7 3 3←36 6 5 3 26 4 5 7 7←33 11 3 3 19 13 11 3 3←20 14 13 3 15 26 2 3 3←16 28 3 3 15 25 3 3 3←16 26 5 3 11 21 11 3 3←12 22 13 3 11 17 7 7 7←41 3 3 3 10 20 5 7 7←17 27 3 3 9 11 15 7 7
4	45 1 1 1←46 2 1 41 5 1 1←42 6 1 39 3 3 3←41 5 3 33 13 1 1←34 14 1 31 11 3 3←33 13 3 27 7 7 7←35 7 7 17 29 1 1←18 30 1 17 13 11 7←43 3 3 15 27 3 3←17 29 3 7 11 15 15	46 1 1 1←48 1 1 35 6 5 3←36 7 7 31 10 5 3←32 11 7 28 7 7 7←34 13 3 19 14 13 3←20 15 15 18 13 11 7←42 5 3 15 28 3 3←16 31 3 15 26 5 3←16 27 7 12 23 7 7←18 29 3 11 22 13 3←12 23 15
3	46 1 1←48 1 45 2 1←46 3 41 6 1←42 7 33 14 1←34 15 17 30 1←18 31	47 1 1←49 1 31 11 7←35 15 19 15 15←43 7 15 27 7←19 31
2	47 1←49	
1		

Stem 48 (filtrations ≦ 7) Stem 49 (filtrations ≦ 7)

10	31 * * 1←34 4 1 1 * 1 29 4 4 1 1 * 1←32 1 2 34411 1 25 8 4 1 1 * 1←26 12 1 1 * 1 23 3 6 2 34411 1←24 8 1 1 24333 22 2..2 45333←27 6 2 34411 1 19 2..2 35733←20 3 6 2 45333 17 16 4 1 1 * 1←18 20 1 1 * 1 17 3 6 2 2 45333←18 5 6 2 45333 13 24333 6653←14 3 6 3 3 6653 11 3 5 6 2 35733←12 5 6 3 3 6653 10 2 3 5 5 3 6653←12 4 7 3 3 6653 7 13 6 2 2 45333←8 15 6 2 45333 3 5 5 6 5 2 3577←9 5 6 5 2 3577	27 5 1 2 34411 1←28 6 2 34411 1 23 2..2 45333←24 4 2 2 45333 21 4 2 2 2 45333←22 6 2 2 45333 20 2..2 35733←25 8 1 1 24333 19 5 8 1 1 24333←20 7 13 1 * 1 17 2 2 2 3 3 6653←18 3 6 2 35733 15 3 5 6 2 45333←16 5 6 2 35733 14 24333 6653←19 5 6 2 45333 11 13 8 1 1 24333←12 15 13 1 * 1 11 2 3 5 5 3 6653←12 3 6 5 2 3577 9 15 8 1 1 24333←10 17 13 1 * 1 7 4 5 5 5 3 6653←10 5 6 5 2 3577 5 7 4 7 3 3 6653←6 11 5 5 3 6653 2 3 5 3 5 9 3577←8 9 5 5 3 6653 1 2 3 5 3 59777←2 4 6 3 59777
9	33 4 1 1 * 1←34 5 1 * 1 31 1 2 34411 1←33 2 34411 1 25 12 1 1 * 1←26 13 1 * 1 23 8 1 1 24333←32 1 1 24333 19 3 6 2 45333←20 6 2 35733 17 20 1 1 * 1←18 21 1 * 1 13 3 6 3 3 6653←14 6 5 2 3577 11 20 1 1 24333←12 21 1 24333 11 5 6 3 3 6653←12 7 12 45333 11 4 7 3 3 6653←16 5 5 3 6653 7 9 6 3 3 6653←8 11 12 45333 6 9 5 5 3 6653←8 15 3 3 6653 5 10 5 5 3 6653←6 11 9 3 6653 4 5 3 5 9 3577 3 6 3 5 9 3577←4 7 8 12 9 3 3 3 3 3 6 5 9 3577←4 7 7 9 3577 2 3 5 3 59777	23 4 2 2 45333←24 6 2 45333 21 6 2 2 45333←27 13 1 * 1 19 7 13 1 * 1←20 10 2 45333 17 3 6 2 35733←18 6 3 3 6653 15 5 6 2 35733←21 6 2 35733 11 15 13 1 * 1←12 18 2 45333 11 3 6 5 2 3577←12 6 9 3 6653 9 17 13 1 * 1←10 20 2 45333 7 9 5 5 3 6653←8 10 9 3 6653 5 5 3 5 9 3577←17 5 5 3 6653 4 6 3 5 9 3577←9 11 12 45333 2 4 5 3 59777←4 8 3 59777
8	37 1 1 * 1←38 2 * 1 33 5 1 * 1←34 6 * 1 31 1 1 24333←32 2 24333 25 13 1 * 1←33 1 24333 17 21 1 * 1←18 22 * 1 15 5 5 3 6653←17 9 3 6653 11 21 1 24333←12 22 24333 11 7 12 45333←12 13 2 3577 11 3 5 7 3577←13 5 9 3577 9 3 5 9 3577←19 12 45333 7 11 12 45333←9 17 3 6653 4 6 3 59777	38 1 1 * 1←40 1 * 1 23 6 2 45333←30 2 45333 19 10 2 45333←20 12 45333 17 6 3 3 6653←24 3 3 6653 13 26 1 * 1←14 27 * 1 12 3 5 7 3577←16 5 7 3577 11 22 1 24333←12 23 24333 11 18 2 45333←12 20 45333 11 6 9 3 6653←12 8 12 9 3 3 3 10 3 5 9 3577←18 9 3 6653 7 10 9 3 6653←8 12 12 9 3 3 3 6 7 5 9 3577←10 17 3 6653 4 7 3 59777←6 7 14 4 5 7 7 3 6 5 59777←4 7 7 13 5 7 7

Stem 48 (filtrations 8 to 10)   Stem 49 (filtrations 8 to 10)

15	22 1 * * * 1←24 * * * 1 21 2 * * * 1←22 4 4 1 1 * * 1 17 6 * * * 1←18 8 4 1 1 * * 1 15 2 * * 24333←16 3 6 2 34411 * 1 12 2.......2 45333←18 * * 24333 9 2.......2 35733←10 3 6 2....2 45333 6 2......2 3 3 6653 5 6 2......2 35733←6 7 6 2....2 45333 3 2....2 4 3 3 3 6653←4 3 5 6 2...2 35733	23 1 * * * 1←25 * * * 1 19 34411 * * 1←20 5 1 2 34411 * 1 13 2.......2 45333←14 4 2.....2 45333 10 2.......2 35733←16 2.......2 45333 7 2......2 3 3 6653←8 3 5 6 2...2 45333 4 2....2 4 3 3 3 6653←10 2.....2 3 3 6653 3 6 2.....2 3 3 6653←4 7 5 6 2...2 45333 1 2 2 2 1 2 24733 6653←2 3 4 1 1 24733 6653 1 1 * 24733 6653
14	23 * * * 1←26 4 1 1 * * 1 21 4 4 1 1 * * 1←24 1 2 34411 * 1 17 8 4 1 1 * * 1←18 12 1 1 * * 1 15 3 6 2 34411 * 1←16 8 1 1 * 24333 14 2......2 45333←19 6 2 34411 * 1 11 2......2 35733←12 3 6 2...2 45333 9 3 6 2....2 45333←10 5 6 2...2 45333 5 7 6 2....2 45333←6 9 6 2...2 45333 5 2...2 4 3 3 3 6653←6 3 6 2..2 3 3 6653 3 3 5 6 2...2 35733←4 5 6 2..2 3 3 6653 2 2 2 1 2 24733 6653←4 4 1 1 24733 6653 1 * 24733 6653	19 5 1 2 34411 * 1←20 6 2 34411 * 1 15 2......2 45333←16 4 2....2 45333 13 4 2.....2 45333←14 6 2....2 45333 12 2......2 35733←17 8 1 1 * 24333 9 2.....2 3 3 6653←10 3 6 2...2 35733 7 3 5 6 2...2 45333←8 5 6 2...2 35733 6 2...2 4 3 3 3 6653←11 5 6 2...2 45333 5 5 5 6 2...2 45333←6 7 6 2...2 35733 3 2 2 1 2 24733 6653←4 3 6 2 24333 6653 2 * 24733 6653
13	25 4 1 1 * * 1←26 5 1 * * 1 23 1 2 34411 * 1←25 2 34411 * 1 17 12 1 1 * * 1←18 13 1 * * 1 15 8 1 1 * 24333←24 1 1 * 24333 11 3 6 2...2 45333←12 6 2...2 35733 5 9 6 2...2 45333←6 12 2...2 35733 5 3 6 2..2 3 3 6653←6 6 2 24333 6653 3 5 6 2..2 3 3 6653←4 8 2 24333 6653 3 4 1 1 24733 6653←8 1 1 24733 6653 1 1 2 2 4 5 5 5 3 6653←2 3 2 4 5 5 5 3 6653	15 4 2....2 45333←16 6 2...2 45333 13 6 2....2 45333←20 2....2 45333 9 3 6 2...2 35733←10 6 2..2 3 3 6653 7 5 6 2...2 35733←13 6 2...2 35733 5 7 6 2...2 35733←6 10 2..2 3 3 6653 3 3 6 2 24333 6653←4 5 2 24733 6653 1 2 2 2 4 5 5 5 3 6653←4 2 2 4 5 5 5 3 6653
12	29 1 1 * * 1←30 2 * * 1 25 5 1 * * 1←26 6 * * 1 23 1 1 * 24333←24 2 * 24333 18 2....2 45333←25 1 * 24333 17 13 1 * * 1←18 14 * * 1 7 10 2...2 35733←8 12 2..2 35733 7 1 1 24733 6653←8 2 24733 6653 5 6 2 24333 6653←9 1 24733 6653 2 2 2 4 5 5 5 3 6653	30 1 1 * * 1←32 1 * * 1 19 2....2 45333←20 4 2..2 45333 15 6 2...2 45333←22 2...2 45333 9 6 2..2 3 3 6653←16 2..2 3 3 6653 5 10 2..2 3 3 6653←6 12 2 2 2 3 3 6653 3 6 1 24733 6653←4 7 24733 6653 3 5 2 24733 6653←4 8 2 3 5 5 3 6653 3 2 2 4 5 5 5 3 6653←6 2 4 5 5 5 3 6653 1 2 3 3 5 6 5 2 3577←2 3 5 5 6 5 2 3577
11	30 1 * * 1←32 * * 1 29 2 * * 1←30 4 4 1 1 * 1 25 6 * * 1←26 8 4 1 1 * 1 23 2 * 24333←24 3 6 2 34411 1 20 2...2 45333←26 * 24333 17 14 * * 1←18 16 4 1 1 * 1 17 2...2 35733←18 3 6 2 2 45333 11 2 24333 6653←12 3 5 6 2 35733 7 12 2..2 35733←8 13 6 2 2 45333 7 2 24733 6653←10 24733 6653 2 3 3 5 6 5 2 3577←4 5 5 6 5 2 3577	31 1 * * 1←33 * * 1 27 34411 * 1←28 5 1 2 34411 1 21 2...2 45333←22 4 2 2 2 45333 19 4 2..2 45333←20 5 8 1 1 24333 18 2...2 35733←24 2..2 45333 15 2..2 3 3 6653←16 3 5 6 2 45333 12 2 24333 6653←18 2 2 2 3 3 6653 11 12 2..2 45333←12 13 8 1 1 24333 9 14 2..2 45333←10 15 8 1 1 24333 5 2 4 5 5 5 3 6653←8 4 5 5 5 3 6653 2 4 3 5 6 5 2 3577

Stem 48 (filtrations 11 to 15) Stem 49 (filtrations 11 to 15)

25		2 4 1 1 * * * * * 1
24	5 1 1 * * * * * 1←6 2 * * * * * 1	6 1 1 * * * * * 1←8 1 * * * * * 1 1 2 34411 * * * * 1
23	6 1 * * * * * 1←8 * * * * * 1 5 2 * * * * * 1←6 4 4 1 1 * * * * 1 2 34411 * * * * 1	7 1 * * * * * 1←9 * * * * * 1 3 34411 * * * * 1←4 5 1 2 34411 * * * 1 1 1 * * * * 24333
22	7 * * * * * 1←10 4 1 1 * * * * 1 5 4 4 1 1 * * * * 1←8 1 2 34411 * * * 1 1 * * * * 24333	3 5 1 2 34411 * * * 1←4 6 2 34411 * * * 1 2 * * * * 24333
21	9 4 1 1 * * * * 1←10 5 1 * * * * 1 7 1 2 34411 * * * 1←9 2 34411 * * * 1 3 4 1 1 * * * 24333←4 5 1 * * * 24333	3 6 2 34411 * * * 1 1 2............2 45333←2 4 2..........2 45333
20	13 1 1 * * * * 1←14 2 * * * * 1 9 5 1 * * * * 1←10 6 * * * * 1 7 1 1 * * * 24333←8 2 * * * 24333 3 5 1 * * * 24333←4 6 * * * 24333 2............2 45333←9 1 * * * 24333	14 1 1 * * * * 1←16 1 * * * * 1 8 1 1 * * * 24333 5 10 1 * * * * 1←6 11 * * * * 1 3 6 1 * * * 24333←4 7 * * * 24333 3 2...........2 45333←4 4 2.........2 45333
19	14 1 * * * * 1←16 * * * * 1 13 2 * * * * 1←14 4 4 1 1 * * * 1 9 6 * * * * 1←10 8 4 1 1 * * * 1 7 2 * * * 24333←8 3 6 2 34411 * * 1 4 2...........2 45333←10 * * * 24333 3 6 * * * 24333←4 7 6 2 34411 * * 1 1 2..........2 35733←2 3 6 2........2 45333	15 1 * * * * 1←17 * * * * 1 11 34411 * * * 1←12 5 1 2 34411 * * 1 5 9 4 4 1 1 * * * 1←6 11 1 2 34411 * * 1 5 2..........2 45333←6 4 2.........2 45333 3 4 2.........2 45333←4 5 8 1 1 * * 24333 2...........2 35733←8 2.........2 45333
18	15 * * * * 1←18 4 1 1 * * * 1 13 4 4 1 1 * * * 1←16 1 2 34411 * * 1 9 8 4 1 1 * * * 1←10 12 1 1 * * * 1 7 3 6 2 34411 * * 1←8 8 1 1 * * 24333 6 2.........2 45333←11 6 2 34411 * * 1 5 10 1 2 34411 * * 1←6 11 2 34411 * * 1 3 2.........2 35733←4 3 6 2.......2 45333	11 5 1 2 34411 * * 1←12 6 2 34411 * * 1 7 9 1 2 34411 * * 1←8 10 2 34411 * * 1 7 2.........2 45333←8 4 2........2 45333 5 4 2........2 45333←6 6 2........2 45333 4 2.........2 35733←9 8 1 1 * * 24333 3 5 8 1 1 * * 24333←4 7 13 1 * * * 1 1 2........2 3 3 6653←2 3 6 2.......2 35733
17	17 4 1 1 * * * 1←18 5 1 * * * 1 15 1 2 34411 * * 1←17 2 34411 * * 1 9 12 1 1 * * * 1←10 13 1 * * * 1 7 14 1 1 * * * 1←8 15 1 * * * 1 7 8 1 1 * * 24333←16 1 1 * * 24333 5 10 1 1 * * 24333←6 11 1 * * 24333 3 3 6 2.......2 45333←4 6 2.......2 35733 2.........2 3 3 6653	7 10 2 34411 * * 1←8 12 34411 * * 1 7 4 2........2 45333←8 6 2.......2 45333 5 6 2........2 45333←11 13 1 * * * 1 3 2........2 3 3 6653←4 4 2......2 3 3 6653
16	21 1 1 * * * 1←22 2 * * * 1 17 5 1 * * * 1←18 6 * * * 1 15 1 1 * * 24333←16 2 * * 24333 9 13 1 * * * 1←17 1 * * 24333 3 6 2.......2 35733 1 2.....2 4 3 3 3 6653←2 4 2...2 4 3 3 3 6653	22 1 1 * * * 1←24 1 * * * 1 7 12 34411 * * 1←8 15 4 4 1 1 * * 1 7 6 2.......2 45333←14 2.......2 45333 3 4 2......2 3 3 6653←4 6 2.....2 3 3 6653 2......2 4 3 3 3 6653←8 2......2 3 3 6653

Stem 48 (filtrations ≧ 16)　　Stem 49 (filtrations ≧ 16)

6	39 4 4 1 1 1←44 4 1 1 1 35 8 4 1 1 1←36 12 1 1 1 35 24333←37 6 2 3 3 33 3 6 2 3 3←34 8 3 3 3 29 35733←33 5 7 3 3 26 23577←28 4 5 7 7 19 24 4 1 1 1←20 28 1 1 1 17 11 3577←18 14 5 7 7 15 59777←19 13 5 7 7 13 23 6 2 3 3←14 28 3 3 3 11 21 5 7 3 3←13 24 6 5 3 11 7 13 5 7 7←13 17 7 7 7 10 18 3577←12 20 5 7 7 9 18 4 5 7 7←10 20 7 7 7 9 9 11 7 7 7←10 10 13 11 7 5 9 15 7 7 7	37 5 1 2 3 3←38 6 2 3 3 33 45333←34 5 7 3 3 31 4 7 3 3 3←32 6653 30 35733←35 8 3 3 3 27 23577←29 4 5 7 7 21 12 9 3 3 3←22 13 11 3 3 18 11 3577←20 13 5 7 7 13 21 8 3 3 3←14 24 6 5 3 13 20 9 3 3 3←14 21 11 3 3 11 22 9 3 3 3←12 23 11 3 3 11 18 3577←13 20 5 7 7 11 7 14 5 7 7←12 11 15 7 7 10 19 3577←12 21 5 7 7 10 9 11 7 7 7←14 17 7 7 7 7 11 14 5 7 7←8 13 13 11 7 6 9 15 7 7 7
5	43 4 1 1 1←44 5 1 1 41 1 2 3 3←48 1 1 1 35 12 1 1 1←36 13 1 1 28 3577←32 5 7 7 27 4 5 7 7←30 7 7 7 19 28 1 1 1←20 29 1 1 17 14 5 7 7←20 13 11 7 14 27 3 3 3←16 29 3 3 11 22 11 3 3←12 23 13 3 11 20 5 7 7←14 23 7 7 10 19 7 7 7←12 21 11 7 9 10 13 11 7←10 11 15 15	42 1 2 3 3←44 2 3 3 33 9 3 3 3←34 10 5 3 31 6653←38 6 5 3 29 3577←33 5 7 7 21 13 11 3 3←22 14 13 3 17 25 3 3 3←18 26 5 3 13 21 11 3 3←14 22 13 3 11 21 5 7 7←13 25 7 7 11 19 7 7 7←13 21 11 7 11 11 15 7 7←21 13 11 7 7 13 13 11 7
4	47 1 1 1←48 2 1 43 5 1 1←44 6 1 42 2 3 3←49 1 1 35 13 1 1←36 14 1 31 5 7 7←37 7 7 29 7 7 7←33 11 7 19 29 1 1←20 30 1 19 13 11 7←21 15 15 15 29 3 3←17 31 3 13 23 7 7←17 27 7 11 21 11 7←13 23 15 9 11 15 15	43 2 3 3←46 3 3 42 3 3 3←44 5 3 37 6 5 3←38 7 7 34 11 3 3←36 13 3 33 10 5 3←34 11 7 21 14 13 3←22 15 15 18 27 3 3←20 29 3 17 26 5 3←18 27 7 13 22 13 3←14 23 15
3	47 2 1←48 3 44 3 3←50 1 43 6 1←44 7 35 14 1←36 15 19 30 1←20 31	45 3 3←49 3 43 5 3←45 7 35 13 3←37 15 19 29 3←21 31
2	47 3←51	
1		
	Stem 50 (filtrations ≦ 6)	Stem 51 (filtrations ≦ 6)

	Stem 50 (filtrations 7 to 9)	Stem 51 (filtrations 7 to 9)
9	35 4 1 1 * 1←36 5 1 * 1 33 1 2 34411 1←40 1 1 * 1 27 12 1 1 * 1←28 13 1 * 1 21 3 6 2 45333←22 6 2 35733 19 20 1 1 * 1←20 21 1 * 1 15 3 6 3 3 6653←16 6 5 23577 13 26 1 1 * 1←14 27 1 * 1 13 5 6 3 3 6653←14 7 12 45333 13 4 7 3 3 6653←19 6 3 3 6653 11 22 1 1 24333←12 23 1 24333 9 15 6 2 45333←10 18 2 35733 4 7 3 5 9 3577←6 7 8 12 9 3 3 3 3 4 5 3 59777←5 8 3 59777	34 1 2 34411 1←36 2 34411 1 25 4 2 2 45333←26 6 2 45333 23 6 2 2 45333←29 13 1 * 1 21 7 13 1 * 1←22 10 2 45333 19 3 6 2 35733←20 6 3 3 6653 13 15 13 1 * 1←14 18 2 45333 13 3 6 5 23577←14 6 9 3 6653 11 17 13 1 * 1←12 20 2 45333 11 5 6 5 23577←17 6 5 23577 9 9 5 5 3 6653←10 10 9 3 6653 5 9 3 5 7 3577 3 4 6 3 59777←4 5 10 11 3577
8	39 1 1 * 1←40 2 * 1 35 5 1 * 1←36 6 * 1 34 2 34411 1←41 1 * 1 33 1 1 24333←34 2 24333 19 21 1 * 1←20 22 * 1 15 6 5 23577←21 12 45333 13 21 1 24333←14 22 24333 13 7 12 45333←14 13 23577 13 3 5 7 3577←17 5 7 3577 11 3 5 9 3577←13 8 12 9 3 3 3 9 18 2 35733←10 20 35733 9 15 3 3 6653←10 17 23577 5 7 8 12 9 3 3 3←6 12 11 3577	35 2 34411 1←38 34411 1 34 1 1 24333←36 1 24333 25 6 2 45333←32 2 45333 21 10 2 45333←22 12 45333 18 5 5 3 6653←20 9 3 6653 14 3 5 7 3577←16 5 9 3577 13 18 2 45333←14 20 45333 13 6 9 3 6653←14 8 12 9 3 3 3 12 3 5 9 3577←18 5 7 3577 11 20 2 45333←12 22 45333 10 11 12 45333←12 17 3 6653 9 10 9 3 6653←10 12 12 9 3 3 3 3 5 10 11 3577
7	39 2 * 1←40 4 4 1 1 1 36 34411 1←42 * 1 35 6 * 1←36 8 4 1 1 1 34 1 24333←36 24333 33 2 24333←34 3 6 2 3 3 27 2 35733←28 3 6653 19 22 * 1←20 24 4 1 1 1 14 5 9 3577←20 9 3577 13 22 24333←14 23 6 2 3 3 13 13 23577←14 14 4 5 7 7 12 3 59777←16 59777 10 19 35733←12 21 5 7 3 3 10 5 59777←12 7 13 5 7 7 9 17 23577←10 18 4 5 7 7 3 6 9 11 7 7 7←4 7 11 15 7 7	37 34411 1←38 5 1 2 3 3 35 1 24333←37 24333 31 2 45333←32 4 7 3 3 3 28 2 35733←34 45333 25 3 3 6653←28 23577 19 9 3 6653←22 12 9 3 3 3 15 5 9 3577←21 9 3577 13 20 45333←14 21 8 3 3 3 13 3 59777←17 59777 11 22 45333←12 23 8 3 3 3 11 17 3 6653←14 20 9 3 3 3 11 5 59777←12 7 14 5 7 7 9 18 23577←10 19 4 5 7 7 9 12 12 9 3 3 3←12 18 3577 7 7 14 4 5 7 7←8 11 14 5 7 7 4 6 9 11 7 7 7

Stem 50 (filtrations 7 to 9) Stem 51 (filtrations 7 to 9)

	Stem 50 (filtrations 10 to 12)	Stem 51 (filtrations 10 to 12)
12	31 1 1 * * 1←32 2 * * 1 27 5 1 * * 1←28 6 * * 1 26 2 34411 * 1←33 1 * * 1 25 1 1 * 24333←26 2 * 24333 19 13 1 * * 1←20 14 * * 1 9 1 1 24733 6653←10 2 24733 6653 7 12 2...2 35733←8 14 2..2 35733 7 6 2 24333 6653←14 2 24333 6653 5 8 2 24333 6653←6 10 24333 6653 1 2 4 3 5 6 5 23577←4 4 3 5 6 5 23577	27 2 34411 * 1←30 34411 * 1 26 1 1 * 24333←28 1 * 24333 21 2....2 45333←22 4 2..2 45333 17 6 2...2 45333←24 2...2 45333 11 22 1 * * 1←12 23 * * 1 10 1 1 24733 6653←12 1 24733 6653 9 18 1 * 24333←10 19 * 24333 7 10 2..2 3 3 6653←8 12 2 2 2 3 3 6653 5 5 2 24733 6653←6 8 2 3 5 5 3 6653 5 2 2 4 5 5 5 3 6653←8 2 4 5 5 5 3 6653 2 2 4 3 5 6 5 23577
11	31 2 * * 1←32 4 4 1 1 * 1 28 34411 * 1←34 * * 1 27 6 * * 1←28 8 4 1 1 * 1 26 1 * 24333←28 * 24333 25 2 * 24333←26 3 6 2 34411 1 19 14 * * 1←20 16 4 1 1 * 1 19 2...2 35733←20 3 6 2 2 45333 13 2 24333 6653←14 3 5 6 2 35733 10 1 24733 6653←12 24733 6653 9 2 24733 6653←16 24333 6653 7 14 2..2 35733←8 15 6 2 2 45333 5 10 24333 6653←6 11 5 6 2 35733 3 5 4 5 5 5 3 6653←4 7 5 6 5 23577 3 4 3 5 6 5 23577←6 5 5 6 5 23577 1 2 3 5 3 5 9 3577←2 4 6 3 5 9 3577	29 34411 * 1←30 5 1 2 34411 1 27 1 * 24333←29 * 24333 23 2...2 45333←24 4 2 2 2 45333 21 4 2..2 45333←22 5 8 1 1 24333 20 2...2 35733←26 2..2 45333 17 2..2 3 3 6653←18 3 5 6 2 45333 11 21 4 4 1 1 * 1←12 23 1 2 34411 1 11 1 24733 6653←13 24733 6653 7 12 2 2 2 3 3 6653←8 13 5 6 2 45333 7 2 4 5 5 5 3 6653←10 4 5 5 5 3 6653 5 8 2 3 5 5 3 6653←6 9 4 7 3 3 6653 3 3 5 5 6 5 23577←5 7 5 6 5 23577 2 2 3 5 3 5 9 3577←4 4 5 3 5 9 3577 1 2 4 5 3 5 9 3577←2 4 7 3 5 9 3577 1 2 2 3 5 3 59777←2 3 4 5 3 59777
10	31 4 4 1 1 * 1←36 4 1 1 * 1 27 8 4 1 1 * 1←28 12 1 1 * 1 27 * 24333←29 6 2 34411 1 25 3 6 2 34411 1←26 8 1 1 24333 21 2..2 35733←22 3 6 2 45333 19 16 4 1 1 * 1←20 20 1 1 * 1 19 3 6 2 2 45333←20 5 6 2 45333 15 24333 6653←16 3 6 3 3 6653 13 3 5 6 2 35733←14 5 6 3 3 6653 12 2 3 5 5 3 6653←17 5 6 2 35733 11 22 1 2 34411 1←12 23 2 34411 1 11 24733 6653←14 4 7 3 3 6653 9 13 6 2 2 45333←10 15 6 2 45333 5 5 5 6 5 23577 2 4 5 3 5 9 3577←4 8 3 5 9 3577 2 2 3 5 3 59777←4 4 5 3 59777 1 2 4 5 3 59777←2 4 7 3 59777	29 5 1 2 34411 1←30 6 2 34411 1 25 2..2 45333←26 4 2 2 45333 23 4 2 2 2 45333←24 6 2 2 45333 22 2..2 35733←27 8 1 1 24333 21 5 8 1 1 24333←22 7 13 1 * 1 19 2 2 2 3 3 6653←20 3 6 2 35733 17 3 5 6 2 45333←18 5 6 2 35733 13 13 8 1 1 24333←14 15 13 1 * 1 13 2 3 5 5 3 6653←14 3 6 5 23577 11 15 8 1 1 24333←12 17 13 1 * 1 9 4 5 5 5 3 6653←12 5 6 5 23577 7 13 5 6 2 45333←8 15 6 2 35733 5 9 4 7 3 3 6653←6 11 6 5 23577 3 6 9 5 5 3 6653←4 7 11 12 45333 3 4 5 3 5 9 3577←5 8 3 5 9 3577 3 2 3 5 3 59777←4 4 6 3 59777

Stem 50 (filtrations 10 to 12)

Stem 51 (filtrations 10 to 12)

17	19 4 1 1 * * * 1←20 5 1 * * * 1 17 1 2 34411 * * 1←24 1 1 * * * 1 11 12 1 1 * * * 1←12 13 1 * * * 1 5 3 6 2.......2 45333←6 6 2.......2 35733 3 5 6 2.......2 45333 1 2......24333 6653←2 4 2....24333 6653	18 1 2 34411 * * 1←20 2 34411 * * 1 9 4 2........2 45333←10 6 2.......2 45333 7 6 2........2 45333←13 13 1 * * * 1 5 7 13 1 * * * 1←6 10 2.......2 45333 3 3 6 2.......2 35733←4 6 2......2 3 3 6653 2.......24333 6653
16	23 1 1 * * * 1←24 2 * * * 1 19 5 1 * * * 1←20 6 * * * 1 18 2 34411 * * 1←25 1 * * * 1 17 1 1 * * 24333←18 2 * * 24333 5 6 2.......2 35733 3 2.....24333 6653←4 4 2...24333 6653 1 1 1 * 24733 6653←2 2 * 24733 6653	19 2 34411 * * 1←22 34411 * * 1 18 1 1 * * 24333←20 1 * * 24333 9 6 2.......2 45333←16 2.......2 45333 5 10 2.......2 45333←6 12 2......2 45333 3 6 2......2 3 3 6653 2 1 1 * 24733 6653←4 1 * 24733 6653 1 2 1 * 24733 6653←2 3 * 24733 6653 1 1 2 * 24733 6653←2 4 2 2 1 2 24733 6653
15	23 2 * * * 1←24 4 4 1 1 * * 1 20 34411 * * 1←26 * * * 1 19 6 * * * 1←20 8 4 1 1 * * 1 18 1 * * 24333←20 * * 24333 17 2 * * 24333←18 3 6 2 34411 * 1 11 2.......2 35733←12 3 6 2....2 45333 5 2....24333 6653←6 3 5 6 2...2 35733 3 4 2...24333 6653←4 5 5 6 2...2 35733 2 1 * 24733 6653←4 * 24733 6653 1 2 * 24733 6653←8 2...24333 6653	21 34411 * * 1←22 5 1 2 34411 * 1 19 1 * * 24333←21 * * 24333 15 2.......2 45333←16 4 2.....2 45333 12 2.......2 35733←18 2......2 45333 9 15 4 4 1 1 * * 1←10 17 1 2 34411 * 1 9 2......2 3 3 6653←10 3 5 6 2...2 45333 6 2....24333 6653 5 6 2.....2 3 3 6653←6 7 5 6 2...2 45333 3 1 * 24733 6653←5 * 24733 6653
14	23 4 4 1 1 * * 1←28 4 1 1 * * 1 19 8 4 1 1 * * 1←20 12 1 1 * * 1 19 * * 24333←21 6 2 34411 * 1 17 3 6 2 34411 * 1←18 8 1 1 * 24333 13 2......2 35733←14 3 6 2...2 45333 11 3 6 2....2 45333←12 5 6 2...2 45333 7 2...24333 6653←8 3 6 2..2 3 3 6653 5 3 5 6 2...2 35733←6 5 6 2..2 3 3 6653 4 2 2 1 2 24733 6653←9 5 6 2...2 35733 3 5 5 6 2...2 35733←4 7 6 2..2 3 3 6653 3 * 24733 6653←6 4 1 1 24733 6653 1 1 2 2 2 4 5 5 5 3 6653←2 3 2 2 4 5 5 5 3 6653	21 5 1 2 34411 * 1←22 6 2 34411 * 1 17 2......2 45333←18 4 2....2 45333 15 4 2.....2 45333←16 6 2....2 45333 14 2......2 35733←19 8 1 1 * 24333 11 2.....2 3 3 6653←12 3 6 2...2 35733 9 17 1 2 34411 * 1←10 18 2 34411 * 1 9 3 5 6 2...2 45333←10 5 6 2...2 35733 5 7 5 6 2...2 45333←6 9 6 2...2 35733 5 2 2 1 2 24733 6653←6 3 6 2 24333 6653 3 3 4 1 1 24733 6653←4 5 6 2 24333 6653 1 2..2 4 5 5 5 3 6653←4 2 2 2 4 5 5 5 3 6653
13	27 4 1 1 * * 1←28 5 1 * * 1 25 1 2 34411 * 1←32 1 1 * * 1 19 12 1 1 * * 1←20 13 1 * * 1 13 3 6 2...2 45333←14 6 2...2 35733 7 9 6 2...2 45333←8 12 2...2 35733 7 3 6 2..2 3 3 6653←8 6 2 24333 6653 5 5 6 2..2 3 3 6653←6 8 2 24333 6653 5 4 1 1 24733 6653←11 6 2..2 3 3 6653 3 6 1 1 24733 6653←4 7 1 24733 6653 2 2 2 2 4 5 5 5 3 6653	26 1 2 34411 * 1←28 2 34411 * 1 17 4 2....2 45333←18 6 2...2 45333 15 6 2....2 45333←22 2....2 45333 11 3 6 2...2 35733←12 6 2..2 3 3 6653 9 18 2 34411 * 1←10 20 34411 * 1 5 9 6 2...2 35733←6 12 2..2 3 3 6653 5 3 6 2 24333 6653←6 5 2 24733 6653 3 5 6 2 24333 6653←4 7 2 24733 6653 3 2 2 2 4 5 5 5 3 6653←6 2 2 4 5 5 5 3 6653 1 1 2 4 3 5 6 5 23577←2 3 4 3 5 6 5 23577
	Stem 50 (filtrations 13 to 17)	Stem 51 (filtrations 13 to 17)

27		1 1 * * * * * * * 1
26	1 * * * * * * * 1	2 * * * * * * * 1 1 1 1 2 34411 * * * * 1←2 2 2 34411 * * * * 1
25	3 4 1 1 * * * * * 1←4 5 1 * * * * * * 1 1 1 2 34411 * * * * 1←8 1 1 * * * * * * 1	4 4 1 1 * * * * * 1 2 1 2 34411 * * * * 1←4 2 34411 * * * * 1 1 2 2 34411 * * * * 1←2 4 34411 * * * * 1
24	7 1 1 * * * * * 1←8 2 * * * * * * 1 3 5 1 * * * * * 1←4 6 * * * * * * 1 2 2 34411 * * * * 1←9 1 * * * * * 1 1 1 1 * * * * 24333←2 2 * * * * 24333	3 6 1 * * * * * 1←4 7 * * * * * * 1 3 2 34411 * * * * 1←6 34411 * * * * 1 2 1 1 * * * * 24333←4 1 * * * * 24333 1 2 1 * * * * 24333←2 3 * * * * 24333
23	7 2 * * * * * 1←8 4 4 1 1 * * * * 1 4 34411 * * * * 1←10 * * * * * 1 3 6 * * * * * 1←4 8 4 1 1 * * * * 1 2 1 * * * * 24333←4 * * * * 24333 1 2 * * * * 24333←2 3 6 2 34411 * * * 1	5 34411 * * * * 1←6 5 1 2 34411 * * * 1 3 5 4 4 1 1 * * * * 1←4 7 1 2 34411 * * * 1 3 1 * * * * 24333←5 * * * * 24333
22	7 4 4 1 1 * * * * 1←12 4 1 1 * * * * 1 3 6 1 2 34411 * * * 1←4 7 2 34411 * * * 1 3 * * * * 24333←5 6 2 34411 * * * 1	5 5 1 2 34411 * * * 1←6 6 2 34411 * * * 1 1 2..............2 45333←2 4 2............2 45333
21	11 4 1 1 * * * * 1←12 5 1 * * * * 1 9 1 2 34411 * * * 1←16 1 1 * * * * 1 5 10 1 1 * * * * 1←6 11 1 * * * * 1 3 6 1 1 * * * 24333←4 7 1 * * * 24333 2..............2 45333	10 1 2 34411 * * * 1←12 2 34411 * * * 1 3 2............2 45333←4 4 2...........2 45333
20	15 1 1 * * * * 1←16 2 * * * * 1 11 5 1 * * * * 1←12 6 * * * * 1 10 2 34411 * * * 1←17 1 * * * * 1 9 1 1 * * * 24333←10 2 * * * 24333 5 5 1 * * * 24333←6 6 * * * 24333 4 2............2 45333 1 2...........2 35733←2 4 2..........2 35733	11 2 34411 * * * 1←14 34411 * * * 1 10 1 1 * * * 24333←12 1 * * * 24333 5 2............2 45333←6 4 2..........2 45333 3 4 2..........2 45333←4 6 2..........2 45333 2.............2 35733←8 2...........2 45333
19	15 2 * * * * 1←16 4 4 1 1 * * * 1 12 34411 * * * 1←18 * * * * 1 11 6 * * * * 1←12 8 4 1 1 * * * 1 10 1 * * * 24333←12 * * * 24333 9 2 * * * 24333←10 3 6 2 34411 * * 1 6 2...........2 45333 5 6 * * * 24333←6 7 6 2 34411 * * 1 3 2...........2 35733←4 3 6 2........2 45333	13 34411 * * * 1←14 5 1 2 34411 * * 1 11 1 * * * 24333←13 * * * 24333 7 2...........2 45333←8 4 2.........2 45333 5 4 2.........2 45333←6 5 8 1 1 * * 24333 4 2...........2 35733←10 2..........2 45333 3 6 2..........2 45333←4 7 8 1 1 * * 24333 1 2..........2 3 3 6653←2 3 5 6 2.......2 45333
18	15 4 4 1 1 * * * 1←20 4 1 1 * * * 1 11 8 4 1 1 * * * 1←12 12 1 1 * * * 1 11 * * * 24333←13 6 2 34411 * * 1 9 3 6 2 34411 * * 1←10 8 1 1 * * 24333 5 7 6 2 34411 * * 1←6 12 1 1 * * 24333 5 2..........2 35733←6 3 6 2.......2 45333 3 3 6 2........2 45333←4 5 6 2.......2 45333 2..........2 3 3 6653	13 5 1 2 34411 * * 1←14 6 2 34411 * * 1 9 2..........2 45333←10 4 2........2 45333 7 4 2........2 45333←8 6 2........2 45333 6 2..........2 35733←11 8 1 1 * * 24333 5 5 8 1 1 * * 24333←6 7 13 1 * * * 1 3 2.........2 3 3 6653←4 3 6 2.......2 35733

Stem 50 (filtrations ≥ 18)

Stem 51 (filtrations ≥ 18)

Bibliography

[A] J. F. Adams, On the structure and applications of the Steenrod algebra, Comm. Math. Helv. **52** (1958), 180-214.

[B] M. G. Barratt, EHP caclulations of homotopy groups of spheres, (unpublished tables), c 1960.

[B,C] A. K. Bousfield, E. B. Curtis, A spectral sequence for the homotopy of nice spaces, Trans. Amer. Math Soc. **151** (1970), 457-479.

[6A] A. K. Bousfield, E. B. Curtis, D. M. Kan, D. G. Quillen, D. L. Rector and J. W. Schlesinger, The mod-p-lower central series and the Adams spectral sequence, Topology **5** (1966), 331-342.

[C1] E. B. Curtis, Lectures on Simplicial homotopy. Matematisk Institut, Aarhus, Denmark, 1968.

[C2] E. B. Curtis, Some non-zero homotopy groups of spheres. Bull. Amer. Math. Soc. **75** (1969), 541-546.

[C3] E. B. Curtis, Simplicial homotopy theory. Advances of Mathematics **6** (1971), 107-209.

[H] W. A. Hansen, Computer calculations of the homology of the lambda algebra. Dissertation, Northwestern University, 1974.

[M,T] M. E. Mahowald, and M. C. Tangora, Some differentials in the Adams spectral sequence, Topology **6**, (1967), 349-369.

[May] J. P. May, The cohomology of the Steenrod algebra; stable homotopy groups of spheres, Bull. Am. Math. Soc. **71** (1967), 377-380.

[T1] M. C. Tangora, On the cohomology of the Steenrod algebra, Math. Z. **116** (1970), 18-64.

[T2] M. C. Tangora, Some remarks on the lambda algebra, Proc. Evanston Conference on algebraic Topology (1971). Lecture Notes in Mathematics **#658**, Springer-Verlag (1978), 476-487.

[T3] M. C. Tangora, Generating Curtis tables, Proc. Canadian Summer Congress on Algebriac Topology, Vancouver B. C. (1977). Lecture Nores in Mathematics **#673**, Springer-Verlag (1978), 243-253.

[T4] M. C. Tangora, Computing the homology of the lambda algebra, Memoirs American Math. Soc. **#337** (1985).

[Tod] H. Toda, Composition methods in homotopy groups of spheres, Princeton, 1962.

[We1] R. J. Wellington, The unstable Adams spectral sequence for free iterated loop spaces. Memoirs Amer. Math. Soc. **#258** (1982).

[We2] R. J. Wellington, The computation of Ext groups for modules over the Steenrod algebra. Preprint, 1982.

[GWW] G. W. Whitehead, (absolutely amazing) unpublished tables, c. 1970.

THE bo-ADAMS SPECTRAL SEQUENCE: SOME CALCULATIONS AND A PROOF OF ITS VANISHING LINE

Donald M. Davis
Lehigh University
Bethlehem, PA 18015

1. INTRODUCTION

Let bo denote the spectrum for connective real K-theory localized at 2. The bo-Adams spectral sequence (bo-ASS) converging to the 2-primary stable homotopy groups of spheres was initiated by Mahowald ([10]) in his study of the image of the J-homomorphism. Its computability was greatly improved in [8].

This spectral sequence is formed by repeatedly smashing with the cofibration $S^0_{(2)} \to bo \to \overline{bo}$ and applying $\pi_*(\ )$. Thus it depends upon the splitting

$$bo \wedge bo \simeq \bigvee_{n\geq 0} (\Sigma^{8n} bo^{<4n-\alpha(n)>} \vee \Sigma^{8n+4} bsp^{<4n-\alpha(n)>}) \vee \bar{H}\ , \qquad (1.1)$$

where $bo^{<m>}$ is formed from bo by killing homotopy classes of Adams filtration less than m , $\bar{H}$ is a wedge of mod 2 Eilenberg-MacLane spectra, and $\alpha(n)$ is the number of 1's in the binary expansion of n .

In [8] it is shown that the E_2-term of this bo-ASS can be approximated by combining 2 calculations--one corresponding to the bo- and bsp-parts of $bo\wedge\ldots\wedge bo$, and the other to the $\bar{H}$-parts. More precisely, there is a long exact sequence

$$\to H^{s,t}(V^*) \to E_2^{s,t} \to H^{s,t}(\mathfrak{C}) \to H^{s+1,t}(V_*) \to\ , \qquad (1.2)$$

where V^* is a cochain complex of graded Z_2-vector spaces with $V^s = \pi_*(H^s)$, with H^s the Eilenberg-MacLane summand of $bo \wedge \overline{bo}^{\wedge s}$. Lellman and Mahowald calculate $H^{s,t}(\mathfrak{C})$ completely in [8]. (See 4.1.) As for computing $H^{s,t}(V^*)$, they say "The differential on V^* is, under a suitable isomorphism, induced from the standard differential of the bar resolution. Since all these formulae are quite manageable, it is possible to pass the problem to a computer. Details will appear elsewhere."

This is an "elsewhere". The purpose of this paper is to describe exactly how these calculations can be performed, and to present the results obtained. The size of these calculations grows (seemingly) exponentially with t , and so we satisfy ourselves with the range $t \leq 25$.

This suffices to complete the calculation of the bo-ASS through the 20-stem, the result of which was presented in [8]. A subsidiary purpose then is to verify the claims made in [8] for $H^{s,t}(V^*)$ in $t-s \leq 20$.

Homotopy classes often have lower filtration in this spectral sequence than in the classical or Novikov ASS's, and unlike those, this one has no nonzero differentials in $t-s \leq 20$. It is observed in [8] that there is a nonzero differential in the bo-ASS in the 30-stem. Whether this is the first is still unresolved. Because the E_2-term of this bo-ASS gives such a good approximation to $\pi_*(S^0)_{(2)}$ at least in the range calculated to date, a third purpose of this paper is to publicize it. While doing this publicizing, it should be mentioned that [8] shows that these techniques can be applied to a large class of spectra, and in view of the success of bo-resolutions in studying geometric dimension qualitatively in [6], the author hopes to return to a more quantitative application of the bo-ASS to obstruction theory in future paper.

Two global results about the E_2-term of the bo-ASS are proved. It is well known that $H^0(\mathfrak{C}) \oplus H^1(\mathfrak{C})$ detects exactly the image of J and Adams μ-elements ([10]). We show that $H^1(V^*) = 0$, so that no new homotopy is detected until the 2-line. A careful study of this 2-line, $E_2^{2,*}$, would certainly seem to be in order. We also present in Section 4 an expanded account of Mahowald's proof that $E_2^{s,t}$ vanishes above a line of slope 1/5 in the usual (t-s,s) chart.

<u>Theorem 1.3</u> (Mahowald [10])

$$E_2^{s,t} = 0 \quad \text{if} \quad t-s < \begin{cases} 3s & s \leq 6 \\ 5s-11 & s \equiv 0,1(4) \\ 5s-13 & s \equiv 2(4) \\ 5s-12 & s \equiv 3(4) \end{cases}.$$

Although the method presented here of calculating $H^*(V^*)$ is effective and conceptually simple, the author feels that a more global approach, analogous to (perhaps) the May spectral sequence, must be developed to carry these calculations much farther.

2. THE METHOD OF CALCULATION

The $E_2^{s,t}$-term of the bo-ASS is the sth cohomology group of the cochain complex

$$\longrightarrow \pi_t(bo \wedge \overline{bo}^{s-1}) \longrightarrow \pi_t(bo \wedge \overline{bo}^{s}) \longrightarrow \pi_t(bo \wedge \overline{bo}^{s+1}) \longrightarrow ,$$

where superscripts indicate iterated smash power, and maps are

$(S^0_{(2)} \longrightarrow bo) \wedge (bo \longrightarrow \overline{bo}) \wedge 1$. These homotopy groups are calculated by the classical ASS with E_2-term $\mathrm{Ext}_{A_1}(\overline{A//A_1}^s, \mathbb{Z}_2)$. Here A_n denotes the subalgebra of the mod 2 Steenrod algebra A generated by $\{Sq^i : i \le 2^n\}$, $A//A_n$ is the quotient by the left ideal generated by the nonunits of A_n , $\overline{A//A_n}$ is its augmentation ideal,
the s in the superscript denotes iterated $\otimes$, and we have used the change-of-rings theorem and $H^*bo \approx A//A_1$. This Ext-group will have a filtration-0 $\mathbb{Z}_2$ dual to each free A_1-summand in the A_1-module $\overline{A//A_1}^s$, and these correspond to $V^s = \pi_*(H^s)$, where H^s is the Eilenberg-MacLane summand of $bo \wedge \overline{bo}^s$. Generators of these free A_1-summands are characterized by $Sq^3Sq^3g \neq 0$, and, dualizing, we obtain V^s as the image of the right action of Sq^3Sq^3 on $\overline{(A//A_1)}{}^s_*$. It was shown in [2] that $A//A_1^* \approx \mathbb{Z}_2[\zeta_1^4, \zeta_2^2, \zeta_3, \zeta_4, \ldots]$, where ζ_i is the conjugate $\chi(\xi_i)$ of the Milnor generator ξ_i of degree $2^i - 1$, and

$$(\zeta_i)(\chi Sq) = \zeta_i + \zeta_{i-1}^2 . \tag{2.1}$$

Note $Sq^3Sq^3 = \chi Sq^3 \cdot \chi Sq^3$.

The vector space dimensions of these graded vector spaces V^s_* can best be calculated (by computer) by using graded dimensions of $A//A_1$, $H^*(bo^{\langle n\rangle})$, and A_1 , and (1.1). Some results are given in Table 1, on the next page. These are used merely as a check that our $\mathrm{im}(Sq^3Sq^3)$ is finding all elements, and as a warning of how large our calculations will be.

$\mathrm{Ext}_A^{**}(\mathbb{Z}_2, \mathbb{Z}_2)$ can be calculated using the exact sequence

$$0 \leftarrow \mathbb{Z}_2 \leftarrow W \xleftarrow{d} W \otimes \bar{W} \xleftarrow{d} W \otimes \bar{W}^2 \xleftarrow{d} \cdots \tag{2.2}$$

where $W = A//A_1$ and $d(a_0)[a_1|\ldots|a_s]) = \varepsilon(a_0)a_1[a_2|\ldots|a_s]$.

TABLE 1

$\dim(V_t^s) = \dim(E_1^{s,t}$ of $H\mathbb{Z}_2$-part of bo-ASS) .

t \ s	1	2	3	4	5	6	7	8	9	10	11
14	0	2	2								
15	0	0	0								
16	0	0	0								
17	0	0	0								
18	0	3	6	3							
19	0	0	0	0							
20	0	1	3	2							
21	0	0	3	3							
22	1	10	18	13	4						
23	0	0	0	0	0						
24	0	2	9	12	5						
25	0	0	6	12	6						
26	1	15	41	46	24	5					
27	0	0	3	8	5	0					
28	0	7	30	48	34	9					
29	0	4	25	42	31	10					
30	2	26	84	126	100	40	6				
:											
45	0	37	406	1685	3751	4955	3962	1849	442	36	
46	6	123	819	2967	6449	8714	7413	3900	1181	174	10

This gives a spectral sequence converging to $\mathrm{Ext}_A^{**}(\mathbb{Z}_2)$ with $E_1^{\sigma,s,t} = \mathrm{Ext}_A^{s,t}(W \otimes \bar{W}^\sigma)$, which will be studied in more detail in Section 4. Here and henceforth we often delete $\mathbb{Z}_2$ from the second component of Ext. We are interested here in a subgroup of the $s = 0$ part. V^σ is the subgroup of $\mathrm{Hom}_A^t(W \otimes \bar{W}^\sigma, \mathbb{Z}_2) \approx \mathrm{Hom}_{A_1}^t(\bar{W}^\sigma, \mathbb{Z}_2)$ corresponding to the A_1-free summand of $\bar{W}^\sigma$, and we desire the cohomology of the complex formed from these.

The chain complex (2.2) is isomorphic to

$$0 \longleftarrow \mathbb{Z}_2 \longleftarrow W \xleftarrow{D} A \otimes_{A_1} \bar{W} \xleftarrow{D} A \otimes_{A_1} \bar{W}^2 \xleftarrow{D} \cdots$$

with $D(a_o[a_1|\ldots|a_s]) = \sum a_o a_i' \cdot \chi(a_i'')[a_2|\ldots|a_s]$, where

$\Psi(a_1) = \sum a_i' \otimes a_i''$. (See [3 ; 2.1].) The associated cochain complex

$$\begin{array}{ccccc} \mathrm{Hom}_{A_1}(\bar{W}, \mathbb{Z}_2) & \xrightarrow{D^*} & \mathrm{Hom}_{A_1}(\bar{W}^2, \mathbb{Z}_2) & \rightarrow & \cdots \\ \cap & & \cap & & \\ \bar{W}^* & \xrightarrow{\delta} & \bar{W}^{*2} & \xrightarrow{\delta} & \end{array}$$

has, if $\Psi(\theta_j) = \sum_{\nu_j} \theta'_{j,\nu_j} \otimes \theta''_{j,\nu_j}$,

$$\delta[\theta_1|\cdots|\theta_s] = \sum_{\nu_1}\cdots\sum_{\nu_s} [\chi(\theta'_{1,\nu_1}\cdots\theta'_{s,\nu_s})|\theta''_{1,\nu_1}|\cdots|\theta''_{s,\nu_s}] . \quad (2.3)$$

This formula appears in [3; p.321] and is easily proved. Indeed, if $\Psi^{(s-1)}a_1 = \sum_i a_i^{(1)} \otimes \cdots \otimes a_i^{(s)}$, then

$$\langle \delta[\theta_1|\cdots|\theta_s] , [a_1|\cdots|a_{s+1}]\rangle = \langle [\theta_1|\cdots|\theta_s] , \chi(a_1)[a_2|\cdots|a_{s+1}]$$

$$= \sum_i \langle\theta_1, \chi(a_i^{(s)})a_2\rangle\cdots\langle\theta_s, \chi(a_i^{(1)})a_{s+1}\rangle$$

$$= \sum_i \sum_{\nu_1}\cdots\sum_{\nu_s} \prod_{k=1}^{s} \langle\chi(\theta'_{k,\nu_k}), a_i^{(s+1-k)}\rangle \langle\theta''_{k,\nu_k}, a_{k+1}\rangle$$

$$= \sum_{\nu_1}\cdots\sum_{\nu_s} \langle\chi(\theta'_{1,\nu_1}\cdots\theta'_{s,\nu_s}), a_1\rangle \prod_{k=1}^{s} \langle\theta''_{k,\nu_k}, a_{k+1}\rangle ,$$

as desired.

Thus the computer uses (2.1) to calculate $\mathrm{im}(Sq^3Sq^3)$ in $\overline{\mathbb{Z}_2[\zeta_1^4,\zeta_2^2,\zeta_3,\ldots]}^{\,s}$, and for each such element it calculates d_1 by (2.3), using the inductive formula $\chi(\zeta_i) = \sum_{j<i} \chi(\zeta_j)\zeta_{i-j}^{2^j}$. Terms in (2.3) with 1 in any position are ignored. The most difficult part is finding $\ker(d_1)/\mathrm{im}(d_1)$, which was done by hand. This step also will have to be computerized if this calculation is to be extended. Advantage was made of the following facts:

i) if $x = \sum [\theta_{1,i}|\cdots|\theta_{s,i}]$ is in E_1 , and σ is a permutation, then $x_\sigma = \sum [\theta_{\sigma(1),i}|\cdots|\theta_{\sigma(s),i}]$ is in E_1 , and $d_1(x_\sigma)$ is obtained from $d_1(x)$ by applying σ to components 2 to $s+1$ of all monomials;

ii) the external product of any element x of E_1 by ζ_1^4 in any component gives an element of E_1 on which d_1 is obtained by a similar external product on $d_1(x)$.

3. STATEMENT OF RESULTS

$H^{s,t}(V^*)$ is the degree t part of

$\ker(\delta_s|\mathrm{im}(Sq^3Sq^3))/\mathrm{im}(\delta_{s-1}|\mathrm{im}(Sq^3Sq^3))$ in

$$\begin{array}{ccc} \mathrm{im}(Sq^3Sq^3) & & \mathrm{im}(Sq^3Sq^3) \\ \cap & & \cap \\ \bar{W}^s & \xrightarrow{\delta_s} & \bar{W}^{s+1} \end{array}$$

where δ_s is as in (2.3). We use the following notation to indicate elements in $\bar{W}^s$: abc means $\zeta_1^a\zeta_2^b\zeta_3^c$, $\alpha \cdot \beta \cdot \gamma$ means $\alpha \otimes \beta \otimes \gamma$ $(= [\alpha|\beta|\gamma])$, and a bar over a group of elements separated by dots means the sum of all distinct permutations of these elements. For example, $\overline{4 \cdot 02} \cdot 8$ means $\zeta_1^4 \otimes \zeta_2^2 \otimes \zeta_1^8 + \zeta_2^2 \otimes \zeta_1^4 \otimes \zeta_1^8$, and $\overline{4 \cdot 4 \cdot 04}$ means $\zeta_1^4 \otimes \zeta_1^4 \otimes \zeta_2^4 + \zeta_1^4 \otimes \zeta_2^4 \otimes \zeta_1^4 + \zeta_2^4 \otimes \zeta_1^4 \otimes \zeta_1^4$.

<u>Theorem 3.1.</u> *If $t \leq 25$, $H^{s,t}(V^*) = 0$ or $\mathbb{Z}_2$, and the only cases in which it is $\mathbb{Z}_2$ are those listed below, where a nonzero element is given (nonuniquely) by Table 2.*

TABLE 2

s	t	element of $H^{s,t}(V^*)$
2	18	$\overline{8 \cdot 42} + \overline{82 \cdot 4} + \overline{\underset{\smile}{12} \cdot 02}$ ($\underset{\smile}{12}$ means ζ_1^{12} rather than $\zeta_1\zeta_2^2$)
3	18	$\overline{4 \cdot 02} \cdot 8$
2	20	$\overline{8 \cdot 04}$
3	20	$\overline{4 \cdot 4 \cdot 04} + \overline{02 \cdot 02 \cdot 8} + \overline{02 \cdot 4 \cdot 42}$
3	21	$\overline{02 \cdot 001 \cdot 8} + \overline{4 \cdot 001 \cdot 42} + \overline{4 \cdot 02 \cdot 401}$
4	21	$4 \cdot \overline{4 \cdot 02 \cdot 001}$
3	22	$4 \cdot 02 \cdot 04 + 4 \cdot 4 \cdot 002 + 4 \cdot 42 \cdot 8 + 02 \cdot 8 \cdot 8$
4	22	$\overline{4 \cdot 02 \cdot 02 \cdot 02} + \overline{42 \cdot 4 \cdot 4 \cdot 4} + 02 \cdot \overline{8 \cdot 4 \cdot 4} + 8 \cdot 02 \cdot 4 \cdot 4$
3	24	$4 \cdot \overline{8 \cdot 04}$
4	24	$8 \cdot \overline{4 \cdot 02 \cdot 02} + 04 \cdot 4 \cdot 4 \cdot 4$
4	25	$8 \cdot \overline{4 \cdot 02 \cdot 001}$
5	25	$4 \cdot 4 \cdot \overline{4 \cdot 02 \cdot 001}$

We illustrate the proof of 3.1 by considering $t = 24$. We begin by listing computer-generated elements of V^s_{24}, and $\delta_s(\ )$.

TABLE 3

s	element of V^s	$= Sq^3Sq^3(\downarrow)$	#	δ_s(this)
2	$8\cdot 44+04\cdot\underline{12}$	$021\cdot 421$	2	$\overline{8\cdot 04}\cdot 4+4\cdot\overline{8\cdot 04}$
3	$4\cdot\overline{8\cdot 04}$	$4\cdot 021\cdot 021$	3	0
3	$4\cdot 8\cdot 04+02\cdot\overline{8\cdot 42}+4\cdot 42\cdot 42$	$02\cdot 401\cdot 021$	3	$4_2+8_3+8_4$
3	$4\cdot 4\cdot 44+\overline{02\cdot 4}\cdot 82+02\cdot 02\cdot\underline{12}$	$02\cdot 001\cdot 421$	3	8_1+8_4
4	$8\cdot 4\cdot 02\cdot 02+04\cdot 4\cdot 4\cdot 4+42\cdot 4\cdot\overline{02\cdot 4}$	$021\cdot 4\cdot 02\cdot 001$	12	$\overline{4\cdot 02}\cdot 4\cdot\overline{4\cdot 02}$
5	$4\cdot\overline{02\cdot 4}\cdot\overline{02\cdot 4}$	$4\cdot 02\cdot 02\cdot 001\cdot 001$	5	0

Remarks

i) # means the number of distinct elements obtained by permuting the given element.

ii) If the second $s = 3$ element is added to the element obtained by interchanging its second and third components, the first $s = 3$ element is obtained.

iii) Numbers a_b in the final column are the sum of the 3 listed $s = 4$ elements whose {8,4,02,02}-term has a in the bth positions.

Row reduction shows that δ_4 is surjective with kernel $\langle 4_1, 4_2, 4_3, 4_4, 8_1, 8_2, 8_3\rangle$ (8_4 is the sum of the others.) Thus $H^{5,24}(V) = 0$ and $H^{4,24}(V)$ has basis $\{8_1\}$. Also, δ_3 is injective on the 6-dimensional subspace of V^3 spanned by elements of the second and third type, while $\ker(\delta_3)/\mathrm{im}(\delta_2)$ has as basis any one of the 3 V^3-elements of the first type.

Table 1 (or [4]) shows that V^1_* eventually becomes large, but we have

Theorem 3.2. $H^{1,*}(V) = 0$

Proof. There is a commutative diagram

$$\begin{array}{ccc} \overline{A//A_1}^* & \xrightarrow{\delta} & \overline{A//A_1}^* \otimes \overline{A//A_1}^* \\ \downarrow & & \downarrow \\ \bar{A}^* & \xrightarrow{\delta'} & \bar{A}^* \otimes \bar{A}^* , \end{array}$$

and $H^{1,*}(V)$ is $\ker(\delta|\mathrm{im}(Sq^3Sq^3))$, while $\ker(\delta') = \mathrm{Ext}_A^{1,*}(\mathbb{Z}_2,\mathbb{Z}_2)$. By [1], $\ker(\delta') = \langle\zeta_1^{2^n} : n \geq 0\rangle$, but these are not in $\mathrm{im}(Sq^3Sq^3)$, since $\mathrm{im}(\chi Sq^3)$ must involve $\zeta_i Sq^1 = \zeta_{i-1}^2$, but odd powers of ζ_2 do not appear in $\overline{A/\!/A_1}^*$. Indeed, in the bo-ASS these elements (h_n in the classical ASS) are in $H^*(\mathfrak{C})$. ∎

4. THE VANISHING LINE

We have been discussing how to calculate the irregular part $H^*(V)$ of the exact sequence (1.2). The other part contributing to the desired $E_2^{*,*}$ is $H^*(\mathfrak{C})$, which was completely calculated in [8]. We restate their result in a form which may be slightly less elucidating in that it does not name the elements, but slightly easier to use in writing abstract groups.

Theorem 4.1. ([8])
$$H^{0,t}(\mathfrak{C}) = \begin{cases} \mathbb{Z} & t = 0 \\ \mathbb{Z}_2 & t \equiv 1,2(8) \\ 0 & \text{otherwise} \end{cases}$$

$$H^{1,t}(\mathfrak{C}) = \begin{cases} \mathbb{Z}/2^{1+\nu(t)} & \text{if } t \equiv 0(4) \\ \mathbb{Z}_2 & \text{if } t \equiv 1,2(8) \\ 0 & \text{otherwise} \end{cases}$$

where $\nu(t)$ is the exponent of 2 in t. For $s \geq 2$, let $\mathcal{S} = \mathcal{S}_{s-1}$ denote the set of nondecreasing sequences of $s-1$ 2-powers with the last not repeated and >1. If $S \in \mathcal{S}$, let $|S|$ denote the sum of the entries of S, and S_1 the smallest entry of S. Let $\mathbb{Z}_2(t)$ denote $\mathbb{Z}_2$ in degree t. Then for $s \geq 2$,

$$H^{s,*}(\mathfrak{C}) = A_* \oplus B_*, \text{ where}$$

$$A_* = \begin{cases} \bigoplus_{S\in\mathcal{S}} \mathbb{Z}_2(8|S|-2s-\{{}^2_1\}) & \text{if } s \equiv \{{}^2_3\} \ (4) \\ 0 & \text{if } s \equiv 0 \text{ or } 1 \ (4) \end{cases}$$

$$B_* = \begin{cases} \bigoplus_{S\in\mathcal{S}} \mathbb{Z}_2(4(|S|-2S_1+m(2S_1,|S|-(2a+2)))) & \text{if } s = 4a+3 \text{ or } 4 \\ \bigoplus_{\substack{S\in\mathcal{S}: \\ |S|-[s/2] \not\equiv 0(2S_1)}} \mathbb{Z}_2(4(|S|-2S_1+m(2S_1,|S|-[s/2]))) & \text{if } s \equiv 1 \text{ or } 2(4) \end{cases}$$

where $m(b,e)$ is the smallest multiple of b greater than e.

The exact sequence (1.2) is not known (or expected) to be short exact, and so it is possible that the desired $E_2^{s,t}$ might be smaller than the sum of $H^{s,t}(V^*)$ and $H^{s,t}(\mathfrak{C}^*)$. In the range calculated so far, short exactness does hold. However the sequence does not split in $E_2^{4,24} \approx \mathbb{Z}_4$, a nontrivial extension of two $\mathbb{Z}_2$'s. We repeat the chart of $E_2^{s,t}$ for $t-s \leq 20$ given in [8], with numbers indicating cyclic summands of the indicated order, dots $\mathbb{Z}_2$'s, and lines the action of $\eta \in \pi_1(S^0)$ or $2 \in \pi_0(S^0)$, depending upon whether they increase the stem. We circle the elements which come from $H^*(V)$. Dashed lines indicate the action of $\nu \in \pi_3(S^0)$.

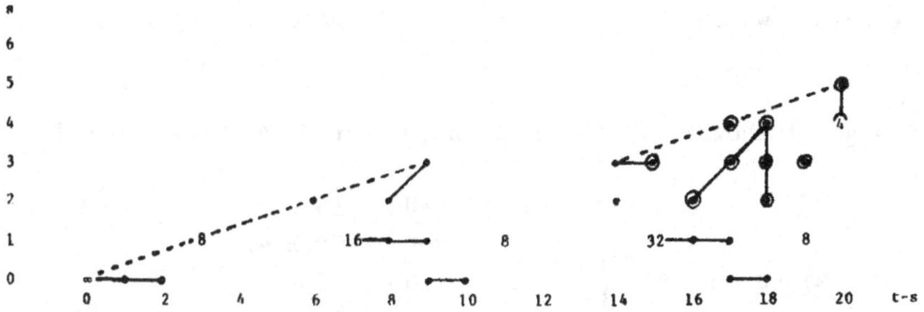

We also point out the following simple consequence of 4.1.

Corollary 4.2. If $0 \leq \varepsilon \leq 3$, then $H^{4a+\varepsilon,t}(\mathfrak{C}^*) = 0$ if $t < 4(6a+\varepsilon)$.

Proof. The element in B_* corresponding to $S = (1,..,1,2)$ is the first possible and occurs in the stated t. ∎

This says that $H^*(\mathfrak{C}^*)$ vanishes above a line of slope 1/5 in the usual (t-s,s)-chart. The same statement is true of the entire $E_2^{*,*}$, although here we do not know the precise edge. This result has played an important role in the work of Mahowald and his collaborators ([10], [6], [7], [12]). We feel that some readers might benefit from an expansion upon Mahowald's proof in [10]. This expansion benefits from many conversations with Mahowald.

We will study the algebraic analogue. For an A-module M, we define $E^{\sigma,s,t}(M)$ to be ker/im in

$$\mathrm{Ext}_A^{s,t}(W\otimes\bar{W}^{\sigma-1}\otimes M)\longrightarrow \mathrm{Ext}_A^{s,t}(W\otimes\bar{W}^{\sigma}\otimes M)\longrightarrow \mathrm{Ext}_A^{s,t}(W\otimes\bar{W}^{\sigma+1}\otimes M) \quad (4.3)$$

where $W = A//A_1$. This is called the E_2-term of the Mahowald spectral sequence in [14;2.2]. If X is a (bo,H)-prime spectrum ([14;3.3], [8;4.4]), then (4.3) with $M = H^*X$ is the Adams-filtration preserving part in filtration s of the homomorphisms

$$\pi_{t-s}(bo\wedge\overline{bo}^{\sigma-1}\wedge X)\longrightarrow \pi_{t-s}(bo\wedge\overline{bo}^{\sigma}\wedge X)\longrightarrow \pi_{t-s}(bo\wedge\overline{bo}^{\sigma+1}\wedge X)$$

whose ker/im define $E_2^{\sigma,t-s}(X)$, and hence $\bigoplus_s E^{\sigma,s,\sigma+u+s}(H^*X)$ provides an upper bound for an associated graded of $E_2^{\sigma,\sigma+u}(X)$.

The result which will be proved, a mild generalization of [11;Cor. 3.5], is

Theorem 4.4. (Mahowald). If M is a connected A_0-free A-module, then

$$E^{\sigma,s,t}(M) = 0 \text{ if } t < \begin{cases} 4\sigma & s=0, \ \sigma\le 6 \\ 6\sigma-10 & s=0, \ \sigma\equiv 0,1(4) \\ 6\sigma-12 & s=0, \ \sigma\equiv 2(4) \\ 6\sigma-11 & s=0, \ \sigma\equiv 3(4) \\ 6\sigma+3s-2 & s>0 \end{cases}$$

Corollary 4.5. *If X is a (bo,H)-prime connected spectrum with H^*X free as an A_0-module, then*

$$E_2^{\sigma,t}(X) = 0 \text{ if } t-\sigma < \begin{cases} 3\sigma & \sigma\le 6 \\ 5\sigma-12 & \sigma\ge 7 \end{cases}.$$

After proving this result, we will comment upon its application to $\pi_*(S^0)$.

When $s > 0$, the theorem follows immediately from

Lemma 4.6. *If M is a connected A_0-free A-module, $s > 0$, and $\sigma \ge 0$, then*

$$\mathrm{Ext}_{A_1}^{s,t}(\bar{W}^\sigma \otimes M) = 0 \text{ if } t-s<6\sigma+2s-\begin{cases} 0 & \\ 2 & \text{if } s-\sigma \equiv 2,3(4) \\ 1 & \end{cases} \quad \begin{matrix} 0\ (4) \\ \\ 1\ (4) \end{matrix}.$$

<u>Proof</u>. It suffices to prove the lemma when $M = A_0$, since short exact sequences of A_1-modules give long exact sequences in Ext. (n.b.: that this is not true for $E^{*,*,*}$ is one reason why Theorem 4.4 is so difficult to prove.)

The stable A_1-structure of $\bar{W}^\sigma$ is completely known (by [13;4.3 and 5.11] or [10;2.5 and 2.7] or [6;3.7 and 3.12]. Only the number of free A_1's is problematic, but this is not relevant to Ext^s when $s > 0$. The calculation of $\mathrm{Ext}_{A_1}(\bar{W}^\sigma \otimes A_0)$ above filtration zero is simpler; it is just a sum of translates of $\mathrm{Ext}_{A_1}(A_0)$, which is well known to have chart

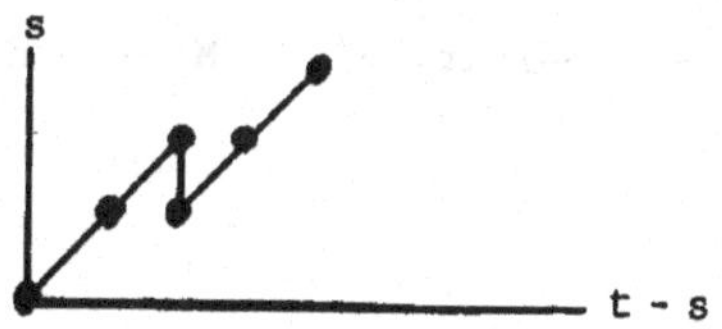

acted on freely by an operator of $(t-s,s) = (8,4)$. Indeed, if $\bar{n} = (n_1,\dots n_\sigma)$ is in the set N^σ of σ-tuples of positive integers, let $|\bar{n}| = \sum n_j$, $\alpha(\bar{n}) = \sum \alpha(n_j) = 4A(\bar{n}) + \varepsilon(\bar{n})$, with $0 \leq \varepsilon(\bar{n}) \leq 3$. Then for $s > 0$

$$\mathrm{Ext}_{A_1}^{s,t}(\bar{W}^\sigma \otimes A_0) \approx \bigoplus_{\bar{n}\in N^\sigma} \mathrm{Ext}_{A_1}^{s+4-\varepsilon(\bar{n}),t+12-\varepsilon(\bar{n})}(\Sigma^{8(|\bar{n}|-A(\bar{n}))}A_0).$$

The first summand corresponds to $\bar{n} = (1,\dots,1)$ and begins

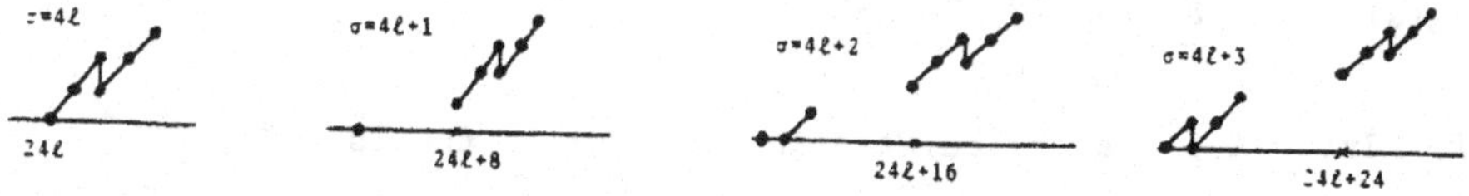

The proof of 4.4 when $s = 0$ will utilize 6 lemmas.

Lemma 4.7. $E^{\sigma,s,t}(\bar{W}\otimes M) = E^{\sigma+1,s,t}(M)$ for $\sigma \geq 0$.

Proof. They have the same definition. ∎

Lemma 4.8. If M is A_1-free, then

$$E^{\sigma,s,t}(M) \approx \begin{cases} \mathrm{Ext}_A^{\sigma,t}(M) & \text{if } s = 0 \\ 0 & \text{if } s > 0 . \end{cases}$$

Proof. The topological proof of [10;5.15] could be translated into an algebraic proof, viewing the modules in (4.3) as part of an A-resolution of M . At a more basic level, this argument can be presented as follows: Since $W\otimes M$ is A-free, $E^{\sigma,s,t}(M) = 0$ if $s > 0$, and there are exact sequences

$$0 \longrightarrow \mathrm{Ext}_A^0(\bar{W}^\sigma\otimes M) \xrightarrow{\alpha_\sigma} \mathrm{Ext}_{A_1}^0(\bar{W}^\sigma\otimes M) \xrightarrow{\beta_\sigma} \mathrm{Ext}_A^0(\bar{W}^{\sigma+1}\otimes M) \longrightarrow \mathrm{Ext}_A^1(\bar{W}^\sigma\otimes M) \longrightarrow 0$$

and isomorphisms

$$\mathrm{Ext}_A^1(\bar{W}^\sigma\otimes M) \approx \mathrm{Ext}_A^{\sigma+1}(M) .$$

Hence

$$E^{\sigma,0,t}(M) = \frac{\ker(\alpha_{\sigma+1}\beta_\sigma)}{\mathrm{im}(\alpha_\sigma\beta_{\sigma-1})} = \frac{\mathrm{im}(\alpha_\sigma)}{\mathrm{im}(\alpha_\sigma\beta_{\sigma-1})} \approx \frac{\mathrm{Ext}_A^0(\bar{W}^\sigma\otimes M)}{\mathrm{im}(\beta_{\sigma-1})} \approx \mathrm{Ext}_A^1(\bar{W}^{\sigma-1}\otimes M) \approx \mathrm{Ext}_A^\sigma(M). \blacksquare$$

Lemma 4.9. *If M is an A_1-free connected A-module, then $\mathrm{Ext}_A^{s,t}(M) = 0$ if $t < g(s)$, where $g(s)$ is given by*

s	1	2	3	4	5	6	≥7
g(s)	4	8	15	22	28	35	6s

Proof. $\mathrm{Ext}_{A_2}(A_1)$ was calculated in W. H. Lin's thesis ([9],[7]). As in [7;p.655] or [3;3.2], an $A//A_2$-resolution of M shows that $\mathrm{Ext}_A(M)$ has the same vanishing line. ∎

Lemma 4.10. *Let $0 \longrightarrow B \xrightarrow{\alpha} C \longrightarrow D \longrightarrow 0$ be a short exact sequence of A-modules*

a) If B or D is A_1-free, then for $s \geq 0$ there is a long exact sequence.

$$\to E^{\sigma-1,s,t}(B) \to E^{\sigma,s,t}(D) \to E^{\sigma,s,t}(C) \to E^{\sigma,s,t}(B) \to E^{\sigma+1,s,t}(D) \to .$$

b) If $Ext^{1,t}_{A_1}(\bar{W}^j \otimes D) \to Ext^{1,t}_{A_1}(\bar{W}^j \otimes C)$ is injective for $j \geq \sigma - 1$, then there is an exact sequence as above for $s = 0$ beginning with $E^{\sigma-1,s,t}(C)$.

Proof.

a) is the algebraic precursor of [11;3.1 and 3.2].

There are long exact sequences $\mathcal{S}_\sigma$

$$\to Ext_A^{s-1,t}(W \otimes \bar{W}^\sigma \otimes B) \to Ext_A^{s,t}(W \otimes \bar{W}^\sigma \otimes D) \to Ext_A^{s,t}(W \otimes \bar{W}^\sigma \otimes C) \to\cdots$$

$$\cdots \to Ext_A^{s,t}(W \otimes \bar{W}^\sigma \otimes B) \to$$

for each σ, and morphisms $\mathcal{S}_\sigma \to \mathcal{S}_{\sigma+1}$. If D is A_1-free, then $Ext_A^{s,t}(W \otimes \bar{W}^\sigma \otimes D) = 0$ for $s > 0$, and so $\langle \mathcal{S}_\sigma \rangle$ reduces to short exact sequences of cochain complexes

$$0 \to \mathcal{E}(D) \to \mathcal{E}(C) \to \mathcal{E}(B) \to 0$$

whose cohomology defines $E^{*,*,*}$ in (4.3). The same conclusion is true when B is A_1-free, but we must be a bit more careful at $s = 0$ and 1. There are exact sequences

$$0 \to Ext_A^0(W \otimes \bar{W}^\sigma \otimes D) \to Ext_A^0(W \otimes \bar{W}^\sigma \otimes C) \xrightarrow{\alpha^*} Ext_A^0(W \otimes \bar{W}^\sigma \otimes B) \to\cdots$$

$$\cdots \to Ext_A^1(W \otimes \bar{W}^\sigma \otimes D) \to Ext_A^1(W \otimes \bar{W}^\sigma \otimes C) \to 0 ,$$

and, since A is an injective A-module ([15]), α^* is surjective. The proof of (b) is the standard argument for the cohomology long exact sequence induced by a short exact sequence of cochain complexes.∎

<u>Lemma 4.11</u>. *Consider the following A-modules:*

$$M_1 = \Sigma^4 \langle 1, Sq^2, Sq^3 \rangle$$

$$M_2 = \Sigma^8 \langle Sq^i : 0 \leq i \leq 7,\ i \neq 1 \rangle \quad (Sq^4 Sq^2 = 0)$$

$$M_3 = \langle 1,\ Sq^1,\ Sq^2,\ Sq^2Sq^1,\ Sq^4 = Sq^3Sq^1\rangle \quad (Sq^3 = 0)\ .$$

There are monomorphisms of A-modules with cokernel A_1-free

$$\Sigma^{10}M_3 \otimes A_0 \longrightarrow M_1 \otimes M_1 \otimes A_0$$

$$\Sigma^{18}M_3 \otimes A_0 \longrightarrow M_1 \otimes M_2 \otimes A_0\ .$$

Proof. The two cases are practically the same; we do the second. One verifies that

$$\Sigma^{18}1 \otimes 1 \mapsto \Sigma^{12}(1 \otimes Sq^6 \otimes 1)$$

$$\Sigma^{18}1 \otimes Sq^1 \mapsto \Sigma^{12}(Sq^2 \otimes Sq^4 \otimes Sq^1 + Sq^3 \otimes Sq^4 \otimes 1 + 1 \otimes Sq^7 \otimes 1)$$

is well-defined on a generating set and injective. (The relations in both the domain module and the submodule of $M_1 \otimes M_2 \otimes A_0$ generated by the indicated elements are $Sq^3g_1 + Sq^2g_2$ and $(Sq^4+Sq^3Sq^1)g_1 + Sq^2Sq^1g_2$, together with those due to the truncation above.

All modules are Q_0-free and have a single Q_1-homology class which is mapped across. Thus by Wall's theorem ([16]), the cokernel is A_1-free. ∎

Corollary 4.12. *If M_i is as in 4.11 and g as in 4.9, then*

$E^{\sigma,0,t}(M_1^4 \otimes A_0)$ is isomorphic to a quotient of $E^{\sigma,0,t}(\Sigma^{24}A_0)$

if $t < g(\sigma) + 16$, with equality if also $t < g(\sigma-1) + 20$,

and $E^{\sigma,0,t}(M_1^3 \otimes M_2 \otimes A_0) \approx E^{\sigma,0,t}(\Sigma^{32}A_0)$ if $t < g(\sigma) + 20$.

Remark. This result (more precisely, its proof) is a corrected version of [11;3.3].

Proof. We first show there are short exact sequences of A-modules with F_j free as A_1-modules

$$
\begin{array}{ccccccccc}
 & & & & 0 & & & & \\
 & & & & \uparrow & & & & \\
 & & & & F_2 & & & & \\
 & & & & \uparrow & & & & \\
 & & 0 & \longrightarrow & L_1 & \xrightarrow{\phi_1} & M\otimes A_0 & \longrightarrow F_1 \longrightarrow & 0 \\
 & & & & \uparrow{\scriptstyle \phi_2} & & & & \\
0 \longrightarrow & F_3 & & \longrightarrow & L_2 & \xrightarrow{\phi_3} & \Sigma^d A_0 & \longrightarrow & 0 \\
 & & & & \uparrow & & & & \\
 & & & & 0 & & & &
\end{array}
$$

where $M = M_1^4$, $d = 24$, and F_i beginning in degree 16,18,20, for $i = 1,2,3$ or $M = M_1^3 \otimes M_2$, $d = 32$, and F_i beginning in degree 20,26,28.

The top horizontal sequence is obtained by applying $\otimes M_1 \otimes M_1$ to the sequences of 4.11. Thus $L_1 = \Sigma^e M_3 \otimes A_0 \otimes M_1 \otimes M_1$, and the vertical sequence is $\Sigma^e M_3 \otimes$ (first sequence of 4.9). $(e = 10$ or $18)$. We have $L_2 = \Sigma^{10+e} M_3 \otimes M_3 \otimes A_0$. Since M_3 is self-dual, there is a homomorphism $M_3 \otimes M_3 \to \Sigma^4 \mathbb{Z}_2$ dual to the identity, and inducing an isomorphism in Q_1-homology.

By Lemma 4.10a these short exact sequences induce long exact sequences in $E^{\sigma,0,t}(\ )$, and we use 4.8 and 4.9 to estimate where ϕ_1, ϕ_2, and ϕ_3 may fail to be isomorphisms. ∎

For the rest of this paper, we let $f(\sigma)$ denote the smallest t such that $E^{\sigma,0,t}(A_0) \neq 0$.

<u>Lemma 4.13</u>. *If $f(\sigma) \leq 6\sigma - 6$, then for every connected A_0-free A-module M*

$$E^{\sigma,0,t}(M) = 0 \quad \text{for} \quad t < f(\sigma) .$$

<u>Proof</u>. We prove it for finite modules M by induction on dim(M). There is a short exact sequence $0 \to M' \to M \to A_0 \to 0$ with M' A_0-free and connected. By 4.6, the assumption that $t < 6\sigma - 6$, and 4.10b there is an exact sequence

$$E^{\sigma,0,t}(A_0) \longrightarrow E^{\sigma,0,t}(M) \longrightarrow E^{\sigma,0,t}(M') ,$$

which implies the desired result since the groups on both ends are 0 by the induction hypothesis.

The passage from finite to infinite M follows by a similar exactness argument.■

<u>Proof of 4.4</u>. We may assume $\sigma \geq 7$, for $f(\sigma) \geq 4\sigma$ is clear. There are short exact sequences of A-modules, with M_i as in 4.11 and K_2 beginning in degree 12.

$$\begin{array}{ccccccccc} & & & & 0 & & \\ & & & & \downarrow & & \\ & & & & K_2 & & \\ & & & & \downarrow & & \\ 0 \leftarrow & M_1 & \leftarrow & \bar{W} & \leftarrow K_1 & \leftarrow 0 \\ & & & & \downarrow & & \\ & & & & M_2 & & \\ & & & & \downarrow & & \\ & & & & 0 & & \end{array}$$

Thus there is a filtration

$$F_5 \subset F_4 \subset F_3 \subset F_2 \subset F_1 \subset \bar{W}^4 \qquad (4.14)$$

so that $\bar{W}^4/F_1 \approx M_1^4$, $F_i/F_{i+1} \approx M_1^3 \otimes K_1$ for $1 \leq i \leq 4$, and F_5 begins in degree 24.
[[F_i is spanned by products $a_1 \otimes a_2 \otimes a_3 \otimes a_4$ which satisfy (i) at least one a_j is in K_1 , and (ii) if a_j is the only factor in K_1, then $j \geq i$]] .

Assume $f(\sigma) \leq 6\sigma - 6$ and $t < f(\sigma) + 24$. By the proof of 4.12, $\mathrm{Ext}_{A_1}^{1,t}(\bar{W}^{\sigma-1} \otimes M_1^3 \otimes M_2 \otimes A_0) \approx \mathrm{Ext}_{A_1}^{1,t}(\bar{W}^{\sigma-1} \otimes \Sigma^{32}A_0)$, and by 4.6 this is 0 in this range. Thus by 4.10b and the vertical sequence above there is an exact sequence.

$$E^{\sigma,0,t}(M_1^3 \otimes M_2 \otimes A_0) \longrightarrow E^{\sigma,0,t}(M_1^3 \otimes K_1 \otimes A_0) \longrightarrow E^{\sigma,0,t}(M_1^3 \otimes K_2 \otimes A_0) \ .$$

The third group is 0 by 4.13 and the first group is 0 by 4.12 and 4.13. Thus $E^{\sigma,0,t}(M_1^3 \otimes K_1 \otimes A_0) = 0$.

Using the filtration 4.14 inductively, we show that $E^{\sigma,0,t}(F_1 \otimes A_0) = 0$ under the same hypothesis $(t < f(\sigma) + 24 \leq 6\sigma + 18)$.
[[It is true for F_5 by 4.13. 4.10b applies to each $(F_{i+1} \longrightarrow F_i \longrightarrow M_1^3 \otimes K_1) \otimes A_0$, since in this range

$\mathrm{Ext}^1_{A_1}(M_1^3 \otimes K_1 \otimes A_0 \otimes \bar{W}^{\sigma-1})$ is no larger than $\mathrm{Ext}^1_{A_1}(M_1^3 \otimes M_2 \otimes A_0 \otimes \bar{W}^{\sigma-1})$ $\approx \mathrm{Ext}^1_{A_1}(\Sigma^{32}A_0 \otimes \bar{W}^{\sigma-1})$. This allows us to work up to F_1 in 4 steps.]]
4.10b also applies to $(0 \longrightarrow F_1 \longrightarrow \bar{W}^4 \xrightarrow{k} M_1^4 \longrightarrow 0) \otimes A_0$ since k is the projection to a direct A_1-summand. Thus, applying 4.12 and 4.13 for $E(M_1^4 \otimes A_0)$, $E^{\sigma,0,t}(\bar{W}^4 \otimes A_0) = 0$ if $t < f(\sigma) + 24 \leq 6\sigma + 18$ and $t < g(\sigma) + 16$. Using 4.7, this says if $f(\sigma) \leq 6\sigma - 6$, then

$$f(\sigma+4) \geq \min(f(\sigma) + 24\ ,\ g(\sigma) + 16)\ . \qquad (4.15)$$

The induction is initialized with $f(\sigma) \geq 4\sigma$ for $\sigma = 3,4,5$ and 6. The recursive formula (4.15) then implies 4.4 for $\sigma \geq 7$. ∎

We add a little more detail to the argument of [11] which deduces the vanishing line for S^0 from that of $S^0 \cup_2 e^1$.

<u>Theorem 1.3</u>. $E_2^{\sigma,t}(S^0) = 0$ if $t - \sigma < \begin{cases} 3\sigma & \sigma \leq 6 \\ 5\sigma - 11 & \sigma \equiv 0,1(4) \\ 5\sigma - 13 & \sigma \equiv 2(4) \\ 5\sigma - 12 & \sigma \equiv 3(4) \end{cases}$

<u>Proof</u>. That the groups are 0 for $t < 4\sigma$ is clear.

As was alluded to in the proof of 4.6, the notation of which we use here, $\mathrm{Ext}_{A_1}(\bar{W}^\sigma)$ was shown in [10] or [6] to be a sum of filtration 0 $\mathbb{Z}_2$'s, plus for every $\bar{n} \in N^\sigma$ a chart of the form

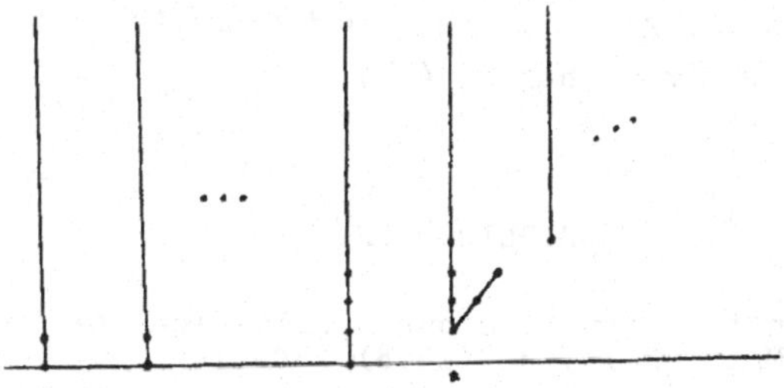

where towers occur every 4 degrees, begin in degree $4|\bar{n}|$, and first leave filtration zero in degree * roughly $8|\bar{n}| - 2\alpha(\bar{n})$. A key point is that this latter degree is greater than those considered in 1.3.

We first argue that the towers form an acyclic subcomplex of

$$\cdots \to \mathrm{Ext}_{A_1}(\bar{W}^{\sigma-1}) \to \mathrm{Ext}_{A_1}(\bar{W}^{\sigma}) \to \mathrm{Ext}_{A_1}(\bar{W}^{\sigma+1}) \to \quad \cdots .$$

This is probably not a consequence of finiteness of $\pi_*(S^0)$, since that would involve a statement about E_∞ , not E_2 . It is certainly a corollary of [8]. At a more basic level, it follows from the correspondence between infinite towers and Q_0-homology. ([5;Lemma 4]). By [2;6.8], $H_*(W^*;Q_0) = \mathbb{Z}_2[\zeta_1^4]$, and the complex

$$\to H_*(W^*\otimes \bar{W}^{*\sigma-1};Q_0) \to H_*(W^*\otimes \bar{W}^{*\sigma};Q_0) \to H_*(W^*\otimes \bar{W}^{*\sigma+1};Q_0) \to$$

is clearly acyclic. Since we are in the range in which all towers begin in filtration 0, we may deduce that they are an acyclic subcomplex. Thus $E^{\sigma,s,t}(\mathbb{Z}_2)$ is 0 in this range for $s > 0$.

In the exact Ext_{A_1}-sequence induced by

$$(0 \to \Sigma\mathbb{Z}_2 \to A_0 \to \mathbb{Z}_2 \to 0) \otimes \bar{W}^{\sigma} ,$$

the boundary homomorphisms $\mathrm{Ext}^0_{A_1}(\Sigma\mathbb{Z}_2 \otimes \bar{W}^{\sigma}) \xrightarrow{\delta} \mathrm{Ext}^1_{A_1}(\mathbb{Z}_2 \otimes \bar{W}^{\sigma})$ correspond to $\cdot h_0$. If we divide out by the towers, or their mod 2 reduction for A_0 , we obtain a short exact sequence of cochain complexes (as σ varies), and hence a long exact sequence

$$\to E^{\sigma,0,t}(\mathbb{Z}_2) \to E^{\sigma,0,t}(A_0) \to E^{\sigma,0,t}(\Sigma\mathbb{Z}_2) \to E^{\sigma+1,0,t}(\mathbb{Z}_2) \to$$

in our range. Thus if $E^{\sigma,0,\sigma+u}(A_0) = 0 = E^{\sigma+1,0,\sigma+u}(A_0)$ for all $\sigma \geq \sigma_0$, then the groups $E^{\sigma,0,\sigma+u-1}(\mathbb{Z}_2)$ are isomorphic for all $\sigma \geq \sigma_0$, and hence are 0 by choosing σ so that $\sigma + u - 1 \leq 4\sigma$. Thus the vanishing line of 1.3 follows from that of 4.4.

BIBLIOGRAPHY

1. J. F. Adams, "On the structure and applications of the Steenrod algebra", Comm. Math. Helv. 32 (1958) 180-214.

2. D. W. Anderson, E. H. Brown, and F. P. Peterson, "The structure of the spin cobordism ring", Annals of Math. 86 (1967) 271-298.

3. D. W. Anderson and D. M. Davis, "A vanishing theorem in homological algebra", Comm. Math. Helv. 48 (1973) 318-327.

4. G. Carlsson, "On the stable splitting of $bo \wedge bo$ and torsion operations in connective K-theory", Pac. Jour. Math. 87 (1980) 283-297.

5. D. M. Davis, "The cohomology of the spectrum bJ", Bol. Soc. Mat. Mex. 20 (1975) 6-11.

6. D. M. Davis, S. Gitler, and M. Mahowald, "The stable geometric dimension of vector bundles over real projective spaces", Trans. Amer. Math. Soc. 268 (1981) 39-61.

7. D. M. Davis and M. Mahowald, "v_1- and v_2-periodicity in stable homotopy theory", Amer. Jour. Math. 103 (1981) 615-659.

8. W. Lellman and M. Mahowald, "The bo-Adams spectral sequence", to appear in Trans. Amer. Math. Soc.

9. W. H. Lin. "An Adams-type spectral sequence for Hopf subalgebras of the Steenrod algebra", thesis, Northwestern Univ., 1974.

10. M. Mahowald, "bo-resolutions", Pac. Jour. Math. 92 (1981) 365-383.

11. __________, "An addendum to bo-resolutions", Pac. Jour. Math. 11 (1984) 117-123.

12. __________, "The image of J in the EHP sequence", Annals of Math. 116 (1982) 65-117.

13. R. J. Milgram, "The Steenrod algebra and its dual for connective K-theory", Reunion sobre homotopia, Soc. Mat. Mex. (1975) 127-158.

14. H. R. Miller, "On relations between Adams spectral sequences with an application to the stable homotopy of the Moore space", J. Pure Appl. Algebra 20 (1981) 287-312.

15. J. Moore and F. P. Peterson, "Nearly Frobenius algebras, Poincare algebras, and their modules", J. Pure Appl. Algebra 3 (1973) 83-93.

16. C. T. C. Wall, "A characterization of simple modules over the Steenrod algebra mod 2", Topology 1 (1962) 249-254.

The rigidity of L(n)

Nicholas J. Kuhn*
Department of Mathematics
Princeton University
Princeton, New Jersey 08544

§1 Introduction

Let $L(n) = \Sigma^{-n} SP^{p^n}(S)/SP^{p^{n-1}}(S)$, localized at a fixed prime p, where $SP^k(S)$ denotes the kth symmetric product of the sphere spectrum S. This family of spectra is known to have many beautiful properties — in particular,

(i) if $M(n) = L(n) \vee L(n-1)$ then $M(n)$ is the stable wedge summand of $B(\mathbb{Z}/p)^n_+$ corresponding to the Steinberg representation of $GL_n(\mathbb{Z}/p)$ [9], and

(ii) the sequence

$$\cdots \longrightarrow L(2) \xrightarrow{\partial_1} L(1) \xrightarrow{\partial_0} L(0) \xrightarrow{\varepsilon} H\mathbb{Z}_{(p)}$$

is "exact" [5,7], thus induces a long exact sequence in homotopy groups.

It is the purpose of this note to prove that $L(n)$ has certain rigidity properties. For $n \geq 1$, the mod p cohomology of $L(n)$ is infinitely generated as a module over the Steenrod algebra A. In spite of this, we show that $H^*(L(n))$ is rigid:

Theorem 1.1 $\mathrm{End}_A(H^*(L(n))) = \mathbb{Z}/p$.

The $n = 1$ case of this theorem is well known [1] - $L(1)$ is the suspension spectrum of $B\Sigma_p$, and $H^*(\Sigma_p)$ can be shown to be rigid over A using only the operations β and P^1. In contrast, we have the following example, presumably typical of the $n > 1$ situation.

Example 1.2 Let $p = 2$, and let B be any finite subalgebra of A. Then

$$\dim_{\mathbb{Z}/2} \mathrm{End}_B(H^*(L(2))) \geq 2.$$

---

*Partially supported by the N.S.F. and the Sloan Foundation.

One immediate consequence of the theorem is

Corollary 1.3 Any self map $f: L(n) \longrightarrow L(n)$ that is nonzero on the bottom cell is a homotopy equivalence.

By "nonzero on the bottom cell" we mean that $H_{2p^n-n-2}(f) \neq 0$.

In [1], Adams used the rigidity of $B\Sigma_p$ to prove a "uniqueness" statement about the Kahn-Priddy Theorem. Corollary 1.4 gives a generalization of this.

Corollary 1.4 Let $E_n = \Sigma^{-n} H\mathbb{Z}_{(p)}/SP^{p^{n-1}}(S)$. If $f: L(n) \longrightarrow E_n$ is any map with $H_{2p^n-n-2}(f) \neq 0$ then $\Omega^\infty f: \Omega^\infty L(n) \longrightarrow \Omega^\infty E_n$ has a homotopy section.

To see that Corollary 1.3 implies this last corollary, we argue as in [6, Example 2.8]: the "projectivity" of $L(n)$ (property (i) above) implies that f lifts

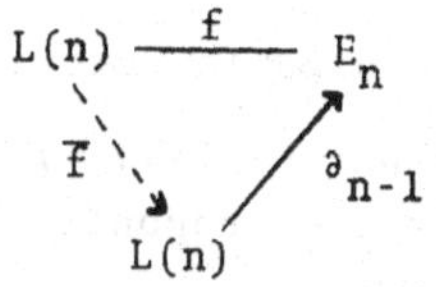

,

where ∂_{n-1} is the map in property (ii). With our assumptions on f, $H_{2p^n-n-2}(\bar{f})$ will be an isomorphism. Thus $\bar{f}$ will be a homotopy equivalence and Corollary 1.4 follows.

Theorem 1.1 is proved by combining (i) with the formula of Adams, Gunawardena, and Miller [2]:

(1.5) $$\mathbb{Z}/p[M_n(\mathbb{Z}/p)] \simeq \mathrm{End}_A(H^*((\mathbb{Z}/p)^n)).$$

Here $\mathbb{Z}/p[M_n(\mathbb{Z}/p)]$ is the semigroup ring with basis the semigroup $M_n(\mathbb{Z}/p)$ of $m \times n$ matrices over $\mathbb{Z}/p$, and the isomorphism is the ring homomorphism induced by the cohomology functor.

Using (1.5), it follows that, if $e \in \mathbb{Z}/p[GL_n(\mathbb{Z}/p)]$ is an idempotent representing the Steinberg module, then

$$\mathrm{End}_A(H^*(M(n))) = e\,\mathbb{Z}/p[M_n(\mathbb{Z}/p)]e.$$

Since $M(n) = L(n) \vee L(n-1)$, Theorem 1.1 thus follows from

Proposition 1.6 $\dim_{\mathbb{Z}/p} e\mathbb{Z}/p[M_n(\mathbb{Z}/p)]e = 2$.

We prove this in §2. The proof is an exercise in the use of row and column operations -- classical elementary linear algebra. Example 1.2 is discussed in §3.

Remarks 1.7

(1) Our proof of Proposition 1.6 yields more: $em = 0$ whenever $m \in M_n(\mathbb{Z}/p)$ has rank $\leq n-2$. From this it follows that

$$\mathrm{Hom}_A(H^*(L(n)),H^*(L(m))) = 0 \quad \text{if} \quad m \neq n.$$

Independently, Nishida [10] has performed a similar calculation and used a geometric version of (1.5) to prove that (when $p = 2$)

$$[L(n),L(m)] = 0 \quad \text{unless} \quad m = n \quad \text{or} \quad n-1,$$
$$[L(n),L(n)] = [L(n),L(n-1)] = \hat{\mathbb{Z}}_2.$$

(2) The isomorphism (1.5) can be used to describe wedge decompositions of $B(\mathbb{Z}/p)^n_+$ into indecomposable pieces [4]. In particular, the splitting of $e \in \mathbb{Z}/p[GL_n(\mathbb{Z}/p)]$ into two orthogonal primitive idempotents in $\mathbb{Z}/p[M_n(\mathbb{Z}/p)]$ (corresponding to the splitting $M(n) = L(n) \vee L(n-1)$) can be shown to be a perfectly generic occurrence -- i.e. has nothing to do with special properties of the Steinberg module.

Some of these results were presented at the University of Washington during a visit in January, 1985. The author wishes to thank the topologists there for the invitation to participate in their topology emphasis year.

§2 Proof of Proposition 1.6

We introduce some notation:

$M = M_n(\mathbb{Z}/p)$,

$G = GL_n(\mathbb{Z}/p)$,

B = Borel subgroup of G consisting of upper triangular matrices,

U = subgroup of B consisting of matrices with 1's on the main diagonal,

W = group of permutation matrices,

E = subset of M consisting of matrices in "echelon form",

C = subset of E consisting of matrices in "canonical form".

Here "echelon matrix" means a matrix reduced by using row operations: the nonzero rows come before the zero rows, the first nonzero term in a nonzero row is a 1, and the leading 1 for the $(i+1)^{st}$ row will be to the right of the leading 1 for the ith row. A "canonical matrix" is an echelon matrix with nonzero rows equal to some subset of the standard basis $\{\vec{e}_1,\ldots,\vec{e}_n\}$ of $(\mathbb{Z}/p)^n$.

In $\mathbb{Z}/p[G]$, let $\overline{B} = \sum_{b\in B} b$, $\overline{W} = \sum_{w\in W} \operatorname{sgn}(w)w$, and $e = \overline{B}\,\overline{W}$. Then $e\,\mathbb{Z}/p[G]$ is the Steinberg module. Let $I_n \in M$ be the identity matrix and let $I_{n-1} \in M$ correspond to the projection onto the first (n-1) coordinates of $(\mathbb{Z}/p)^n$.

Lemma 2.1

(1) $e\,\mathbb{Z}/p[G] = e\,\mathbb{Z}/p[U]$.

(2) $\mathbb{Z}/p[U]e = \mathbb{Z}/p\, e$.

Proof (1) is well known, and (2) is clear: $u\overline{B} = \overline{B}$ for all $u \in U$.

We remark, en passant, that (1) and (2) imply that $e\,\mathbb{Z}/p[G]e \simeq \mathbb{Z}/p$, from which it follows that the Steinberg module is absolutely irreducible (since a group ring is a symmetric algebra [8]).

The next lemma generalizes Lemma 2.1.

Lemma 2.2

(1) $e\,\mathbb{Z}/p[M] = e\,\mathbb{Z}/p[E]$.

(2) $\mathbb{Z}/p[E]e = \mathbb{Z}/p[C]e$.

Proof Given $m \in M$, there exists $g \in G$ and $h \in E$ such that $m = gh$. Then $em = egh \in \mathbb{Z}/p[UE]$, by Lemma 2.1(1). But $UE = E$, so (1) follows. To show (2), given $h \in E$, there exists $c \in C$ and $u \in U$ such that $h = cu$. Then $he = cue \in \mathbb{Z}/p[C]e$, using Lemma 2.1(2).

Lemma 2.3

(1) If $h \in E$ and the rank of $h \le n-2$, then $\overline{W}h = 0$ (and thus $eh = 0$).

(2) If $c \in C$, c has rank $n-1$, and $c \neq I_{n-1}$, then p divides $c\overline{B}$ (and thus $ce = 0$ in $\mathbb{Z}/p[C]e$).

Proof of (1) The matrices will cancel in pairs: if $w_{n-1} \in W$ interchanges rows (n-1) and n then, for all $w \in W$, $ww_{n-1}h = wh$ and

$\mathrm{sgn}(ww_{n-1}) = -\mathrm{sgn}(w)$.

Proof of (2) If $c \in C$ has rows $\vec{e}_1,\dots,\vec{e}_{i-1},\vec{e}_{i+1},\dots,\vec{e}_n,\vec{0}$ then left multiplication by c deletes the ith rows and adds a zero row at the bottom of any matrix. It follows that, given $b \in B$ there will be $(p-1)p^{n-i}$ matrices $b' \in B$ with $cb' = cb$. Thus p^{n-i} will divide $c\bar{B}$.

From the last two lemmas we conclude

Corollary 2.4 $e\,\mathbb{Z}/p[M]e$ is spanned by $\{e, eI_{n-1}e\}$.

Proposition 1.3 then follows from the observation that $\dim e\,\mathbb{Z}/p[M]e \geq 2$, a consequence of either $M(n) = L(n) \vee L(n-1)$ or Remark 1.7(2).

Remark 2.5 With a bit more calculation, one can actually check that $eI_{n-1}e$ is itself idempotent (and thus corresponds to $L(n-1)$). To do this, one can quickly reduce to the case when $n = 2$ by using Corollary 2.8 of [7]. The $n = 2$ case can be checked explicitly.

§3 Proof of Example 1.2

Let $A(n) \subset A$ be the subalgebra generated by $Sq^1,\dots,Sq^{2^n}$. For each n, we construct an element $f_n \in \mathrm{End}_{A(n)}(H^*(L(2)))$ that is neither the identity nor the zero map.

Recall [9] that the filtration

$$S \subset SP^2(S) \subset SP^4(S) \subset \dots \subset H\mathbb{Z}_{(2)}$$

realizes the length filtration on admissible sequences in $\bar{A} \equiv A/ASq^1$:

$$H^*(SP^{2^n}(S)) = \bar{A}/\{Sq^I \in \bar{A} \mid I \text{ is admissible and } \ell(I) > n\}.$$

It follows that $H^*(L(2))$ has a basis given by elements $\bar{Sq}^i\,\bar{Sq}^j$ with $i \geq 2j$ and $j \geq 2$. It is also known [3] that the elements $\bar{Sq}^{2^m}\,\bar{Sq}^{2^n}$ with $m > n \geq 1$ are A module indecomposables.

Lemma 3.1 If $k < 2^n$, $Sq^k\,Sq^{2^n a+1} = \sum_{j=1}^{[k/2]} \binom{2^n a-j}{k-2j} Sq^{2^n a+k+1-j}\,Sq^j$.

The point here is that when $j = 0$, $\binom{2^n a-j}{k-2j} = \binom{2^n a}{k} = 0$, since

$k < 2^n$.

Lemma 3.2 $\quad Sq^{2^i}\ \overline{S}q^{3\cdot 2^n+1}\ \overline{S}q^{2^n+1} = 0$ for $i < n$.

Proof We just need to check that, in A, $Sq^{2^i}\ Sq^{3\cdot 2^n+1}\ Sq^{2^n+1}$ can be written as a sum of admissible sequences of length 3. We have:

$$Sq^{2^i}\ Sq^{3\cdot 2^n+1}\ Sq^{2^n+1} = \sum_{k=1}^{2^{i-1}} \sum_{j=1}^{2^{i-2}} \binom{3\cdot 2^{n-k}}{2^i-2k}\binom{2^n-j}{k-2j} Sq^{3\cdot 2^n+2^i+1-k}\ Sq^{2^n+k+1-j}\ Sq^j.$$

Here the limits of summation follow from the last lemma. Now we claim that each term is admissible. Thus we need to verify that $3\cdot 2^n + 2^i + 1-k \geq 2(2^n+k+1-j)$, i.e. that $2^n + 2^i \geq 3k + 1-j$. But since $j \geq 1$ and $k \leq 2^{i-1}$, we have $3k + 1-j \leq 3\cdot 2^{i-1} = 2^{i-1} + 2^i \leq \leq 2^n + 2^i$.

We now define $f_n\colon H^*(L(2)) \longrightarrow H^*(L(2))$ by

$f_n\left(\overline{S}q^{2^{n+3}}\ \overline{S}q^2\right) = \overline{S}q^{3\cdot 2^{n+1}+1}\ \overline{S}q^{2^{n+1}+1}$, and $f_n\left(\overline{S}q^i\ \overline{S}q^j\right) = 0$ otherwise. By Lemma 3.2 it follows that $f_n \in \mathrm{End}_{A(n)}(H^*(L(2)))$.

References

1. J. F. Adams, The Kahn-Priddy Theorem, Proc. Camb. Phil. Soc. 73 (1973), 45-55.

2. J. F. Adams, J. H. Gunawardena, and H. R. Miller, The Segal conjecture for elementary abelian p-groups I, Topology, to appear.

3. R. Cohen, lecture given at the Midwest topology conference, Chicago, 1983.

4. J. C. Harris and N. J. Kuhn, Stable decompositions of classifying spaces of finite abelian p-groups, preprint, 1985.

5. N. J. Kuhn, A Kahn-Priddy sequence and a conjecture of G. W. Whitehead, Math. Proc. Camb. Phil. Soc. 92 (1982), 467-483.

6. N. J. Kuhn, Spacelike resolutions of spectra, Proc. of the Northwestern Homotopy Theory Conference, A.M.S. Cont. Math. Series 19 (1983), 153-165.

7. N. J. Kuhn and S. B. Priddy, The Transfer and Whitehead's conjecture, Math. Proc. Camb. Phil. Soc., to appear.

8. P. Landrock, Finite Group Algebras and their Modules, London Math. Soc. Lecture Note Series no. 84, Cambridge Univ. Press, Cambridge, 1983.

9. S. A. Mitchell and S. B. Priddy, Stable splittings derived from the Steinberg module, Topology 22 (1983), 285-298.

10. G. Nishida, On the spectra $L(n)$ and a theorem of Kuhn, preprint, 1985.

Thom complexes and the spectra bo and bu

BY MARK MAHOWALD[1]

Recent work has shown that many interesting spectra can be constructed as Thom complexes of rather easily constructed bundles. In particular the Eilenberg-Mac Lane spectra $K(Z/2)$, $K(Z)$,etc. have all been constructed as Thom complexes, [M]. It should be noted that, if $Z_{(2)}$ is the integers localized at 2, then $K(Z_{(2)})$ is a Thom complex of a bundle classified by BSO. In order to construct $K(Z)$ itself as a Thom complex we needed to consider spherical fibrations. This paper considers the problem of constructing the spectra in the title in this fashion. First we have the observation:

PROPOSITION 1. *The spectra bo and bu, as spectra over the integers, can not be Thom spectra of bundles classified by BU or BO.*

PROOF: At odd primes BO and BU are the same and so if we could construct such a Thom complex we would have maps

$$bo \to MU$$

$$bu \to MU$$

Each of these maps has degree one in dimension 0 and clearly this is not possible at odd primes because of the presence of Bocksteins in $H_*(bo)$ and $H_*(bu)$.

For the rest of the note we wish to consider only the prime 2, that is, all spaces are 2-local. We have the following where BG is the classifying space for spherical fibrations.

THEOREM 2. *There does not exist an H-space Y with a bundle $\xi : Y \to BG$ which is an H-map such that*

$$Y^\xi = bo.$$

PROOF: Suppose Y exists. Let X be the seven skeleton of Y. Then X has cells in dimension 4, 6 and 7 with Sq^3 non zero. Also the Pontryjagin ring structure of $H_*(Y)$ is given by $H_*(bo)$ and so we have

$$H_*(Y) = Z/2[x_4, x_6, x_j;\, j = 2^i - 1,\, i > 2].$$

Using the H-space structure there is a map

$$f : \Omega\Sigma X \to Y$$

[1]Supported by the National Science Foundation

which has degree one in dimension 4. The fiber of f, F, is 9-connected and $H_{10}(F;Z) =\cdot$ $Z/2$. The map of the fiber into $\Omega\Sigma X$ induces a non-trivial map in homology in dimension 10. The 11-skeleton of the fiber, after adjoining, defines a map $k : P^{12}(2) \to \Sigma X$. The cofiber, Z, has a nontrivial cup product structure as follows:

$$\begin{aligned}
H^5 Z &= Z/2 \quad \text{generated by} \quad \alpha \\
H^7 Z &= Z/2 \quad \text{generated by} \quad \beta = Sq^2\alpha \\
H^8 Z &= Z/2 \quad \text{generated by} \quad Sq^1\beta \\
H^{12} Z &= Z/2 \quad \text{generated by} \quad \alpha\cup\beta \\
H^{13} Z &= Z/2 \quad \text{generated by} \quad \alpha\cup Sq^1\beta
\end{aligned}$$

We wish to show that Z, with this cohomology algebra, cannot exist. This is equivalent to showing that there is no map $P^{11}(2) \to \Omega\Sigma X$ which is non-zero in homology in dimension 10. The strategy will be to construct a map

$$f : Z \to P^8(2)$$

This map will exist if and only if the 12-cell is attached trivially to the 8-cell. Clearly, the 12-cell is attached to the 8-cell by η^3 or 0. We need to show that if a space with the cohomology algebra above exists then we can construct a space, which we will also call Z, with the same cohomology algebra and with the 12-cell attached trivially to the 8-cell. The following lemma will allow us to do this.

LEMMA 3. *There is a map $k : P^{12}(2) \to \Sigma X$ whose cofiber, Z', has a trivial cohomology ring. In addition, the 12-cell of Z' is attached to the 8-cell by η^3.*

Before we prove the lemma we will complete the proof of the theorem. The lemma gives a map $a : P^{12}(2) \to \Sigma X$ so that the cofiber has a trivial comohology ring but the composite

$$P^{12}(2) \to \Sigma X \to S^8$$

is η^3. Having zero cup products in the cofiber is equivalent, in this case, to the adjoint

$$b : P^{11}(2) \to \Omega\Sigma X$$

being zero in homology. Thus if the desired map from Z to $P^8(2)$ does not exist then replacing the map k, which defines the orginal Z, by the map $k + a$ will give a cofiber with the desired cohomology ring and the desired map $f : Z \to P^8(2)$. The fiber of this map is just S^5. This spherical fibration has a classifying map

$$g : P^8(2) \to BG(5)$$

The Sq^2 in $H_*(Z)$ implies that this bundle does not have a section over the 7 skeleton. On the other hand, we have the following lemma which will complete the proof of the theorem.

Lemma 4. *Every spherical 5-sphere bundle over S^7 has a section.*

Proof: Let $G(5)$ be the space of all homotopy equivalences of S^5. We have the fibration

$$\Omega^5 S^5 \to G(5) \to S^5$$

The projection map is evaluation at the base point. This corresponds to

$$S^5 \to BF(5) \to BG(5)$$

There is a map

$$G(5) \to \Omega^6 S^6$$

This gives a commutative diagram

$$\begin{array}{ccccc} & & A & \to & A \\ & & \downarrow & & \downarrow \\ \Omega^5 S^5 & \to & G(5) & \to & S^5 \\ \downarrow & & \downarrow & & \downarrow \\ \Omega^5 S^5 & \to & \Omega^6 S^6 & \to & \Omega^6 S^{11} \end{array}$$

The calculations in [T] show that the suspension map

$$\pi_{11}(S^5) \to \pi_{12}(S^6)$$

is an isomorphism. The EHP sequence applied to the right hand column implies that A is 8-connected. Thus

$$\pi_6(G(5)) = \pi_6(\Omega^6 S^6)$$

Therefore

$$\pi_6(g(5)) = \pi_6(\Omega^5 S^5)$$

This implies the lemma.

The proof of theorem 2 will be complete once we prove lemma 3.

Proof of Lemma 3: The 7-skeleton of Z, $\Sigma^3 CP^2$, is a stable complex through homotopy dimension 8. The Adams spectral sequence E_2 term is given as follows:

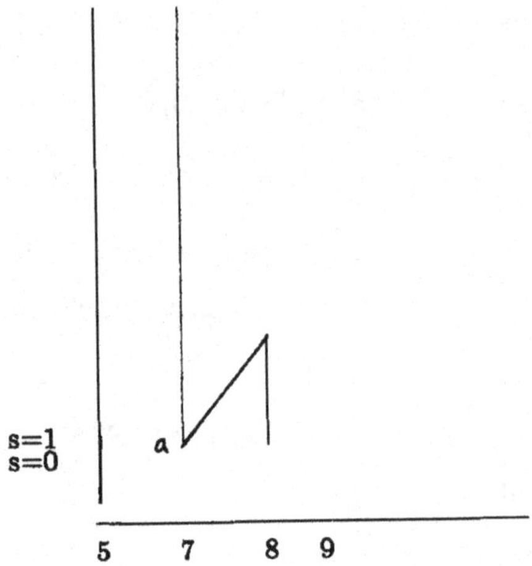

The class of (s,t) filtration (1,8), which we will label a, is the attaching map of the 8-cell in Z (or Z'). Clearly composition with η is divisible by 2 and so composition with η^2 is zero. Thus we can construct Z' as the cofiber of the composite

$$S^7 \cup_{\eta^3} e^{11} \cup e^{12} \to S^7 \cup_{\eta^2} e^{10} \to \Sigma^3 CP^2.$$

The fact that the attaching map is a composite forces the cup products in Z' to be zero. This proves the lemma.

Lemma 3 can be paraphrased by the observation that there is a null homotopy of

$$< \eta, 2, \eta^3 >= 2\nu\eta^2$$

on S^5.

We would like to stress the fact that a space Y, which is a likely candidate for the base space of the bundle to construct bo as a Thom complex, does not exist. The question of constructing a map into either BO or BG does not occur. Also the question of how nice the Pontryjagin ring of Y is does not enter the discussion. If bo is the Thom spectrum of a bundle classified by an H-map, the ring structure of bo determines the Pontryjagin ring of Y. The ring structure of bo defines a commutative and associative multiplication in $H_*(bo)$. Thus $H_*(Y)$ would also have such a multiplication.

Finally, we have the somewhat simpler theorem.

THEOREM 5. *There does not exit a space Y with a bundle ξ over ΩY classified by an H-map such that*

$$\Omega Y^{\xi} = bu.$$

PROOF: If Y exists then

$$H_* \Omega Y = Z/2[x_2, x_6, x_j | j = 2^k - 1;\ k > 2] = H_*(bu).$$

This implies that the first three cells of Y must be in dimensions 3, 7 and 8. In addition the cells in dimension 7 and 8 must be related by a Sq^1. There are two classes in dimension 6, x_2^3 and x_6. They can be distinguished by the action of Sq^4 in $H_*(bo)$

$$x_6 Sq^4 = 0 \quad and \quad x_2^3 Sq^4 \neq 0.$$

This implies that $x_6 Sq^2 = x_2^2$. Thus, in order that the Steenrod algebra be correct in ΩY^ξ the Hopf invariant of the attaching map of the 7 cell must carry a Sq^2. There is an element, usually designated ν', of order 4 in $\pi_6(S^3)$ with Hopf invariant η, [T]. This will carry the Sq^2 but the 8 cell cannot be attached since ν' has order 4 and not 2. Therefore, the 8 skeleton of Y cannot exist and this completes the proof.

Again, the conclusion is arrived at by just considering the possible base space ΩY together with Y. The problem of construction a map never occurs.

The next question we would like to consider is the possibility of there being a space Y with a bundle ξ so that Y^ξ is either bo or bu. If such a Y exists which is not an H space, then several envisioned applications would not follow. D. Waggnor has constructed a space X so that $H_*\Omega X = Z/2[x_6, x_j | j = 2^k - 1;\ k > 2]$. If there is a map

$$f : \Omega X \to S^5$$

so that the composite

$$S^6 \to \Omega X \to S^5$$

is η, then, if Y is the fiber of f, there should be a bundle $\xi : Y \to BO$ so that y^ξ is bo.

The evidence for this is essentially that the homology of the fiber would force the correct Steenrod algebra action on the Thom complex. Constructing the map f seems to be hard. It is rather easy to check that the extension can be made through the 15 skeleton but beyond that there is no easy way to proceed.

[M] Mark Mahowald, Ring spectra which are Thom complexes, Duke Math. J. 46 (1979) 549-559.

[T] Hirosi Toda, *Composition Methods in Homotopy Groups of Spheres*, Ann. of Math Studies #49 (1962).

A commentary on the "Image of J in the EHP Sequence"

Mark Mahowald and Robert Thompson

The second named author found some ambiguities in the discussion contained in sections 3, 4, 5, and 7 of [M1]. Our purpose here is to describe these ambiguities and to complete the discussion.

We will first discuss the material from sections 3, 4, and 5. Let $W(n)$ denote the fiber of the double suspension map $E : S^{2n-1} \to \Omega^2 S^{2n+1}$. Proposition 4.4 of [M1] states that the homotopy of $W(n)$ is isomorphic to the homotopy of $\Omega^4 W(n+1)$ above a "1/5 line." An algebraic version of this was proved in [M2]. The first step in the proof of 4.4 is to construct a map $\sigma : W(n) \to \Omega^4 W(n+1)$. Secondly, certain resolutions of $W(n)$ and $\Omega^4 W(n+1)$ are constructed. Finally, propositon 4.10 states that σ can be extended to a map between these resolutions and the proof of 4.4 is completed by comparing the E_2-terms of the resolutions above the $1/5$ line. We wish to discuss the proof of 4.10. The starting point is the idea of a resolution of a space. In this context we mean a tower of fibrations $(X_i, p_{i-1} : X_i \to X_{i-1})$ with fibers F_i which satisfy:

1) Each F_i is a generalized Eilenberg-MacLane space $K(V_i)$ where V_i is a graded $\mathbf{Z}/2$ vector space (if V^i is the homogeneous part of V in degree i then $K(V) = \prod_i K(V^i, i)$).

2) The fibration

$$F_i \to X_i \to X_{i-1}$$

is principal with

$$g_{i-1} : X_{i-1} \to BF_i$$

as the clasifying map.

3) There are maps

$$f_i : X \to X_i$$

such that the induced map

$$f : X \to \operatorname{holim} X_i$$

is a homotopy equivalence if X is 2-adically complete.

We will use the symbol $X^\bullet$ to represent a resolution of X.

An Adams resolution, yielding an unstable Adams spectral sequence, requires, in addition, that

$$\ker p_i^* = \ker f_i^* \quad \text{and } f_0^* \text{ is onto.}$$

In general we wish to drop this requirement. A useful example of a resolution which is not an Adams resolution is obtained by applying Ω^k to an Adams resolution. The resulting tower of fibrations will be a resolution but the cohomology condition is no longer satisfied.

Definition. A map of resolutions $f^\bullet : X^\bullet \to Y^\bullet$ covering a map $f : X \to Y$ is a sequence of maps $f_s : X_s(X) \to X_s(Y)$ such that the following diagram commutes:

$$\begin{array}{ccc} X_s(X) & \stackrel{f_s}{\to} & X_s(Y) \\ \downarrow & & \downarrow \\ X_{s-1}(X) & \stackrel{f_{s-1}}{\to} & X_{s-1}(Y) \\ \uparrow & & \uparrow \\ X & \stackrel{f}{\to} & Y \end{array} \tag{0.1}$$

The following proposition is immediate.

Proposition 1. If $f : X \to Y$ is some map and $X^\bullet$ is an Adams resolution of X then, for any resolution $Y^\bullet$ of Y, there exists a map of resolutions

$$f^\bullet : X^\bullet \to Y^\bullet$$

which covers f.

Thus Adams resolutions play a special role in this theory.

For each space X with a resolution $X^\bullet$ there is an obviously defined spectral sequence:

$$E_1^{s,t} = \pi_{t-s}(F_s) = V_s^{t-s}$$

which converges to $\pi_*(X)$. If $X^\bullet$ is an Adams resolution then the E_2-term has been identified as an appropriate Ext group. In the generality that we have here, the E_2-term does not have a nice description. In what follows we will be using the functorial notation of [M1].

Proposition 2. If $f^\bullet : X^\bullet \to Y^\bullet$ is a map of resolutions covering a map $f : X \to Y$ then there is a resolution of the fiber of f, $F(f)^\bullet$, so that we have a long exact sequence

$$\cdots \to E_2^{s,t}(F(f)^\bullet) \to E_2^{s,t}(X^\bullet) \stackrel{\partial}{\to} E_2^{s,t}(Y^\bullet) \to E_2^{s+1,t}(F(f)^\bullet) \to \cdots$$

This is proposition 3.3 of [M1] and is proved in some detail there. The following proposition is used implicitly in [M1].

Proposition 3. The resolution $F(f)^\bullet$ can be constructed in such a way that the boundary map in the above long exact sequence coincides with the map of E_2-terms induced by $f^\bullet$.

Proof: We must construct $F(f)^\bullet$ in such a way that we can compute d_1. Recall that $X_s(f)$ is defined as the fiber of the composite $X_s(X) \xrightarrow{f_s} X_s(Y) \xrightarrow{p} X_{s-1}(Y)$. Then it is proved that $F_s(f) = F_s(X) \times \Omega F_{s-1}(Y)$. We will define a map $g_s(f) : X_s(f) \to BF_{s+1}(f) = BF_{s+1}(X) \times F_s(Y)$ which yields the desired formula for d_1 and show that it has $X_{s+1}(f)$ as its fiber.

Define $g_{s,1}(f)$ by the composite $X_s(f) \to X_s(X) \xrightarrow{g_s(X)} BF_{s+1}(X)$. Then the composite $F_s(X) \times \Omega F_{s-1}(Y) \to X_s(f) \xrightarrow{g_{s,1}} BF_{s+1}(X)$ induces a map

$$E_1^{s,t}(X) \oplus E_1^{s-1,t}(Y) \to E_1^{s+1,t}(X)$$

given by $(x, y) \to d_1(x)$.

Now let $g_{s,2}(f)$ be the induced map of fibers in the following diagram:

$$\begin{array}{ccccccc} \Omega X_{s-1}(Y) & \longrightarrow & X_s(f) & \longrightarrow & X_s(X) & \longrightarrow & X_{s-1}(Y) \\ \downarrow{=} & & \downarrow{g_{s,2}(f)} & & \downarrow{f_s} & & \downarrow{=} \\ \Omega X_{s-1}(Y) & \longrightarrow & F_s(Y) & \longrightarrow & X_s(Y) & \longrightarrow & X_{s-1}(Y) \end{array} \quad . \qquad (3.1)$$

It is straightforward to check that the composite

$$\Omega F_{s-1}(Y) \to F_s(f) \to X_s(f) \xrightarrow{g_{s,2}} F_s(Y)$$

just induces d_1 for the resolution $Y^\bullet$.

It is also easy to check that the composite

$$F_s(X) \to F_s(f) \to X_s(f) \xrightarrow{g_{s,2}} F_s(Y)$$

is just the map of fibers induced by the top half of diagram (0.1). Thus if we define $g_s(f) : X_s(f) \to BF_{s+1}(X) \times F_s(Y)$ by $g_s(f) = (g_{s,1}(f), g_{s,2}(f))$ then g_s induces a homomorphism

$$d_1 : E_1^{s,t}(X) \oplus E_1^{s-1,t}(Y) \to E_1^{s+1,t}(X) \oplus E_1^{s,t}(Y)$$

satisfying the formula $(x, y) \to (d_1(x), f_*(x) + d_1(y))$. This formula shows that the boundary homomorphism in the long exact sequence of E_2-terms is that which is induced by $f^\bullet$.

What remains is to show that $X_{s+1}(f)$ is the fiber of $g_s(f)$. This will follow from the following lemma:

lemma 4. Consider a diagram

$$\begin{array}{ccccc} F & \longrightarrow & D & \longrightarrow & B \\ \downarrow & & \downarrow & & \downarrow{=} \\ E & \longrightarrow & A & \xrightarrow{\beta} & B \\ \downarrow & & \downarrow{\gamma} & & \downarrow \\ C & \xrightarrow{=} & C & \longrightarrow & * \end{array} \tag{4.1}$$

in which all the rows and columns are fiber sequences. Then F is also the fiber of the map $(\beta, \gamma) : A \to B \times C$.

This follows immediately from, for example, the proof of lemma 2.1 of [CMN]. To apply the lemma to the situation at hand, set diagram 4.1 equal to the diagram

$$\begin{array}{ccccc} F & \longrightarrow & D & \longrightarrow & F_s(Y) \\ \downarrow & & \downarrow & & \downarrow{=} \\ E & \longrightarrow & X_s(f) & \xrightarrow{g_{s,2}(f)} & F_s(Y) \\ \downarrow & & \downarrow{g_{s,1}(f)} & & \downarrow \\ BF_{s+1}(X) & \xrightarrow{=} & BF_{s+1}(X) & \longrightarrow & * \end{array}$$

Diagram 3.1 shows that E is just the fiber of $X_s(X) \xrightarrow{f_s} X_s(Y)$. By considering the following diagram

$$\begin{array}{ccccc} X_{s+1}(f) & \longrightarrow & X_{s+1}(X) & \longrightarrow & X_s(Y) \\ \downarrow & & \downarrow & & \downarrow{=} \\ E & \longrightarrow & X_s(X) & \longrightarrow & X_s(Y) \\ \downarrow & & \downarrow & & \downarrow \\ BF_{s+1}(X) & \xrightarrow{=} & BF_{s+1}(X) & \longrightarrow & * \end{array}$$

we see that $F = X_{s+1}(f)$. By lemma 4, F is the fiber of $g_s(f)$, as claimed.

We now begin the discussion of the particular material from [M1]. Let $(S^{2n-1})^{\bullet}$ be an Adams resolution for S^{2n-1}. Then the map $E : S^{2n-1} \to \Omega^2 S^{2n+1}$ lifts to a map

$$E^{\bullet} : (S^{2n-1})^{\bullet} \to \Omega^2[(S^{2n+1})^{\bullet}].$$

Thus we have a resolution of $W(n)$ constructed from the fiber of E. We also have an algebraic resolution of $W(n)$ constructed from the lambda algebra [M2]. This gives $W(n)$ as the fiber of the "d_1" map in the EHP spectral sequence as follows:

$$W(n) \to \Omega^3 S^{4n+1} \to \Omega S^{4n-1}$$

Both of these resolutions fit into long exact sequences

$$\cdots \to \mathrm{E}_2^{s,t}(F(E)^\bullet) \to \mathrm{E}_2^{s,t}(S^{2n-1}) \xrightarrow{\partial} \mathrm{E}_2^{s,t}(S^{2n+1}) \to \cdots$$

$$\cdots \to \mathrm{E}_2^{s,t}(\Lambda(W(n))) \to \mathrm{E}_2^{s,t}(S^{2n-1}) \to \mathrm{E}_2^{s,t}(S^{2n+1}) \to \cdots$$

In the second sequence the map $\mathrm{E}_2^{s,t}(S^{2n-1}) \to \mathrm{E}_2^{s,t}(S^{2n+1})$ is induced by the inclusion $\Lambda(2n-1) \to \Lambda(2n+1)$ and by proposition 3 we see that this is the same map as ∂ in the first sequence. Since all the terms involved are vector spaces there are no extension problems and we get an isomorphism $\mathrm{E}_2^{s,t}(F(E)^\bullet) \simeq \mathrm{E}_2^{s,t}(\Lambda(W(n)))$ as $\mathbf{Z}/2$ vector spaces.

Unfortunately, in the discussion of this material in [M1], the first resolution is constructed but the proof of proposition 4.10 uses properties of the E_2-term of the second one. It is rather easy to see that the conclusion of proposition 4.10 requires only information about the E_2-term . We formulate this in the following way. We will call $X^\bullet$ a minimal resolution if

$$\mathrm{E}_1^{s,t}(X^\bullet) = \mathrm{E}_2^{s,t}(X^\bullet)$$

Note that since we are not dealing with Adams resolutions a minimal resolution need not be unique.

Theorem 5. For any resolution $X^\bullet$, there is a minimal resolution of ΩX, denoted by $\Omega X^\bullet$, such that there are lifts of the identity map of ΩX to

$$f^\bullet : \Omega(X^\bullet) \to \Omega X^\bullet$$

and

$$g^\bullet : \Omega X^\bullet \to \Omega(X^\bullet)$$

with $f^\bullet$ and $g^\bullet$ inducing isomorphisms in $\mathrm{E}_2^{s,t}$.

Proof: Consider d_1 for the resolution $\Omega(X^\bullet)$. Recall that this d_1 can be thought of as the map in homotopy induced by the composite

$$\Omega F_i \to \Omega X_i \to \Omega BF_{i+1}.$$

If d_1 is non-zero on $\Sigma^{-1}V_i^j \to \Sigma^{-1}V_{i+1}^{j-1}$ then we can find a non-zero vector subspace A and a non-zero quotient space B such that the composite

$$A \to \Sigma^{-1}V_i^j \to \Sigma^{-1}V_{i+1}^{j-1} \to B$$

is an isomorphism. This allows us to define maps a and b so that

$$K(A) \xrightarrow{a} \Omega F_i \to \Omega X_i \to \Omega BF_{i+1} \xrightarrow{b} K(B)$$

is a homotopy equivalence. Since ΩX_i is an H-space we can conclude that

$$\Omega X_i = X_i' \times K(A).$$

The resolution $\Omega X^\bullet$ is constructed by removing from ΩF_i, ΩX_i and ΩBF_{i+1} the $K(A)$ factor. Thus the resulting resolution $\Omega X^\bullet$ will have a trivial d_1. This gives the theorem.

This result, then, completes the proof of the proposition 4.10. Indeed, the conclusion depends on the E_2-term and the argument given in [M1] is already valid for one of the two possible E_2-terms , that is, the one constructed from the lambda algebra.

Secondly, we wish to elaborate on the proof of Theorem 1.5. Consider the composite

$$\pi_j(\Omega^{2n+1}S^{2n+1}) \to \pi_j(QP^{2n}) \to J_j(P^{2n})$$

where the first map is induced by the Snaith map and the second map is the J-homology Hurewicz homomorphism. Theorem 1.5 states that the above composite is surjective if $j \geq 2n+1$ and $j \not\equiv -2\ (8)$ or if $j \geq 2n+8+2i$ and $j \equiv -2\ (8)$. The following material is related to the discussion which is in section 7 of [M1] and, in particular, the results beginning with lemma 7.18. We wish to discuss the case $j \equiv -2\ (8)$. A straightforward modification of the proof of lemma 7.18 yields the following.

Proposition 6. Let $j \equiv 2^i - 2\ (2^{i+1})$ and $k \geq 0$. If $j \geq 2n+8k+8+2i$, then the Hurewicz map

$$\pi_j^S(P_{8k+1}^{2n+8k}) \to J_j(P_{8k+1}^{2n+8k})$$

is surjective.

Now, for each $k \geq 0$, we have the diagram

$$\Omega^{2n+1}\Sigma^{2n+1}P_{8k+1}^{2n+8k} \to \Omega^{2n+1}\Sigma^{2n+1}P^{2n} \to \Omega^{2n+1}S^{2n+1} \to QP^{2n}$$

If $j \geq 2n+8k+8+2i$ (condition 1) for some k, then let $\alpha : S^j \to QP_{8k+1}^{2n+8k}$ be a class whose Hurewicz image is a generator. Consider the following diagram:

$$\begin{array}{ccc} & \overset{\alpha}{\nearrow} & Q\Sigma P_{8k+1}^{2n+8k} \\ & & \uparrow \\ S^{j+1} & \xrightarrow{\alpha'} & QP_{2n+8k+1} \end{array}$$

The proof of proposition 6 shows that α factors through $QP_{2n+8k+1}$. Now consider the diagram:

$$\begin{array}{ccccc} S^{j+1} & \xrightarrow{\hat{\alpha}} & \Omega^{2n}\Sigma^{2n}P_{2n+8k+1} & \to & \Omega^{2n}\Sigma^{2n+1}P_{8k+1}^{2n+8k} \\ \downarrow & & \downarrow & & \downarrow \\ S^{j+1} & \xrightarrow{\alpha'} & QP_{2n+8k+1} & \to & Q\Sigma P_{8k+1}^{2n+8k} \end{array}$$

If α' has a lifting $\hat{\alpha}$ then the proof of 1.5 would be complete. But α' would be in the range to have such a lifting if $j \leq 16k + 6n$ (condition 2). One can now easily check that for all but a few small values of j and n satisfying the hypothesis of 1.5 there exists a k simultaneously satisfying conditions 1 and 2. The remaining cases are handled separately using the methods presented at the end of section 7 of [M1].

[CMN] F. Cohen, J. C. Moore, J. Neisendorfer, The double suspension homomorphism and exponents of the homotopy groups of spheres. Ann. of Math., 110(1979), 549-565.

[M1] Mark Mahowald, The Image of J in the EHP Sequence,Ann. of Math., 116(1982),65-112.

[M2] Mark Mahowald, The Double Suspension Homomorphism,Trans. AMS 214(1975), 169-178

ON THE Λ-ALGEBRA AND THE HOMOLOGY OF SYMMETRIC GROUPS

William M. Singer*
Department of Mathematics
Fordham University
Bronx, New York 10458

1. Introduction.

The lambda algebra was introduced and studied in [1]. This algebra is an E_1-term for the Adams spectral sequence for the stable homotopy groups of the sphere, completed at the prime two. It arises naturally if one obtains the Adams spectral sequence from the lower central series filtration of a simplicial free group, as in [2], [9], and [12].

In this paper we establish a relationship between the lambda algebra and the homology theory of the symmetric groups Σ_r. Our object is to begin work on the following:

Conjecture 1. Let $\Lambda = \oplus_{s,t}\Lambda^{s,t}$ be the bigraded lambda algebra of [1]. Then for each integer $s \geq 1$ there is a finite module N_s over the group-ring $F_2(\Sigma_{2^s})$, and an isomorphism of graded F_2-vector spaces:

$$\Lambda^{s,*} = H_{*-s}(\Sigma_{2^s};N_s) \tag{1.1}$$

Here by $H_*(\Sigma_{2^s};N_s)$ we mean $\mathrm{Tor}_*^{F_2(\Sigma_{2^s})}(F_2,N_s)$.

We will prove this conjecture in the cases $s = 1$ and $s = 2$. We will show:

Theorem 2. The graded vector spaces are isomorphic:

$$\Lambda^{1,*} = H_{*-1}(\Sigma_2;F_2) \tag{1.2}$$

$$\Lambda^{2,*} = H_{*-2}(\Sigma_4;S^{(2,2)}\otimes S^{(2,2)}) \tag{1.3}$$

Here $S^{(2,2)}$ is the classical "Specht module" associated with the partition (2,2) of the integer 4. It is a two-dimensional represen-

*Research partially supported by National Science Foundation grants DMS 8503335 and MCS 8101702.

tation of Σ_4.

The case $s = 1$ of this theorem is easy; but some work is required to give a useful proof for the case $s = 2$. The truth of Conjecture 1 in this latter case gives us good cause to believe it is true in general. In fact, all relations that define Λ as an algebra are present in filtration degree $s = 2$. Work on the general case is underway.

Our hope is that (1.1) will lead to new information about the Adams spectral sequence. In order that this be so it is necessary to derive (1.1)-(1.3) directly from the geometrical definition of Λ, without using the explicit generators and relations for that algebra that were worked out in [1]. We recall now the geometrical definition of Λ, and reformulate Conjecture 1 and Theorem 2 in geometrical terms.

Let A be the functor that to each pointed set T assigns the F_2-vector space spanned by the members of T: one imposes in AT the single relation that the basepoint of T be zero. Let L^{res} be the functor that to each F_2-vector space V assigns the free restricted Lie algebra generated by V. Then $L^{res}V$ is graded: $L^{res}V = \oplus L_r^{res}V$, where $L_r^{res}V$ is the span of the elements of weight r. By the process of "prolongation" both A and L^{res} become functors defined on categories of semisimplicial spectra ([7]): one simply applies them dimensionwise. If S is the semisimplicial sphere spectrum ([10, p. 241, or (2.4) below), then the definition of Λ given in [1] is:

$$\Lambda^{s,t} = \pi_{t-s}(L_{2^s}^{res}AS) \tag{1.4}$$

In view of (1.4), Conjecture 1 and Theorem 2 are reformulated:

<u>Conjecture 1'</u>. For each integer $s \geq 1$ there is a finite module N_s over the group ring $F_2(\Sigma_{2^s})$, and an isomorphism of graded F_2-vector spaces:

$$\pi_*(L_{2^s}^{res}AS) = H_*(\Sigma_{2^s};N_s) \tag{1.5}$$

<u>Theorem 2'</u>. The graded vector spaces are isomorphic:

$$\pi_*L_2^{res}AS = H_*(\Sigma_2;F_2) \tag{1.6}$$

$$\pi_*L_4^{res}AS = H_*(\Sigma_4;S^{(2,2)}\otimes S^{(2,2)}) \tag{1.7}$$

The statement of our conjecture in the form (1.5) immediately raises a further question. What can be said about the homotopy groups $\pi_*(L_r^{res}AS)$ if r is not a power of two? In [1] it is shown that:

$$\pi_*(L_r^{res}AS) = 0 \quad \text{if} \quad r \quad \text{not a power of two} \tag{1.8}$$

Is there a group-homological explanation of (1.8)?

A moment's thought about what constructions are needed to prove (1.5)-(1.7) will show that similar constructions should give (1.8) as well. Indeed, suppose that for each integer r we could construct a functor J_r from the category of pointed sets to the category of $F_2(\Sigma_r)$-modules, having the following properties:

i) For each pointed set T, J_rT is a projective $F_2(\Sigma_r)$-module.

ii) For each pointed set T, there is a natural isomorphism of F_2-vector spaces:

$$F_2 \otimes_{\Sigma_r} J_rT = L_r^{res}AT \tag{1.9}$$

iii) If J_r is applied dimensionwise to the simplicial sphere spectrum S, there results a simplicial spectrum of $F_2(\Sigma_r)$-modules, J_rS, for which:

$$\pi_iJ_rS = 0 \quad \text{if} \quad i > 0 \tag{1.10}$$

The existence of such a sequence of functors J_r would at once imply Conjecture 1'. Indeed, J_rS could be regarded as simply a chain complex of $F_2(\Sigma_r)$-modules, with differential ∂ given by the alternating sum of the face operators. Since the homotopy groups of J_rS are just its homology groups under this differential (see (2.1) below), properties i) and iii) imply that we can regard J_rS as a projective resolution of the $F_2(\Sigma_r)$-module π_0J_rS. From (1.9) we would then have at once

$$\pi_*L_r^{res}AS = H_*(\Sigma_r;\pi_0J_rS) \tag{1.11}$$

Making the identification $\pi_0J_{2^s}S = N_s$ then gives Conjecture 1'. Suppose in addition it should happen that, when r is not a power of two, π_0J_rS is a projective module over $F_2(\Sigma_r)$, satisfying $F_2 \otimes_{\Sigma_r} \pi_0J_rS = 0$.

Then (1.11) would give us at once the desired group-homological interpretation of (1.8). We therefore formulate our conjecture and our theorem in their final forms:

__Conjecture 1"__. For each integer $r \geq 2$ there is a functor J_r from the category of pointed sets to the category of $F_2(\Sigma_r)$-modules, satisfying conditions i)-iii) above. If r is not a power of two, π_0J_rS

is projective over $F_2(\Sigma_r)$, and $F_2 \underset{\Sigma_r}{\otimes} \pi_0 J_r S = 0$.

We will prove:

<u>Theorem 2"</u>. The above conjecture is true if $r = 2,3,4$. Further one has $\pi_0 J_2 S = F_2$; $\pi_0 J_3 S = 0$; and $\pi_0 J_4 S = P \oplus (S^{(2,2)} \otimes S^{(2,2)})$ as modules over $F_2(\Sigma_2)$, $F_2(\Sigma_3)$, and $F_2(\Sigma_4)$, respectively; where P is a certain projective satisfying $F_2 \underset{\Sigma_4}{\otimes} P = 0$.

Theorem 2' is an immediate corollary, as is the group-homological interpretation of (1.8) in the case $r = 3$.

Work on Conjecture 1" in the general case is in progress.

It is a pleasure to acknowledge assistance I have had from several persons in the course of this work. The idea of using the representations M_r of Sections 6 and 7 as "universal examples" of the various weights of a free Lie algebra grew out of a conversation with Gunnar Carlsson. This has turned out to be a key idea, and I thank Carlsson for his contribution to it. For information on the homologies of groups related to this project I thank Stewart Priddy, Fred Cohen, and Shaun Bullett. For assistance with modular representation theory I am indebted to Charles Curtis, Nick Kuhn, and Steve Mitchell. For a suggestion simplifying the formulation of Conjecture 1" and Theorem 2" I thank Daniel Kan.

2. <u>Semisimplicial Spectra</u>

We review here the few elementary facts we need about semisimplicial spectra. The basic references are the papers of Kan [7] and Kan-Whitehea [10]. We will prove a result (Lemma 3 below) about the dimensionwise tensor product of Eilenberg-Maclane spectra that we will need in computin $\pi_* J_r S$.

The (very simple) definition of a semisimplicial spectrum is given by Kan [7], p. 467. Suppose R is a ring, and X a semisimplicial spectrum for which each X_q is an R-module, and all face and degeneracy operators are R-linear. Then we will call X a simplicial spectrum of R-modules. In this case the homotopy groups of X can be computed by the formula:

$$\pi_*(X) = H_*(CX;\partial) \tag{2.1}$$

where (CX,∂) is the chain complex of R-modules defined by $(CX)_q = X_q$;

$\partial = \Sigma d_i$. It is easy to derive (2.1) from the general definition of the homotopy groups of a semisimplicial spectrum that is given by Kan in Section 10 of [7]. One uses the theory of simplicial abelian groups, as developed by Kan [8], Moore [11], and Dold [5]. For our purposes it would suffice to take (2.1) as the definition of $\pi_*(X)$, since all the spectra we will work with are spectra of modules.

The notion of a fibration of semisimplicial spectra is defined by Kan and Whitehead in Section 5 of [10]. To a fibration they associate a long exact homotopy sequence: see Definition 5.2 of [10].

Suppose

$$0 \to X \to Y \to Z \to 0 \tag{2.2}$$

is a diagram of spectra of R-modules, which is a short-exact sequence of R-modules in each dimension. Then (Proposition 5.4 of [10]) (2.2) is a fibration of semisimplicial spectra. Of course, the corresponding diagram of chain complexes

$$0 \to CX \to CY \to CZ \to 0 \tag{2.3}$$

is also short-exact, and the associated long-exact sequence in homology coincides, under the identifications (2.1), with the homotopy sequence of the original fibration. In particular, all maps in this homotopy sequence are R-linear.

The semisimplicial sphere spectrum S ([10], Example 2.2) has in dimension q the basepoint, and all q-fold iterated degeneracy operators applied to the zero-dimensional cell σ:

$$S_q = \{*\} \cup \{s_{j_q} \cdots s_{j_2} s_{j_1} \sigma \mid 0 \leq j_1 < j_2 < \dots < j_q\} \tag{2.4}$$

Applying the functor A dimensionwise to S gives a spectrum AS of vector spaces for which $\pi_0 AS = F_2$; $\pi_i AS = 0$ if $i > 0$. For each integer $k \geq 1$ we define a spectrum $(AS)^{\otimes k}$ of vector spaces by writing $[(AS)^{\otimes k}]_q = [(AS)_q]^{\otimes k}$; face and degeneracy operators are the k-fold tensor products $d_i^{\otimes k}$, $s_i^{\otimes k}$. For later use we prove:

<u>Lemma</u> <u>3</u>. If $k \geq 2$ then $\pi_q(AS)^{\otimes k} = 0$ for all q.

<u>Proof</u>. Let S^n be the semisimplicial n-sphere. Applying A dimensionwise gives a simplicial vector space AS^n for which $\pi_q(AS^n) = 0$ if $q \neq n$ and $\pi_n(AS^n) = F_2$. For integers $n \geq 0$, $k \geq 1$, define a chain complex $C^{n,k}$ by setting

$$(C^{n,k})_q = [(AS^n)_{n+q}]^{\otimes k}$$

with differential $\partial:(C^{n,k})_q \to (C^{n,k})_{q-1}$ given by the sum of the face operators $d_i^{\otimes k}$. Then the Eilenberg-Zilber and Kunneth theorems give:

$$H_q(C^{n,k}) = 0 \quad \text{unless} \quad q = (k-1)n \tag{2.5}$$

Now for each integer q, define a map $\lambda:(S^n)_{n+q} \to (S^{n+1})_{n+q+1}$ of pointed sets by setting $\lambda(*) = *$ and

$$\lambda(s_{j_q}\ldots s_{j_2}s_{j_1}\sigma_n) = s_{j_q}\ldots s_{j_2}s_{j_1}\sigma_{n+1}$$

Then

$$\lambda d_i x = d_i \lambda x \quad (i \geq 0,\ x \in S^n) \tag{2.6}$$

where (2.6) is to be interpreted with the convention $d_i y = *$ if $i > \dim y$. The pointed set S_q of (2.4) is the direct limit of the system defined by the maps λ:

$$S_q = \varinjlim_n (S^n)_{n+q} \tag{2.7}$$

By virtue of (2.6) and (2.7), the maps λ induce chain maps $\lambda^{\otimes k}:C^{n,k} \to C^{n+1,k}$, and (2.7) extends to an isomorphism of chain complexes:

$$C((AS)^{\otimes k}) = \varinjlim_n C^{n,k} \tag{2.8}$$

where the "C" on the left of (2.8) is as in (2.1). Hence for each q:

$$\pi_q((AS)^{\otimes k}) = \varinjlim_n H_q(C^{n,k}) \tag{2.9}$$

But (2.5) implies that if $k \geq 2$ then this direct limit is zero. So the lemma is proved.

3. A Strategy for the Proof of Conjecture 1"

If V is an F_2-vector space we write $L^{res}V$ for the free restricted Lie algebra generated by V, and LV for the free Lie algebra generated by V. Both $L^{res}V$ and LV are graded by weight. If r is even let $F: L_{r/2}^{res}V \to L_r^{res}V$ be the Frobenius map $F(x) = x^2$; and let $i: L_rV \to L_r^{res}V$ be the natural inclusion. Then F and i induce a direct-sum decomposition

$$L_r^{res}V = L_{r/2}^{res}V \oplus L_rV \quad (r \text{ even}) \tag{3.1}$$

On the other hand, if r is odd one has $L_r^{res}V = L_rV$. So in general:

$$L_r^{res}V = \bigoplus_{j=0}^{\nu(r)} L_{r/2^j}V \tag{3.2}$$

where $\nu(r)$ is the largest integer ν for which 2^ν divides r.

Equation (3.2) suggests an approach to the construction of functors J_r satisfying conditions i)-iii) of Section 1. We should first construct for each integer $r \geq 2$ a functor $J_{r,0}$ from the category of pointed sets to the category of modules over $F_2(\Sigma_r)$, and satisfying

$$F_2 \otimes_{\Sigma_r} J_{r,0}T = L_rAT \tag{3.3}$$

More generally, for each pair of integers r, k with $r \geq 2$ and $k \leq \nu(r)$ we aim to construct a functor $J_{r,k}$ from the category of pointed sets to the category of modules over $F_2(\Sigma_r)$, satisfying

$$F_2 \otimes_{\Sigma_r} J_{r,k}T = \bigoplus_{j=0}^{k} L_{r/2^j}AT \tag{3.4}$$

The construction is to begin with $k = 0$, and proceeds by induction on k. Finally, J_r of (1.9) is obtained by setting

$$J_r = J_{r,\nu(r)} \tag{3.5}$$

Equation (3.4) suggests how $J_{r,k}$ should be obtained from $J_{r,k-1}$. In fact, (3.3) and (3.4) imply there is a short-exact sequence of vector spaces, functorial in T and naturally split:

$$0 \to F_2 \otimes_{\Sigma_{r/2^k}} (J_{r/2^k,0}T) \to F_2 \otimes_{\Sigma_r} J_{r,k}T \to F_2 \otimes_{\Sigma_r} J_{r,k-1}T \to 0 \tag{3.6}$$

Then it seems reasonable to seek $J_{r,k}$ as the middle term of a short-exact sequence of $F_2(\Sigma_r)$-modules, of the form:

$$0 \to F_2(\Sigma_r) \otimes_{\Sigma_{r/2^k}} J_{r/2^k,0}T \to J_{r,k}T \to J_{r,k-1}T \to 0 \tag{3.7}$$

The reader will naturally ask: since (3.6) will be naturally split over F_2, shouldn't (3.7) be naturally split over $F_2(\Sigma_r)$? It will turn out we will not want to do this! In fact, for any given T, (3.7) will be split over $F_2(\Sigma_r)$ ... each module in (3.7) will be projective. But we will want to arrange things so that there is no splitting that is natural with respect to morphisms of the variable T. The reason is that for low values of k, $\pi_i(J_{r,k-1}S)$ will turn out to be non-zero for certain $i > 0$. As we increase k we want these "higher" homotopy groups to be killed off, so by the time we have reached $J_rS = J_{r,\nu(r)}S$, we will have $\pi_i(J_rS) = 0$ for all $i > 0$, as required by (1.10). In order that this be so, the fibration:

$$0 \to F_2(\Sigma_r) \otimes_{\Sigma_{r/2^k}} (J_{r/2^k,0}S) \to J_{r,k}S \to J_{r,k-1}S \to 0 \tag{3.8}$$

must be twisted. Thus, we cannot allow (3.7) to be naturally split ... not even over the category of F_2-vector spaces.

4. Invariant Subspaces of an Iterated Tensor Product

In our construction of the functors J_r, it will be useful for us, for each integer k, to be able to pick out certain Σ_k-invariant subspaces of the k-fold tensor product $(AT)^{\otimes k}$. In fact, let $\mu = (\mu_1,\mu_2,\ldots\mu_\ell)$ be a partition of the integer k. We agree to write partitions so that $\mu_1 \geq \mu_2 \geq \cdots \geq \mu_\ell \geq 1$. Suppose $x_1,x_2,\ldots x_k$ members of T. We say that $x_1 \otimes x_2 \otimes \ldots \otimes x_k$ is a basic tensor of type μ if there is a function $f:\{1,2,\ldots,k\} \to \{1,2,\ldots,\ell\}$ having the properties:

i) $f^{-1}(j)$ consists of exactly μ_j elements, for each $j \leq \ell$.

ii) If $a,b \leq k$ are such that $f(a) = f(b)$, then $x_a = x_b$.

For example, if $x,y,z \in T$ then $y \otimes x \otimes y \otimes z \otimes x \otimes y$ is a basic tensor of type $(3,2,1)$. It is also of types $(2,2,1,1)$ and $(2,1,1,1,1)$ and $(1,1,1,1,1,1)$.

If μ is a partition of the integer k we will write $(AT)^\mu$ for the subspace of $(AT)^{\otimes k}$ that is spanned by all basic tensors of type μ.

If Σ_k acts on $(AT)^{\otimes k}$ by permuting the factors, then $(AT)^{\mu}$ is clearly an $F_2(\Sigma_k)$-submodule. A special case is: $(AT)^{(1,1,\ldots 1)} = (AT)^{\otimes k}$

The submodules $(AT)^{\mu}$ satisfy certain containment relations. In fact, let $\mu = (\mu_1,\ldots,\mu_\ell)$ be a partition of k, and let $(\nu_1,\ldots,\nu_{\ell-1})$ be a partition of k having one less "part". Of course μ and ν are assumed given in descending order; but suppose it is possible to reorder the parts ... say, $\mu = (\mu_1',\ldots,\mu_\ell')$ and $\nu = (\nu_1',\ldots,\nu_{\ell-1}')$ ... in such a way that $\nu_1' = \mu_1'$, $\nu_2' = \mu_2',\ldots,\nu_{\ell-1}' = \mu_{\ell-1}' + \mu_\ell'$. Then we will say that "ν is obtained from μ by an elementary change". Now we introduce a partial order on the set of partitions of k by writing $\nu < \mu$ if ν is obtainable from μ by a sequence of elementary changes. Clearly:

$$\text{If } \nu < \mu \text{ then } (AT)^{\nu} \subseteq (AT)^{\mu} \tag{4.1}$$

We observe finally that the subspaces $(AT)^{\mu}$ of $(AT)^{\otimes k}$ can be used to define subspaces of L_kAT. In fact, for $x_1,x_2,\ldots,x_k \in T$ let us write $[x_1,\ldots,x_{k-1},x_k]$ for the "simple" commutator in L_kAT that is defined inductively by

$$[x_1,x_2,\ldots,x_k] = [[x_1,\ldots,x_{k-1}],x_k] \tag{4.2}$$

Define the F_2-linear $\pi:(AT)^{\otimes k} \to L_kAT$ by $\pi(x_1\otimes x_2\otimes\ldots\otimes x_k) = [x_1,x_2,\ldots x_k]$. Then for each partition μ of k we define the F_2-subspace $L_k^{\mu}AT \subseteq L_kAT$ by writing

$$L_k^{\mu}AT = \pi(AT)^{\mu} \tag{4.3}$$

Of course from (4.1) we find that if $\nu < \mu$ then $L_k^{\nu}AT \subseteq L_k^{\mu}AT$.

5. Construction of J_2

In this section we prove Conjecture 1" in the case $r = 2$, and compute π_0J_2S. We will use the approach suggested in Section 3.

If T is a pointed set, define a "relations module" $R_{2,0}T \subseteq (AT)^{\otimes 2}$ by setting:

$$R_{2,0}T = (AT)^{(2)} = \text{Span}\{x\otimes x \mid x \in T\} \tag{5.1}$$

Here the notation $(AT)^{(2)} \subseteq (AT)^{\otimes 2}$ refers to the conventions of Section 4. Now we define $J_{2,0}T$ by:

$$J_{2,0}T = \frac{(AT)^{\otimes 2}}{R_{2,0}T} = \frac{(AT)^{\otimes 2}}{(AT)^{(2)}} \tag{5.2}$$

as a module over $F_2(\Sigma_2)$. Define $\phi : J_{2,0}T \to L_2AT$ by

$$\phi(x_1 \otimes x_2) = [x_1, x_2] \tag{5.3}$$

This map is Σ_2-equivariant (where Σ_2 acts trivially on the right). Clearly:

<u>Proposition 4</u>. $J_{2,0}T$ is a free module over $F_2(\Sigma_2)$, and the map ϕ of (5.3) passes to a natural, F_2-linear isomorphism:

$$F_2 \underset{\Sigma_2}{\otimes} J_{2,0}T \xrightarrow{\phi} L_2AT \tag{5.4}$$

We observe in passing that it is easy to compute the homotopy groups of $J_{2,0}S$. In fact $(AS)^{(2)}$ is clearly isomorphic to the spectrum AS, whose homotopy groups are zero in all positive dimensions. So from (5.2) and Lemma 3 we have

$$\pi_* J_{2,0}S = \begin{cases} F_2 & \text{if } * = 1 \\ 0 & \text{if } * \neq 1. \end{cases} \tag{5.5}$$

Now for any pointed set T we define the $F_2(\Sigma_2)$-module $J_{2,1}T$ to be the cokernel in the short-exact sequence:

$$0 \to R_{2,1}T \xrightarrow{i} [F_2(\Sigma_2) \otimes AT] \oplus [(AT)^{\otimes 2}] \to J_{2,1}T \to 0 \tag{5.6}$$

Here $F_2(\Sigma_2)$ acts on $F_2(\Sigma_2)\otimes AT$ by way of its action on the left-hand factor, and $F_2(\Sigma_2)$ acts on $(AT)^{\otimes 2}$ by permuting the factors. The "relations module" $R_{2,1}T$ is a submodule of the middle term that we define by:

$$R_{2,1}T = \mathrm{Span}\{e \otimes x + (1,2) \otimes x + x \otimes x \mid x \in T\} \tag{5.7}$$

<u>Proposition 5</u>. $J_2T = J_{2,1}T$ is a free module over $F_2(\Sigma_2)$, and there is a natural isomorphism of vector spaces

$$F_2 \underset{\Sigma_2}{\otimes} J_2T = L_1AT \oplus L_2AT = L_2^{res}AT \tag{5.8}$$

Proof. The inclusion of $F_2(\Sigma_2)\otimes AT$ into the direct sum in (5.6) gives an inclusion of $F_2(\Sigma_2)\otimes AT$ into $J_{2,1}T$, and so a short-exact sequence over $F_2(\Sigma_2)$:

$$0 \to F_2(\Sigma_2)\otimes AT \to J_{2,1}T \to J_{2,0}T \to 0 \tag{5.9}$$

Since both ends of (5.9) are free over $F_2(\Sigma_2)$, so is $J_{2,1}T$. Now divide out the Σ_2-action from (5.9). The resulting short-exact sequence of vector spaces is naturally split by the map $s: F_2 \underset{\Sigma_2}{\otimes} J_{2,0}T \to F_2 \underset{\Sigma_2}{\otimes} J_{2,1}T$ given by $s(x_1\otimes x_2) = x_1\otimes x_2$ for all $x_1, x_2 \in T$. So (5.8) follows from (5.4).

Notice that (5.9) realizes the program (3.7) in the case $r = 2$, $k = 1$, if by $J_{1,0}T$ we understand AT.

To complete the proof of Conjecture 1" in the case $r = 2$ we must compute $\pi_* J_{2,1}S$. For this purpose we consider the fibration that arises from (5.6) when T is replaced by S. Clearly $R_{2,1}S$ is isomorphic to AS, with trivial Σ_2-action. This fact, together with Lemma 3, imply that both the left-hand and middle terms of our fibration have vanishing homotopy in positive dimensions, while $\pi_0 R_{2,1}S = F_2$, and π_0 of the middle term is $F_2(\Sigma_2)$. If i_* is the map induced by i on zero-dimensional homotopy then $i_*(1) = e + (1,2)$, so the long-exact homotopy sequence of our fibration gives:

$$\pi_* J_2 S = \pi_* J_{2,1} S = \begin{cases} \operatorname{coker} i_* = F_2 & \text{if } * = 0 \\ 0 & \text{if } * \neq 0 \end{cases} \tag{5.10}$$

as modules over $F_2(\Sigma_2)$. This completes the proof of Conjecture 1" in the case $r = 2$, and of all statements in Theorem 2" concerning the case $r = 2$.

We note that the sequence (5.9) illustrates the remarks we made at the end of Section 3. There is no natural splitting of (5.9). One proves this by replacing T by the spectrum S. In the resulting fibration, the connecting morphism $\partial: \pi_1 J_{2,0}S \to \pi_0(F_2(\Sigma_2)\otimes AS)$ is monic. In this way we obtain another proof of (5.10), while showing the value of having constructed $J_{2,1}$ in such a way that (5.9) is not naturally split. By passing from $J_{2,0}$ to $J_{2,1}$ we are able to kill the unwanted homotopy group π_1.

6. Construction of J_3

To construct J_3 we have to introduce a kind of "universal example" of the third weight of a free Lie algebra. In fact, let V_3 be the three-dimensional F_2-vector space spanned by basis vectors e_1, e_2, e_3. Let Σ_3 act on the left of V_3 by permuting the basis. This action extends to an action on the free Lie algebra LV_3. Let $M_3 \subset L_3V_3$ be the $F_2(\Sigma_3)$-submodule defined by:

$$M_3 = \mathrm{Span}\{[e_{\pi(1)}, e_{\pi(2)}, e_{\pi(3)}] \mid \pi \in \Sigma_3\} \tag{6.1}$$

M_3 is our "universal example", or model space. Using anticommutation and Jacobi relations, one checks easily that M_3 is a two-dimensional representation of Σ_3. In fact, M_3 is isomorphic to the classical "Specht module" $S^{(2,1)}$. If we identify Σ_3 with $GL_2(F_2)$, then M_3 is the Steinberg module. Then by Steinberg's work [13], M_3 is projective over $F_2(\Sigma_3)$.

If T is a pointed set, define the $F_2(\Sigma_3)$-module J_3T by:

$$J_3T = M_3 \otimes (AT)^{\otimes 3} \tag{6.2}$$

Here Σ_3 acts on $(AT)^{\otimes 3}$ by permuting the factors, and Σ_3 acts diagonally on the tensor product with M_3.

Proposition 6. J_3T is a projective module over $F_2(\Sigma_3)$, and

$$\pi_i J_3 S = 0 \quad (i \geq 0) \tag{6.3}$$

Proof. The projectivity of J_3T follows from that of M_3. (6.3) follows at once from Lemma 3.

So our functor J_3 certainly satisfies conditions i) and iii) of the Introduction, with $\pi_0 J_3 S = 0$, a projective! It remains to check condition ii), equation (1.9).

Define an F_2-linear $\phi: J_3T \to L_3AT$ by:

$$\phi([e_{\pi(1)}, e_{\pi(2)}, e_{\pi(3)}] \otimes x_1 \otimes x_2 \otimes x_3) = [x_{\pi(1)}, x_{\pi(2)}, x_{\pi(3)}]$$

for each $\pi \in \Sigma_3$; and $x_1, x_2, x_3 \in T$. A moment's reflection will convince the reader that ϕ is well-defined. Indeed, if s is a formal sum of commutators $[e_{\pi(1)}, e_{\pi(2)}, e_{\pi(3)}]$ that happens to represent zero in M_3, then ϕ carries $s \otimes x_1 \otimes x_2 \otimes x_3$ to a formally identical sum in L_3AT that also represents zero.

A quick calculation shows that ϕ is linear over the groupring $F_2(\Sigma_3)$, where Σ_3-action on L_3AT is taken to be trivial. So ϕ passes to an F_2-linear map:

$$F_2 \underset{\Sigma_3}{\otimes} J_3T \xrightarrow{\phi} L_3AT \tag{6.4}$$

We wish to show:

<u>Proposition 7</u>. ϕ of (6.4) is an isomorphism.

To aid in the proof of Proposition 7, observe that the submodules $(AT)^\mu \subseteq (AT)^{\otimes 3}$ defined in Section 4 determine submodules $J_3^\mu T$ of J_3T, according to the formula

$$J_3^\mu T = M_3 \otimes (AT)^\mu \tag{6.5}$$

for each partition μ of 3. It is also clear that ϕ of (6.3) satisfies $\phi(J_3^\mu T) \subseteq L_3^\mu AT$; and in particular, $\phi(J_3^{(3)}T) = 0$.

Consider the short-exact sequence of $F_2(\Sigma_3)$-modules:

$$0 \to J_3^{(3)}T \to J_3T \to \frac{J_3T}{J_3^{(3)}T} \to 0 \tag{6.6}$$

An easy computation shows that $F_2 \underset{\Sigma_3}{\otimes} M_3 = 0$. Hence

$$F_2 \underset{\Sigma_3}{\otimes} J_3^{(3)}T = F_2 \underset{\Sigma_3}{\otimes} (M_3 \otimes AT) = (F_2 \underset{\Sigma_3}{\otimes} M_3) \otimes AT = 0$$

So application of the right-exact functor $F_2 \underset{\Sigma_3}{\otimes} (\quad)$ to (6.6) gives an isomorphism $F_2 \otimes_{\Sigma_3} (J_3T) = F_2 \otimes_{\Sigma_3} (J_3T/J_3^{(3)}T)$. So to prove Proposition 7, it will suffice to show:

<u>Lemma 8</u>. The module $J_3T/J_3^{(3)}T$ is free over $F_2(\Sigma_3)$, and

$$F_2 \underset{\Sigma_3}{\otimes} (J_3T/J_3^{(3)}T) \xrightarrow{\phi} L_3AT \tag{6.7}$$

is an isomorphism.

In order to prove Lemma 8, we consider the diagram:

$$\begin{array}{ccccccccc} 0 & \to & \dfrac{J_3^{(2,1)}T}{J_3^{(3)}T} & \to & \dfrac{J_3T}{J_3^{(3)}T} & \to & \dfrac{J_3T}{J_3^{(2,1)}T} & \to & 0 \\ & & \downarrow{\scriptstyle \phi'} & & \downarrow{\scriptstyle \phi} & & \downarrow{\scriptstyle \phi''} & & \\ 0 & \to & L_3^{(2,1)}AT & \to & L_3AT & \to & \dfrac{L_3AT}{L_3^{(2,1)}AT} & \to & 0 \end{array} \tag{6.8}$$

where ϕ' is the restriction of ϕ, and ϕ'' is the induced map on quotients.

<u>Lemma 9</u>. For each pointed set T, $J_3^{(2,1)}T/J_3^{(3)}T$ is free over $F_2(\Sigma_3)$, and ϕ' of (6.8) induces an isomorphism of vector spaces when Σ_3-action is divided out.

<u>Proof</u>. Each pair of distinct elements $x_1, x_2 \in T$ (with $x_1 \neq *$ and $x_2 \neq *$) determine a submodule of $J_3^{(2,1)}T/J_3^{(3)}T$: that generated over Σ_3 by the elements $\alpha \otimes x_1 \otimes x_1 \otimes x_2$ and $\alpha \otimes x_2 \otimes x_2 \otimes x_1$, as α runs through M_3. Further, $J_3^{(2,1)}T/J_3^{(3)}T$ is the direct sum of the subspaces corresponding to distinct pairs. Similarly, the pair $x_1, x_2 \in T$ determine a subspace of $L_3^{(2,1)}AT$ ... that spanned by $[x_1, x_2, x_1]$ and $[x_2, x_1, x_2]$ .. and $L_3^{(2,1)}AT$ is the direct sum of all such subspaces. The map ϕ' respects the splitting we have just described. So it suffices to prove Lemma 9 in the case $T = \{*, x_1, x_2\}$. But in this case it is easy to check that $J_3^{(2,1)}T/J_3^{(3)}T$ is the free module over $F_2(\Sigma_3)$ on the two generators $[e_1, e_2, e_3] \otimes x_1 \otimes x_2 \otimes x_1$ and $[e_1, e_2, e_3] \otimes (x_2 \otimes x_1 \otimes x_2)$. So the proof is complete.

Continuing our analysis of (6.8) we have:

<u>Lemma 10</u>. For each pointed set T, $J_3T/J_3^{(2,1)}T$ is free over $F_2(\Sigma_3)$, and ϕ'' induces an isomorphism of vector spaces when Σ_3-action is divided out.

Proof. By an argument similar to that used in the proof of Lemma 9, we reduce Lemma 10 to the case in which $T = \{*, x_1, x_2, x_3\}$. In this case, $(AT)^{\otimes 3}/(AT)^{(2,1)}$ is obviously free over $F_2(\Sigma_3)$ on the single generator $x_1 \otimes x_2 \otimes x_3$, so that $M_3 \otimes ((AT)^{\otimes 3}/(AT)^{(2,1)}) = J_3 T / J_3^{(2,1)} T$ is free on the two generators $[e_1, e_2, e_3] \otimes (x_1 \otimes x_2 \otimes x_3)$ and $[e_2, e_3, e_1] \otimes (x_1 \otimes x_2 \otimes x_3)$. On the other hand, an F_2-basis for $L_3 AT / L_3^{(2,1)} AT$ is the set $\{[x_1, x_2, x_3], [x_2, x_3, x_1]\}$. So the proof is complete.

Lemma 8 now follows from Lemmas 9 and 10, by way of diagram (6.8).

As we have remarked, Proposition 7 follows from Lemma 8.

We have shown the functor J_3 has properties i), ii), and iii) of the Introduction. This completes the proof of Conjecture 1" in the case $r = 3$, and of all parts of Theorem 2" for this case.

7. Construction of $J_{4,0}$

In this section we construct a functor $J_{4,0}$ in such a way that (3.3) is satisfied, and so that $J_{4,0}S$ has few non-vanishing homotopy groups.

As in the previous section, we will need a "model space". Let V_4 be the F_2-vector space with basis $\{e_1, e_2, e_3, e_4\}$. Let Σ_4 act on the left of V_4 by permuting the basis. This action of Σ_4 extends to an action on LV_4. Let $M_4 \subseteq L_4 V_4$ be the $F_2(\Sigma_4)$-submodule defined by:

$$M_4 = \mathrm{Span}\{[e_{\pi(1)}, e_{\pi(2)}, e_{\pi(3)}, e_{\pi(4)}] \mid \pi \in \Sigma_4\} \tag{7.1}$$

This is our model space. It will serve as a universal example of the relations that define the fourth weight of a free Lie algebra. In fact, one has in M_4, and in the fourth weight of any free Lie algebra:

$$\begin{gathered} [w,x,y,z] + [x,w,y,z] = 0 \\ [w,x,y,z] + [x,y,w,z] + [y,w,x,z] = 0 \\ [w,x,y,z] + [w,x,z,y] = [y,z,w,x] + [y,z,x,w]. \end{gathered} \tag{7.2}$$

Use of these relations is sufficient to show that M_4 is spanned by the elements

$$\{[e_4, e_{\sigma(1)}, e_{\sigma(2)}, e_{\sigma(3)}] \mid \sigma \in \Sigma_3\} \tag{7.3}$$

On the other hand, use of M. Hall's famous basis ([6]) shows that M_4 is 6-dimensional. So we have:

Lemma 11. M_4 is a 6-dimensional representation of Σ_4, and (7.3) is a basis.

A more detailed analysis of M_4 will be given in Section 12, where we will write it as a tensor product.

If T is a pointed set, define the $F_2(\Sigma_4)$-module $J_{4,0}T$ by:

$$J_{4,0}T = \frac{M_4 \otimes (AT)^{\otimes 4}}{R_{4,0}T} \tag{7.4}$$

Here Σ_4 acts on $(AT)^{\otimes 4}$ by permuting the factors, and Σ_4 acts diagonally on $M_4 \otimes (AT)^{\otimes 4}$. $R_{4,0}T$ is a submodule of the tensor product, that we define by:

$$\begin{aligned} R_{4,0}T &= \mathrm{Span}\{\alpha \otimes x^{\otimes 4} \mid \alpha \in M_4,\ x \in T\} \\ &\quad + \Sigma_4\text{-}\mathrm{Span}\{([e_1,e_2,e_3,e_4]+[e_4,e_3,e_2,e_1]) \otimes x_1 \otimes x_2 \otimes x_1 \otimes x_2 \mid x_1, x_2 \in T\} \end{aligned} \tag{7.5}$$

This "module of relations" $R_{4,0}T$ has an interesting structure, that we will explore shortly.

Define an F_2-linear map $\phi : M_4 \otimes (AT)^{\otimes 4} \to L_4AT$ by analogy with (6.4):

$$\begin{aligned} &\phi([e_{\pi(1)},e_{\pi(2)},e_{\pi(3)},e_{\pi(4)}] \otimes x_1 \otimes x_2 \otimes x_3 \otimes x_4) \\ &\qquad = [x_{\pi(1)},x_{\pi(2)},x_{\pi(3)},x_{\pi(4)}] \end{aligned} \tag{7.6}$$

for each $\pi \in \Sigma_4$ and $x_i \in T$. By an argument similar to that given after (6.3), one sees that ϕ is well defined. It is also easy to check that ϕ is linear over $F_2(\Sigma_4)$, where we understand trivial Σ_4 action on L_4AT. Finally, we observe that as a consequence of the identities $[x,x,x,x] = 0$ and $[x_1,x_2,x_1,x_2] + [x_2,x_1,x_2,x_1] = 0$ in L_4AT, we have $\phi(R_{4,0}T) = 0$. So ϕ passes to an $F_2(\Sigma_4)$-linear map

$$J_{4,0}T \xrightarrow{\phi} L_4AT \tag{7.7}$$

The main result of this section is:

Proposition 12. $J_{4,0}T$ is free as a module over $F_2(\Sigma_4)$, and ϕ induces a natural isomorphism of vector spaces:

$$F_2 \underset{\Sigma_4}{\otimes} J_{4,0}T \xrightarrow{\phi} L_4AT. \qquad (7.8)$$

The proof of Proposition 12 will occupy us for the rest of this section. We must first learn something about the structure of the "relations module" $R_{4,0}T$. In particular, let

$$d: M_4 \otimes AT \to R_{4,0}T$$

be the natural map of $F_2(\Sigma_4)$-modules defined by

$$d(\alpha\otimes x) = \alpha\otimes x\otimes x\otimes x\otimes x \qquad (7.9)$$

for each $\alpha \in M_4$ and $x \in T$. We will investigate the structure of the cokernel of d, and so consider the second term in (7.5).

Given $x_1 \neq x_2$ in T-*, let us determine the subspace $K \subset M_4$ for which

$$K\otimes x_1\otimes x_2\otimes x_1\otimes x_2 \subset R_{4,0}T \qquad (7.10)$$

It is clear from (7.5) that $K = F_2(\Lambda)\circ([e_1,e_2,e_3,e_4]+[e_4,e_3,e_2,e_1])$, where $\Lambda \subset \Sigma_4$ is the stabilizer of the tensor $x_1\otimes x_2\otimes x_1\otimes x_2$. Thus, $\Lambda = \{e,(13),(24),(13)(24)\}$, and a short computation using the relations (7.2) shows that K is two-dimensional:

$$\begin{aligned} K &= F_2(\Lambda)\circ([e_1,e_2,e_3,e_4]+[e_4,e_3,e_2,e_1]) \\ &= \mathrm{Span}\{[e_1,e_2,e_3,e_4]+[e_4,e_3,e_2,e_1],[e_3,e_2,e_1,e_4]+[e_4,e_1,e_2,e_3]\} \end{aligned} \qquad (7.11)$$

So (7.10) holds. What else lies in $R_{4,0}T$? From (7.5) it is clear we must also consider the subspaces of $R_{4,0}T$ given by

$$\tau\circ(K\otimes x_1\otimes x_2\otimes x_1\otimes x_2) = (\tau\circ K)\otimes\tau\circ(x_1\otimes x_2\otimes x_1\otimes x_2) \qquad (7.12)$$

as τ runs through Σ_4. But the subspace (7.12) depends only on which of the left cosets of Λ the element τ belongs to. Further, as τ runs through the six cosets, $\tau\circ(x_1\otimes x_2\otimes x_1\otimes x_2)$ will run through the six "tensors of type (2.2)" (in the sense of Section 4) that involve the elements x_1 and x_2. Summarizing: we see that (7.5) implies a direct sum decomposition of vector spaces:

$$\frac{R_{4,0}T}{d(M_4\otimes AT)} = \bigoplus_{x_1\neq x_2} \bigoplus_{\tau\in\Sigma_4/\Lambda} \tau\circ(K\otimes(x_1\otimes x_2\otimes x_1\otimes x_2)) \qquad (7.13)$$

where the first sum is extended over all unordered pairs $x_1 \neq x_2$ in $T-*$, and the second sum is over a set of coset representatives τ.

Equation (7.13) is an isomorphism of vector spaces only. It says nothing about the structure of coker d as a module over $F_2(\Sigma_4)$; nor is the isomorphism natural with respect to maps of the pointed set T. In fact, at the end of this section we will use (7.13) as a stepping stone to a better result. Meanwhile, we will find (7.13) adequate for a proof of Proposition 12; and to this task we now turn our attention.

Our proof is analogous to the proof of Lemma 8, and proceeds as follows. For each partition μ of the integer 4, we introduce a submodule $J^{\mu}_{4,0}T \subseteq J_{4,0}T$, defined by analogy with (6.5):

$$J^{\mu}_{4,0}T = \frac{M_4 \otimes (AT)^{\mu}}{R_{4,0}T \cap (M_4 \otimes (AT)^{\mu})} \tag{7.14}$$

Clearly ϕ of (7.7) satisfies $\phi(J^{\mu}_{4,0}T) \subseteq L^{\mu}_4 AT$. We will prove Proposition 12 by induction on the partitions μ, and begin with:

<u>Lemma</u> <u>13</u>. For each pointed set T, the module $J^{(3,1)}_{4,0}T$ is free over $F_2(\Sigma_4)$, and ϕ induces an isomorphism of vector spaces:

$$F_2 \underset{\Sigma_4}{\otimes} J^{(3,1)}_{4,0}T \xrightarrow{\phi} L^{(3,1)}_4 T \tag{7.15}$$

<u>Proof</u>. By an argument like that used in the proof of Lemma 9, we find it is enough to consider the case $T = \{*, x_1, x_2\}$. But then it is easily checked that $J^{(3,1)}_{4,0}T$ is the free module over $F_2(\Sigma_4)$ on the two generators $[e_1,e_2,e_3,e_4]\otimes(x_2\otimes x_1\otimes x_1\otimes x_1)$ and $[e_1,e_2,e_3,e_4]\otimes(x_1\otimes x_2\otimes x_2\otimes x_2)$ (use Lemma 11). On the other hand, an F_2-basis for $L^{(3,1)}_4 AT$ consists of the two elements $[x_2,x_1,x_1,x_1]$ and $[x_1,x_2,x_2,x_2]$. So the lemma is proved.

<u>Lemma</u> <u>14</u>. For each pointed set T, the module $J^{(2,2)}_{4,0}T$ is free over $F_2(\Sigma_4)$, and ϕ induces an isomorphism of vector spaces

$$F_2 \underset{\Sigma_4}{\otimes} J^{(2,2)}_{4,0}T \xrightarrow{\phi} L^{(2,2)}_4 AT \tag{7.16}$$

<u>Proof</u>. It suffices to consider the case $T = \{*, x_1, x_2\}$. Then we have from (7.13):

$$R_{4,0}T = (M_4 \otimes x_1^{\otimes 4}) \oplus (M_4 \otimes x_2^{\otimes 4}) \oplus \sum_{\tau \in \Sigma_4/\Lambda} \tau\circ(K \otimes x_1 \otimes x_2 \otimes x_1 \otimes x_2) \qquad (7.17)$$

Then it is not hard to see that the $F_2(\Sigma_4)$-module

$$J_{4,0}^{(2,2)}T = \frac{M_4 \otimes (AT)^{(2,2)}}{R_{4,0}T} \qquad (7.18)$$

is free on the single generator

$$[e_1,e_2,e_3,e_4] \otimes (x_1 \otimes x_2 \otimes x_1 \otimes x_2) \qquad (7.19)$$

In fact, the elements of the subgroup $\Lambda \subset \Sigma_4$ fix the tensor $x_1 \otimes x_2 \otimes x_1 \otimes x_2$, and when applied to $[e_1,e_2,e_3,e_4]$ generate a four-dimensional subspace of M_4 that is complementary to K (an easy computation using (7.2) and (7.11) is needed to verify this). It follows that when the four elements of any fixed left coset $\tau\Lambda$ are applied to $[e_1,e_2,e_3,e_4]$, they generate a subspace of M_4 that is complementary to τK. Combining this remark with (7.17) and (7.18), we see at once that $J_{4,0}^{(2,2)}T$ is free over $F_2(\Sigma_4)$, with (7.19) as generator. We now complete the proof of Lemma 14 by observing that $L_4^{(2,2)}AT$ is one-dimensional on the element $[x_1,x_2,x_1,x_2] = \phi([e_1,e_2,e_3,e_4] \otimes (x_1 \otimes x_2 \otimes x_1 \otimes x_2))$.

Preparing for the next lemma, we observe that $J_{4,0}^{(2,2)}T \cap J_{4,0}^{(3,1)}T = 0$, so that $J_{4,0}^{(2,2)}T \oplus J_{4,0}^{(3,1)}T$ is a submodule of $J_{4,0}T$.

Lemma 15. For each pointed set T, the module

$$\frac{J_{4,0}^{(2,1,1)}T}{J_{4,0}^{(2,2)}T \oplus J_{4,0}^{(3,1)}T} = M_4 \otimes \frac{(AT)^{(2,1,1)}}{(AT)^{(2,2)} + (AT)^{(3,1)}} \qquad (7.20)$$

is free over $F_2(\Sigma_4)$, and ϕ induces an isomorphism of vector spaces:

$$F_2 \underset{\Sigma_4}{\otimes} \left[\frac{J_{4,0}^{(2,1,1)}T}{J_{4,0}^{(2,2)}T \oplus J_{4,0}^{(3,1)}T}\right] \xrightarrow{\phi} \frac{L_4^{(2,1,1)}AT}{L_4^{(2,2)}AT \oplus L_4^{(3,1)}AT} \qquad (7.21)$$

Proof. It suffices to consider the case $T = \{*,x_1,x_2,x_3\}$. For each i, $1 \leq i \leq 3$, let $X^i \subset (AT)^{(2,1,1)}$ be the 12-dimensional subspace spanned by those basic tensors in which x_i appears twice, and the two other x's appear once each. Then from (7.20) we have

$$\frac{J_{4,0}^{(2,1,1)}T}{J_{4,0}^{(2,2)}T \oplus J_{4,0}^{(3,1)}T} = \bigoplus_{i=1}^{3} (M_4 \otimes X^i) \tag{7.22}$$

as a direct sum of modules over $F_2(\Sigma_4)$. On the other hand, we define subspaces $L_4^i \subset L_4 AT$, by writing $L_4^i = \pi(X^i)$, where $\pi: (AT)^{\otimes 4} \to L_4 AT$ is the map defined in Section 4. Then as vector spaces:

$$\frac{L_4^{(2,1,1)}AT}{L_4^{(2,2)}AT \oplus L_4^{(3,1)}AT} = \bigoplus_{i=1}^{3} L_4^i \tag{7.23}$$

Consequently, Lemma 15 will be proved if we can show that each $M_4 \otimes X^i$ is free over $F_2(\Sigma_4)$, and that for each i, ϕ induces an isomorphism:

$$F_2 \otimes_{\Sigma_4} (M_4 \otimes X^i) \xrightarrow{\phi} L_4^i \tag{7.24}$$

In fact it is easily checked that $M_4 \otimes X^1$ is free over $F_2(\Sigma_4)$ on the three generators $[e_1,e_2,e_3,e_4] \otimes x_1 \otimes x_2 \otimes x_1 \otimes x_3$; $[e_1,e_2,e_3,e_4] \otimes (x_1 \otimes x_3 \otimes x_1 \otimes x_2)$; $[e_1,e_2,e_3,e_4] \otimes (x_1 \otimes x_2 \otimes x_3 \otimes x_1)$. But by using Hall's basis [6] and the relations (7.2), one finds that L_4^1 is 3-dimensional, with basis $[x_1,x_2,x_1,x_3]$, $[x_1,x_3,x_1,x_2]$, $[x_1,x_2,x_3,x_1]$. So (7.24) is an isomorphism if $i = 1$. The cases $i = 2,3$ are similar.

Lemma 16. For each pointed set T, the module $J_{4,0}^{(2,1,1)}T$ is free over $F_2(\Sigma_4)$, and the map ϕ induces an isomorphism of vector spaces:

$$F_2 \otimes_{\Sigma_4} J_{4,0}^{(2,1,1)}T \xrightarrow{\phi} L_4^{(2,1,1)}AT \tag{7.25}$$

Proof. This now follows from Lemmas 13, 14 and 15, together with the Five Lemma.

Lemma 17. For each pointed set T, the module

$$\frac{J_{4,0}T}{J_{4,0}^{(2,1,1)}T} = M_4 \otimes \frac{(AT)^{\otimes 4}}{(AT)^{(2,1,1)}} \tag{7.26}$$

is free over $F_2(\Sigma_4)$, and ϕ induces an isomorphism of vector spaces:

$$F_2 \underset{\Sigma_4}{\otimes} \left(\frac{J_{4,0}T}{J_{4,0}^{(2,1,1)}T}\right) \xrightarrow{\phi} \frac{L_4AT}{L_4^{(2,1,1)}AT} \tag{7.27}$$

Proof. It suffices to consider the case $T = \{*,x_1,x_2,x_3,x_4\}$. Then $(AT)^{\otimes 4}/(AT)^{(2,1,1)}$ is clearly the free module over $F_2(\Sigma_4)$ on the generator $x_1\otimes x_2\otimes x_3\otimes x_4$. Then from (7.26) we have that $J_{4,0}T/J_{4,0}^{(2,1,1)}T$ is free over $F_2(\Sigma_4)$ on the generators $\alpha\otimes x_1\otimes x_2\otimes x_3\otimes x_4$, where α runs through a basis for M_4. Let us take the basis (7.3). By (7.6):

$$\begin{aligned}\phi([e_4,e_{\sigma(1)},e_{\sigma(2)},e_{\sigma(3)}]\otimes x_1\otimes x_2\otimes x_3\otimes x_4)& \\ = [x_4,x_{\sigma(1)},x_{\sigma(2)},x_{\sigma(3)}]&\end{aligned} \tag{7.28}$$

But as σ runs through Σ_3, the right hand side of (7.28) runs through a basis for $L_4AT/L_4^{(2,1,1)}AT$. So the proof is complete.

Proof of Proposition 12. Proposition 12 now follows easily from Lemmas 16, 17, and the Five Lemma.

We close this section by giving a more detailed analysis of the relations module $R_{4,0}T$ of (7.5). This analysis will be helpful in Sections 9, 10 and 11, when we construct the functors $J_{4,1}$ and $J_{4,2}$.

Let $\Gamma \subset \Sigma_4$ be the subgroup $\Gamma = \{e,(13)(24)\}$. Embed Σ_2 in Σ_4 by letting Σ_2 be the subgroup $\Sigma_2 = \{e,(12)(34)\}$. With $d:M_4\otimes AT \to R_{4,0}T$ as in (7.9), our aim is to establish for each pointed set T a natural isomorphism of $F_2(\Sigma_4)$-modules:

$$j:F_2(\Sigma_4/\Gamma)\underset{\Sigma_2}{\otimes} ((AT)^{\otimes 2}/(AT)^{(2)}) \to \frac{R_{4,0}T}{d(M_4\otimes AT)} \tag{7.29}$$

Here the right action of Σ_2 upon $F_2(\Sigma_4/\Gamma)$ is given by $(\sigma\Gamma)\circ(12)(34) = \sigma(12)(34)\Gamma$ for all $\sigma \in \Sigma_4$. The reader will recognize $(AT)^{\otimes 2}/(AT)^{(2)}$ as $J_{2,0}T$ (see (5.2)); we will exploit this identity later (see (9.1) and (9.2)). We define j by:

$$\begin{aligned}j(\sigma\Gamma\underset{\Sigma_2}{\otimes} (x_1\otimes x_2))& \\ = \sigma\circ(([e_1,e_2,e_3,e_4]+[e_4,e_3,e_2,e_1])\otimes x_1\otimes x_2\otimes x_1\otimes x_2)&\end{aligned} \tag{7.30}$$

for all $\sigma \in \Sigma_4$ and $x_1,x_2 \in T$. Our choice of embedding of Σ_2 into

Σ_4 was dictated by the requirement that (7.30) make j a well defined map: the reader will verify that (7.30) gives the same result for $j(\sigma(12)(34)\otimes_{\Sigma_2}(x_1\otimes x_2))$ as it does for $j(\sigma\otimes_{\Sigma_2}(x_2\otimes x_1))$. To verify that j is an isomorphism, it is enough to consider the case $T = \{*,x_1,x_2\}$. That j is onto then follows from (7.11) and (7.13). That j is one-one then follows from a dimension count. In fact, the left hand side of (7.29) has dimension twelve. One uses (7.11) and (7.13) to verify that the right hand side also has dimension twelve.

We have shown the existence of a natural, short-exact sequence of $F_2(\Sigma_4)$-modules:

$$0 \to M_4\otimes AT \xrightarrow{d} R_{4,0}T \xrightarrow{p} F_2(\Sigma_4/\Gamma)\otimes_{\Sigma_2}((AT)^{\otimes 2}/(AT)^{(2)}) \to 0 \qquad (7.31)$$

where d is the diagonal of (7.9). We close this section by observing that (7.31) can be used to compute $\pi_*J_{4,0}S$. We will not make explicit use of the result, so we will only sketch its derivation. From (7.4) and Lemma 3 we have $\pi_*J_{4,0}S = \pi_{*-1}R_{4,0}S$. Since (7.31) is natural, we can look at the fibration that results from replacing T everywhere by the spectrum S, and use the associated long-exact homotopy sequence to compute $\pi_*R_{4,0}S$. The reader will easily check that the base of the fibration has non-vanishing homotopy only in dimension 1, while the fiber has non-vanishing homotopy only in dimension 0. The connecting morphism between the two non-vanishing homotopy groups is easily computed; it is neither monic nor epic. The end result is:

$$\pi_iJ_{4,0}S = \pi_{i-1}R_{4,0}S = \begin{cases} S^{(2,2)} & \text{if } i = 1,2 \\ 0 & \text{otherwise} \end{cases} \qquad (7.32)$$

as modules over $F_2(\Sigma_4)$. Here $S^{(2,2)}$ is the Specht module whose definition is recalled in the next section.

We will not use this result. However, (7.32) does suggest the sense in which the program sketched in Section 3 will be carried out. We will construct $J_{4,1}S$ as a fibration over $J_{4,0}S$; and $J_{4,2}S$ as a fibration over $J_{4,1}S$. In the process, the "higher" homotopy groups present in (7.32) will get killed off. $J_{4,1}S$ will have only one non-vanishing homotopy group: in dimension one. $J_{4,2}S = J_4S$ will have only zero-dimensional homotopy, as required by Conjecture 1" and Theorem 2".

8. Some Representations of Σ_4

Our constructions of $J_{4,1}$ and $J_{4,2}$ in the following sections, and especially the computations of the homotopy groups $\pi_* J_{4,1} S$ and $\pi_* J_{4,2} S$, are based on a certain four-term exact sequence of Σ_4-modules: (8.11) below. Our object now is to define the modules that appear in this sequence, and to specify the first homomorphism. In order not to impede the exposition, we have postponed the construction of the other homomorphisms, and the proof of exactness, to Section 12.

The Specht modules are of course well-known, and are usually defined in the context of the general representation theory of Σ_4. We will need only the Specht modules $S^{(3,1)}$ and $S^{(2,2)}$; and in order to make this paper self-contained, will give independent definitions of them.

Let $V_4 = \mathrm{Span}\{e_1,e_2,e_3,e_4\}$ be the $F_2(\Sigma_4)$-module introduced in the previous section. The map $\varepsilon: V_4 \to F_2$ given by $\varepsilon(e_i) = 1$ for each $i \leq 4$ is linear over $F_2(\Sigma_4)$. We take

$$S^{(3,1)} = \ker(\varepsilon: V_4 \to F_2) \tag{8.1}$$

The Specht module $S^{(2,1)}$ is a module over $F_2(\Sigma_3)$. It can be obtained by identifying Σ_3 with $GL_2(F_2)$. Then $S^{(2,1)}$ is the two-dimensional vector space on which $GL_2(F_2)$ naturally acts. To define the $F_2(\Sigma_4)$-module $S^{(2,2)}$, let Δ_2 be the "diagonal" elementary abelian 2-group in Σ_4:

$$\Delta_2 = \{e,(12)(34),(13)(24),(14)(23)\} \tag{8.2}$$

Then Δ_2 is normal in Σ_4, and $\Sigma_4/\Delta_2 = \Sigma_3$. By composing the resulting projection $\Sigma_4 \to \Sigma_3$ with the action of Σ_3 upon $S^{(2,1)}$, we obtain an action of Σ_4 upon that two-dimensional space. The resulting $F_2(\Sigma_4)$-module is called $S^{(2,2)}$. It is easy to see that $S^{(2,2)}$ has an F_2-basis $\{\alpha,\beta\}$ for which

$$\begin{aligned} (1,2)\alpha &= (3,4)\alpha = \alpha+\beta; \quad (2,3)\alpha = \beta \\ (1,2)\beta &= (3,4)\beta = \beta \ ; \quad (2,3)\beta = \alpha \end{aligned} \tag{8.3}$$

Note that if Σ_3 is allowed to act on $S^{(2,2)}$ through the canonical inclusion $\Sigma_3 \to \Sigma_4$, the resulting representation of Σ_3 is identical with $S^{(2,1)}$:

$$\mathrm{Res}_{\Sigma_3}(S^{(2,2)}) = S^{(2,1)} \tag{8.4}$$

Finally we need the projective cover P of $S^{(2,2)}$, and (8.4) will be useful in constructing it. We set P equal to the induced module

$$P = \mathrm{Ind}_{\Sigma_3}^{\Sigma_4}(S^{(2,1)}) = F_2(\Sigma_4) \otimes_{\Sigma_3} S^{(2,1)} \tag{8.5}$$

In view of (8.4) we can just as well write

$$P = F_2(\Sigma_4) \otimes_{\Sigma_3} S^{(2,2)} \tag{8.6}$$

By [13], $S^{(2,1)}$ is projective over $F_2(\Sigma_3)$. So from (8.5) we have that P is projective over $F_2(\Sigma_4)$. We wish to show further that P maps onto $S^{(2,2)}$. For this purpose we use (8.6) to establish an isomorphism of Σ_4-modules:

$$(P = F_2(\Sigma_4) \otimes_{\Sigma_3} S^{(2,2)}) \to (F_2(\Sigma_4/\Sigma_3) \otimes S^{(2,2)} = V_4 \otimes S^{(2,2)}). \tag{8.7}$$

Here Σ_4 acts diagonally on the tensor products on the right-hand side, and the map is given by $\sigma \otimes_{\Sigma_3} \gamma \to \sigma\Sigma_3 \otimes \gamma$ for each $\sigma \in \Sigma_4$ and $\gamma \in S^{(2,2)}$

Finally, with $\varepsilon: V_4 \to F_2$ as in (8.1) we form $\varepsilon \otimes \mathrm{id}: V_4 \otimes S^{(2,2)} \to S^{(2,2)}$; this provides us with a Σ_4-linear map of P onto $S^{(2,2)}$.

<u>Lemma 18</u>. The $F_2(\Sigma_4)$-module P defined by (8.5) or (8.6) is a projective that maps onto $S^{(2,2)}$, and satisfies

$$F_2 \otimes_{\Sigma_4} P = 0 \tag{8.8}$$

<u>Proof</u>. All statements have been proved except (8.8). But it is easy to check that $F_2 \otimes_{\Sigma_3} S^{(2,1)} = 0$, so from (8.5) we have $F_2 \otimes_{\Sigma_4} P =$

$F_2 \otimes_{\Sigma_4} (F_2(\Sigma_4) \otimes_{\Sigma_3} S^{(2,1)}) = F_2 \otimes_{\Sigma_3} S^{(2,1)} = 0.$

We are now ready to state the main result of this section. With

Σ_2 embedded in Σ_4 as in Section 7, we define $i:F_2(\Sigma_4/\Delta_2)\to F_2(\Sigma_4/\Sigma_2)$ to be the unique Σ_4-linear map for which

$$i(e\Delta_2) = e\Sigma_2 + (13)(24)\Sigma_2 \tag{8.9}$$

i is monic. Let $j:F_2(\Sigma_4/\Delta_2)\to M_4$ be the unique Σ_4-linear map for which

$$j(e\Delta_2) = [e_1,e_2,e_3,e_4] + [e_4,e_3,e_2,e_1] \tag{8.10}$$

Proposition 19. There exist homomorphisms f, g of $F_2(\Sigma_4)$-modules such that the sequence is exact:

$$0 \to F_2(\Sigma_4/\Delta_2) \xrightarrow{(i,j)} F_2(\Sigma_4/\Sigma_2)\oplus M_4 \xrightarrow{f} F_2(\Sigma_4) \xrightarrow{g} P\oplus(S^{(2,2)}\otimes S^{(2,2)}) \longrightarrow 0 \tag{8.11}$$

Further, the map f will satisfy:

$$\text{im } f \subset IF_2(\Sigma_4) \tag{8.12}$$

where $IF_2(\Sigma_4)$ is the augmentation ideal.

The proof will be given in Section 12. In fact, we will prove the somewhat stronger:

Proposition 20. There exist homomorphisms h, k, g of $F_2(\Sigma_4)$-modules such that the sequences are exact:

$$0 \to F_2(\Sigma_4/\Delta_2) \xrightarrow{(i,j)} F_2(\Sigma_4/\Sigma_2)\oplus M_4 \xrightarrow{h} S^{(3,1)}\otimes S^{(2,2)}\otimes S^{(2,2)} \to 0 \tag{8.13}$$

$$0 \to S^{(3,1)}\otimes S^{(2,2)}\otimes S^{(2,2)} \xrightarrow{k} F_2(\Sigma_4) \xrightarrow{g} P\oplus(S^{(2,2)}\otimes S^{(2,2)}) \to 0 \tag{8.14}$$

We will obtain (8.11) from (8.12) and (8.13) by setting $f = kh$. Propositions 19 and 20 will be used in Sections 10 and 11; to compute the homotopy groups of $J_{4,1}S$ and $J_{4,2}S$.

9. Construction of $J_{4,1}$

In this section we will construct the functor $J_{4,1}$, obtaining it as a fibration over $J_{4,0}$ in the manner sketched out in Section 3. Fo each pointed set T we define the $F_2(\Sigma_4)$-module $J_{4,1}T$ by setting

$$J_{4,1}T = \frac{(F_2(\Sigma_4)\underset{\Sigma_2}{\otimes} J_{2,0}T)\oplus(M_4\otimes AT^{\otimes 4})}{S_{4,1}T} \tag{9.1}$$

Here Σ_2 is embedded in Σ_4 as in Section 7, and $S_{4,1}T$ is a submodule of the numerator that we are going to describe.

For this purpose we first define a Σ_4-linear map theta:

$$\theta : R_{4,0}T \to (F_2(\Sigma_4)\underset{\Sigma_2}{\otimes} J_{2,0}T) = (F_2(\Sigma_4)\underset{\Sigma_2}{\otimes} ((AT)^{\otimes 2}/(AT)^{(2)})) \tag{9.2}$$

as follows. With $\Gamma \subset \Sigma_4$ as in Section 7, let $\eta : F_2(\Sigma_4/\Gamma) \to F_2(\Sigma_4)$ be the Σ_4-linear map defined by $\eta(e\Gamma) = e+(13)(24)$. Then define

$$\bar{\eta} : F_2(\Sigma_4/\Gamma)\underset{\Sigma_2}{\otimes} ((AT)^{\otimes 2}/(AT)^{(2)}) \to F_2(\Sigma_4)\underset{\Sigma_2}{\otimes} ((AT)^{\otimes 2}/(AT)^{(2)})$$

by $\bar{\eta} = \eta\otimes_{\Sigma_2}$ (identity). Finally, define θ in (9.2) by $\theta = \bar{\eta}p$, where p is the projection of (7.31). Thus, in terms of the description of $R_{4,0}T$ given in (7.5) we have:

$$\theta(\alpha\otimes x^{\otimes 4}) = 0 \quad (\alpha \in M_4,\ x \in T) \tag{9.3}$$

$$\begin{aligned} &\theta(([e_1,e_2,e_3,e_4]+[e_4,e_3,e_2,e_1])\otimes x_1\otimes x_2\otimes x_1\otimes x_2) \\ &\qquad = (e+(13)(24))\underset{\Sigma_2}{\otimes}(x_1\otimes x_2) \end{aligned} \tag{9.4}$$

for any $x_1, x_2 \in T$. These equations determine θ.

We can now define the denominator $S_{4,1}T$ in (9.1):

$$S_{4,1}T = \{\theta(\gamma)\oplus\gamma \mid \gamma \in R_{4,0}T \subset M_4\otimes(AT)^{\otimes 4}\} \tag{9.5}$$

This completes our definition of $J_{4,1}T$.

We will find it useful to write descriptions of $S_{4,1}T$ analogous to the descriptions of $R_{4,0}T$ given by (7.5) and (7.31). The analogue of (7.5) is:

$$S_{4,1}T = \text{Span}\{\alpha \otimes x^{\otimes 4} \mid \alpha \in M_4,\ x \in T\}$$

$$+\ \Sigma_4 \text{Span}\{(e+(13)(24)) \underset{\Sigma_2}{\otimes} (x_1 \otimes x_2) + ([e_1,e_2,e_3,e_4]+[e_4,e_3,e_2,e_1]) \otimes x_1 \otimes x_2 \otimes x_1 \otimes x_2 \mid x_1, x_2 \in T\} \tag{9.6}$$

To get the analogue of (7.31), note that the projection of the numerator of (9.1) onto $M_4 \otimes (AT)^{\otimes 4}$ induces an isomorphism of Σ_4-modules $S_{4,1}T \simeq R_{4,0}T$. So, diagram (7.31) can be extended to:

$$\begin{array}{ccccccccc} 0 & \longrightarrow & M_4 \otimes AT & \xrightarrow{d} & S_{4,1}T & \longrightarrow & F_2(\Sigma_4/\Gamma) \underset{\Sigma_2}{\otimes} ((AT)^{\otimes 2}/(AT)^{(2)}) & \longrightarrow & 0 \\ & & & \searrow^{d} & \downarrow \simeq & \nearrow_{p} & & & \\ & & & & R_{4,0}T & & & & \end{array} \tag{9.7}$$

Our main result about $J_{4,1}T$ is:

<u>Proposition 21</u>. $J_{4,1}T$ is a free module over $F_2(\Sigma_4)$, and there is a natural isomorphism of vector spaces:

$$F_2 \underset{\Sigma_4}{\otimes} J_{4,1}T = L_2AT \oplus L_4AT. \tag{9.8}$$

<u>Proof</u>. The inclusion $F_2(\Sigma_4) \otimes_{\Sigma_2} J_{2,0}T \to J_{4,1}T$ gives rise to a short exact sequence

$$0 \to F_2(\Sigma_4) \underset{\Sigma_2}{\otimes} J_{2,0}T \to J_{4,1}T \to J_{4,0}T \to 0 \tag{9.9}$$

as envisioned in (3.7).. That $J_{4,1}T$ is free now follows from Propositions 4 and 12. We claim further that when the Σ_4-action is divided out of (9.9), the resulting short-exact sequence of vector spaces:

$$0 \to F_2 \underset{\Sigma_2}{\otimes} J_{2,0}T \to F_2 \underset{\Sigma_4}{\otimes} J_{4,1}T \to F_2 \underset{\Sigma_4}{\otimes} J_{4,0}T \to 0 \tag{9.10}$$

is naturally split. In fact the map θ, as determined by (9.3) and (9.4), satisfies:

$$\text{im}\ \theta \subseteq IF_2(\Sigma_4) \underset{\Sigma_2}{\otimes} J_{2,0}T$$

where $IF_2(\Sigma_4) \subset F_2(\Sigma_4)$ is the augmentation ideal. So θ becomes the zero homomorphism when Σ_4-action is divided out of its domain and range

It follows that the inclusion of $M_4 \otimes (AT)^{\otimes 4}$ into the numerator of (9.1) can be used to induce a natural F_2-linear map $s: F_2 \underset{\Sigma_4}{\otimes} J_{4,0}T \to F_2 \underset{\Sigma_4}{\otimes} J_{4,1}T$ that splits (9.10). (9.8) now follows from Propositions 4 and 12.

10. Homotopy Groups of $J_{4,1}S$

In this section we show $\pi_* J_{4,1}S = 0$ unless $* = 1$, and we compute π_1. For this purpose it is useful to express $J_{4,1}S$ as a quotient of an acyclic complex. In fact, for any pointed set T the projection $(AT)^{\otimes 2} \to J_{2,0}T$ induces a projection of $(F_2(\Sigma_4) \otimes_{\Sigma_2} (AT)^{\otimes 2}) \oplus (M_4 \otimes (AT)^{\otimes 4})$ onto the numerator of (9.1). Consequently we get the projection π:

$$0 \to R_{4,1}T \to [F_2(\Sigma_4) \underset{\Sigma_2}{\otimes} (AT)^{\otimes 2}] \oplus [M_4 \otimes (AT)^{\otimes 4}] \xrightarrow{\pi} J_{4,1}T \to 0 \qquad (10.1)$$

where we have defined $R_{4,1}T$ to be the kernel. (10.1) is analogous to (5.6). If in (10.1) we replace T by the sphere spectrum S, the middle term has vanishing homotopy groups in all dimensions. This follows easily from Lemma 3 and the fact that $F_2(\Sigma_4)$ is free as a right module over $F_2(\Sigma_2)$. So (10.1) gives:

$$\pi_* J_{4,1}S = \pi_{*-1} R_{4,1}S \qquad (10.2)$$

Then our next task must be to get a useful description of $R_{4,1}$. We deduce easily from (9.1) and (10.1) an exact sequence of Σ_4-modules:

$$0 \to F_2(\Sigma_4) \underset{\Sigma_2}{\otimes} AT \xrightarrow{d'} R_{4,1}T \to S_{4,1}T \to 0 \qquad (10.3)$$

where d' is the "diagonal map" $d'(\sigma \otimes_{\Sigma_2} x) = \sigma \otimes_{\Sigma_2} (x \otimes x)$ for all $\sigma \in \Sigma_4$, $x \in T$, and the action of Σ_2 upon AT is trivial. Further, the diagonal $d: M_4 \otimes AT \to S_{4,1}T$ of (9.7) clearly lifts to a map into $R_{4,1}T$, so using (9.7) we get a diagram of exact sequences:

$$
\begin{array}{ccccccccc}
 & & & & 0 & & 0 & & \\
 & & & & \downarrow & & \downarrow & & \\
 & & & & M_4 \otimes AT & = & M_4 \otimes AT & & \\
 & & & & \downarrow d & & \downarrow d & & \\
0 & \rightarrow & F_2(\Sigma_4) \underset{\Sigma_2}{\otimes} AT & \xrightarrow{d'} & R_{4,1}T & \rightarrow & S_{4,1}T & \rightarrow & 0 \\
 & & & & & & \downarrow & & \\
 & & & & & & F_2(\Sigma_4/\Gamma) \underset{\Sigma_2}{\otimes} ((AT)^{\otimes 2}/(AT)^{(2)}) & & \\
 & & & & & & \downarrow & & \\
 & & & & & & 0 & &
\end{array}
\tag{10.4}
$$

But it is easily verified that the images of d and d' in $R_{4,1}T$ intersect only in the zero vector, so (10.4) gives a short-exact sequence of Σ_4-modules:

$$0 \rightarrow (F_2(\Sigma_4) \underset{\Sigma_2}{\otimes} AT) \oplus (M_4 \otimes AT) \xrightarrow{(d',d)} R_{4,1}T \rightarrow F_2(\Sigma_4/\Gamma) \underset{\Sigma_2}{\otimes} ((AT)^{\otimes 2}/(AT)^{(2)}) \rightarrow 0 \tag{10.5}$$

This sequence is sufficient for the computation of $\pi_* R_{4,1}S$. In fact, $(AS)^{\otimes 2}/(AS)^{(2)} = J_{2,0}S$ has non-vanishing homotopy only in dimension 1, and $\pi_1 J_{2,0}S = F_2$, as in (5.5). So $J_{2,0}S$ can be regarded as a free resolution of F_2 as a left $F_2(\Sigma_2)$-module, and:

$$\pi_*(F_2(\Sigma_4/\Gamma) \underset{\Sigma_2}{\otimes} ((AS)^{\otimes 2}/(AS)^{(2)})) = \mathrm{Tor}_{*-1}^{F_2(\Sigma_2)}(F_2(\Sigma_4/\Gamma), F_2) \tag{10.6}$$

But $F_2(\Sigma_4/\Gamma)$ is free as a right module over $F_2(\Sigma_2)$, and $F_2(\Sigma_4/\Gamma) \otimes_{\Sigma_2} F_2 = F_2(\Sigma_4/\Delta_2)$, where $\Delta_2 \lhd \Sigma_4$ is as in (8.2). So from (10.6) we get

$$\pi_*(F_2(\Sigma_4/\Gamma) \underset{\Sigma_2}{\otimes} (AS^{\otimes 2}/AS^{(2)})) = \begin{cases} F_2(\Sigma_4/\Delta_2) & \text{if } * = 1 \\ 0 & \text{if } * \neq 1 \end{cases} \tag{10.7}$$

as Σ_4-modules. On the other hand, we have easily:

$$\pi_*[(F_2(\Sigma_4) \underset{\Sigma_2}{\otimes} AS) \oplus (M_4 \otimes AS)] = \begin{cases} F_2(\Sigma_4/\Sigma_2) \oplus M_4 & \text{if } * = 0 \\ 0 & \text{if } * \neq 0 \end{cases} \tag{10.8}$$

as Σ_4-modules. Hence, if in (10.5) we replace T by S, the long exact homotopy sequence of the resulting fibration reduces to:

$$0 \to F_2(\Sigma_4/\Delta_2) \xrightarrow{\partial} F_2(\Sigma_4/\Sigma_2)\oplus M_4 \xrightarrow{(d',d)_*} \pi_0 R_{4,1}S \to 0 \tag{10.9}$$

where ∂ is the connecting morphism from π_1 to π_0. But a quick computation shows that

$$\partial(e\Delta_2) = e\Sigma_2 + (13)(24)\Sigma_2 + [e_1,e_2,e_3,e_4] + [e_4,e_3,e_2,e_1] \tag{10.10}$$

That is,

$$\partial = (i,j) \tag{10.11}$$

with i and j as in (8.9) and (8.10). So we have at once from (10.9) and (8.13) an isomorphism of Σ_4-modules:

$$\pi_* R_{4,1}S = \begin{cases} \operatorname{coker}(i,j) \overset{h}{=} S^{(3,1)}\otimes S^{(2,2)}\otimes S^{(2,2)} & \text{if } * = 0 \\ 0 & \text{if } * \neq 0 \end{cases} \tag{10.12}$$

Then from (10.2):

$$\pi_* J_{4,1}S = \begin{cases} S^{(3,1)}\otimes S^{(2,2)}\otimes S^{(2,2)} & \text{if } * = 1 \\ 0 & \text{if } * \neq 1 \end{cases} \tag{10.13}$$

We have achieved the main goals of this section. For use in the next section, we are going to give now another description of $R_{4,1}$. The functorial picture in (10.5) is analogous to (9.7). We need a (non-functorial) picture of $R_{4,1}$ in terms of elements. Such a description is analogous to (9.6), and is easily obtained from (10.5). We find that as a submodule of $[F_2(\Sigma_4)\otimes_{\Sigma_2}(AT)^{\otimes 2}]\oplus[M_4\otimes(AT)^{\otimes 4}]$, $R_{4,1}T$ is given by:

$$R_{4,1}T = F_2(\Sigma_4)\otimes_{\Sigma_2} \operatorname{Span}\{x\otimes x \mid x \in T\} + \operatorname{Span}\{\alpha\otimes x^{\otimes 4} \mid \alpha \in M_4,\ x \in T\}$$

$$+ \Sigma_4\operatorname{Span}\{(e+(13)(24))\otimes_{\Sigma_2}(x_1\otimes x_2)+([e_1,e_2,e_3,e_4]+[e_4,e_3,e_2,e_1])\otimes x_1\otimes x_2\otimes x_1\otimes x_2\} \tag{10.14}$$

where the last line is taken over all pairs $x_1,x_2 \in T$.

11. Construction of $J_{4,2} = J_4$

In this section we construct the functor J_4, and show it has properties i)-iii) of the Introduction. We will also compute $\pi_0 J_4 S$, so completing the proof of Theorem 2". Our definition of $J_4 = J_{4,2}$ has the form:

$$J_{4,2}T = \frac{(F_2(\Sigma_4)\otimes AT)\oplus(F_2(\Sigma_4)\otimes_{\Sigma_2}(AT)^{\otimes 2})\oplus(M_4\otimes(AT)^{\otimes 4})}{R_{4,2}T} \tag{11.1}$$

Here $R_{4,2}T$ is a submodule of the numerator of the form:

$$R_{4,2}T = \{\lambda(\gamma)\oplus\gamma \mid \gamma \in R_{4,1}T\} \tag{11.2}$$

where $\lambda : R_{4,1}T \to F_2(\Sigma_4)\otimes AT$ is an $F_2(\Sigma_4)$-linear map that we are going to define. We use the description of $R_{4,1}T$ given in (10.14), defining λ on the first two summands of the right-hand side by:

$$\begin{aligned} \lambda(\sigma\otimes_{\Sigma_2}(x\otimes x)) &= f(\sigma\Sigma_2)\otimes x \\ \lambda(\alpha\otimes x\otimes x\otimes x\otimes x) &= f(\alpha)\otimes x \end{aligned} \tag{11.3}$$

for each $\sigma \in \Sigma_4$, $\alpha \in M_4$, $x \in T$; where f is the homomorphism of (8.11) Finally, we define λ on the third summand of (10.14) by setting

$$\lambda[(e+(13)(24))\otimes_{\Sigma_2}(x_1\otimes x_2)+([e_1,e_2,e_3,e_4]+[e_4,e_3,e_2,e_1])\otimes x_1\otimes x_2\otimes x_1\otimes x_2)]=0 \tag{11.4}$$

for all $x_1, x_2 \in T$. The third summand has non-zero intersection with the sum of the first two: we must check that (11.4) agrees with (11.3) in case $x_1 = x_2$. But this fact follows at once from the exactness of (8.11). Having defined λ, we have completed the definition of $J_{4,2}$.

Proposition 22. For each pointed set T, $J_{4,2}T$ is a free module over $F_2(\Sigma_4)$, and there is a natural isomorphism of vector spaces:

$$F_2 \underset{\Sigma_4}{\otimes} J_{4,2}T = L_1AT \oplus L_2AT \oplus L_4AT = L_4^{res}AT \tag{11.5}$$

Proof. The inclusion of $F_2(\Sigma_4)\otimes AT$ into the numerator of (11.1) gives a short-exact sequence of Σ_4-modules:

$$0 \to F_2(\Sigma_4)\otimes AT \to J_{4,2}T \to J_{4,1}T \to 0 \tag{11.6}$$

as envisioned in (3.7). That $J_{4,2}T$ is free follows from (11.6) and Proposition 21. Dividing out the Σ_4-action from (11.6) we obtain a short-exact sequence of vector spaces:

$$0 \to AT \to F_2 \underset{\Sigma_4}{\otimes} J_{4,2}T \to F_2 \underset{\Sigma_4}{\otimes} J_{4,1}T \to 0 \qquad (11.7)$$

that we claim is naturally split. In fact, from (8.12), (11.3) and (11.4) we have that $\operatorname{im} \lambda \subset IF_2(\Sigma_4)\otimes AT$. From (11.1) and (11.2) it follows that the elements of $R_{4,1}T$ represent zero in $F_2 \otimes_{\Sigma_4} J_{4,2}T$. Consequently, the inclusion of $(F_2(\Sigma_4)\otimes_{\Sigma_2}(AT)^{\otimes 2})\oplus(M_4\otimes(AT)^{\otimes 4})$ into the numerator of (11.1) induces a map of $F_2 \otimes_{\Sigma_4} J_{4,1}T$ into $F_2 \otimes_{\Sigma_4} J_{4,2}T$. This is our natural splitting of (11.7). (11.5) now follows from (9.8).

Our final task is the computation of $\pi_* J_{4,2}S$. We will use the following commutative diagram of $F_2(\Sigma_4)$-modules.

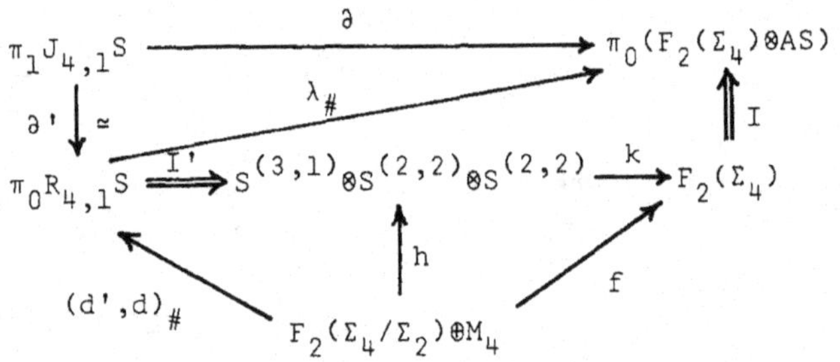

(11.8)

We describe each map in (11.8) and justify commutativity briefly. h and k are the homomorphisms of Proposition 20, and the relation $f = kh$ is the definition of f. The maps d and d' are as in (10.4) and (10.5), and $(d',d)_\#$ is the induced map on zero-dimensional homotopy, as in (10.9). The identification I' of $\pi_0 R_{4,1}S$ with $S^{(3,1)}\otimes S^{(2,2)}\otimes S^{(2,2)}$ that we made in (10.12) is clearly such that $(d',d)_\# \circ I' = h$ (use (10.9), (10.11), (10.12)). The map ∂ of (11.8) is the connecting morphism associated to the fibration that results from replacing T by S in (11.6). Similarly, ∂' is the connecting morphism associated with (10.1). That $\partial = \lambda_\# \partial'$ in (11.8) results from the naturality of the connecting morphism: one need only consider the map of fibrations:

$$\begin{array}{ccccccccc}
0 & \longrightarrow & R_{4,1}S & \longrightarrow & (F_2(\Sigma_4)\otimes_{\Sigma_2}(AS)^{\otimes 2})\oplus(M_4\otimes(AS)^{\otimes 4}) & \longrightarrow & J_{4,1}S & \longrightarrow & 0 \\
 & & \lambda\downarrow & & \downarrow & & \| & & \\
0 & \longrightarrow & F_2(\Sigma_4)\otimes AS & \longrightarrow & J_{4,2}S & \longrightarrow & J_{4,1}S & \longrightarrow & 0
\end{array}$$

Continuing our analysis of (11.8) we assert that

$$\lambda_{\#}(d',d)_{\#} = If \tag{11.9}$$

In fact, this follows at once from (11.3). That $\lambda_{\#} = IkI'$ now follows from (11.9) and the fact that $(d',d)_{\#}$ is onto. We have completed our proof of the commutativity of (11.8).

The determination of $\pi_* J_{4,2}S$ is now easy. From (11.8) and the fact that k is monic, we deduce that ∂ is monic. We now consider the long-exact homotopy sequence of the fibration that results from replacing T by S in (11.6). Since ∂ is monic we have from (10.13)

$$\pi_* J_{4,2}S = \begin{cases} \operatorname{coker} \partial & \text{if } * = 0 \\ 0 & \text{if } * > 0 \end{cases} \tag{11.10}$$

But again from (11.8) we have $\operatorname{coker} \partial = \operatorname{coker} k$, so from (8.14) we have finally:

$$\pi_* J_4 S = \pi_* J_{4,2} S = \begin{cases} P\oplus(S^{(2,2)}\otimes S^{(2,2)}) & \text{if } * = 0 \\ 0 & \text{if } * > 0. \end{cases} \tag{11.11}$$

This result, in conjunction with Proposition 22 and equation (8.8), completes the proof of Theorem 2" in the case $r = 4$. We have now proved Theorem 2" in all the cases for which we have claimed it.

12. Appendix: Proofs of Propositions 19 and 20

This section serves as a kind of appendix, in which we actually construct the exact sequences of $F_2(\Sigma_4)$-modules on which our proofs have depended. We begin with (8.13). In order to define $h|M_4$ it is useful to have a decomposition of M_4 as a tensor product. With V_4 as in Section 8, let $\eta: F_2 \to V_4$ be the Σ_4-linear map defined by $\eta(1) = e_1+e_2+e_3+e_4$. Write $\hat{S}^{(3,1)} = \operatorname{coker} \eta$; this is the dual of the representation $S^{(3,1)}$. If we write g_i for the image of e_i under the projection $V_4 \to \hat{S}^{(3,1)}$, then $\hat{S}^{(3,1)}$ is spanned by $\{g_1,g_2,g_3,g_4\}$, subject to the single relation $\Sigma g_i = 0$. Now it is not hard to verify the existence of a unique isomorphism of $F_2(\Sigma_4)$-modules

$$M_4 \xrightarrow{\simeq} \hat{S}^{(3,1)} \otimes S^{(2,2)} \tag{12.1}$$

for which $[e_1,e_2,e_3,e_4] \to g_1\otimes(\alpha+\beta)+g_2\otimes\alpha$. The decomposition (12.1) enables us to define $h|M_4$: if a suitable homomorphism $\ell:\hat{S}^{(3,1)} \to S^{(3,1)}\otimes S^{(2,2)}$ be chosen, we define $h|M_4$ to be the tensor product $\ell\otimes(\mathrm{id}): \hat{S}^{(3,1)}\otimes S^{(2,2)} \to S^{(3,1)}\otimes S^{(2,2)}\otimes S^{(2,2)}$. In fact, a Σ_4-linear map ℓ is determined by the formula

$$\ell(g_1) = (e_1+e_2)\otimes\beta + (e_1+e_3)\otimes\alpha + (e_1+e_4)\otimes(\alpha+\beta) \tag{12.2}$$

The resulting homomorphism $h|M_4$ satisfies

$$\begin{aligned} h[e_1,e_2,e_3,e_4] &= (e_1+e_2)\otimes(\beta\otimes\alpha+\beta\otimes\beta) + (e_1+e_3)\otimes(\alpha\otimes\beta+\beta\otimes\alpha) \\ &\quad + (e_1+e_4)\otimes(\alpha\otimes\beta+\beta\otimes\alpha+\beta\otimes\beta) \end{aligned} \tag{12.3}$$

and is determined by this equation. To complete the construction of (8.13) we must define $h|F_2(\Sigma_4/\Sigma_2)$. We set

$$h(e\Sigma_2) = (e_1+e_2)\otimes(\alpha\otimes\beta+\beta\otimes\beta) \tag{12.4}$$

There is in fact a unique Σ_4-linear $h:F_2(\Sigma_4/\Sigma_2) \to S^{(3,1)}\otimes S^{(2,2)}\otimes S^{(2,2)}$ that satisfies (12.4), since the right-hand side is invariant under action of (12)(34), the generator of Σ_2. We claim that

$$h(i,j) = 0 \tag{12.5}$$

as required by the exactness of (8.13). In view of (8.9) and (8.10), it will suffice to show that:

$$(e+(13)(24))\circ h(e\Sigma_2) = (e+(14)(23))\circ h[e_1,e_2,e_3,e_4]$$

But this follows easily from (12.3) and (12.4). We next claim:

<u>Lemma</u> <u>23</u>. h of (8.13) is onto.

<u>Proof</u>. Let R_2 be the unique 2-Sylow subgroup of Σ_4 that contains the transposition $(1,2)$. We begin by observing that the map ℓ of

(12.2) fits into a short-exact sequence of Σ_4-modules:

$$0 \to \hat{S}^{(3,1)} \xrightarrow{\ell} S^{(3,1)} \otimes S^{(2,2)} \xrightarrow{p} F_2(\Sigma_4/R_2) \to 0 \qquad (12.6)$$

This is most easily seen by defining first the Σ_4-linear isomorphism $q: F_2(\Sigma_4/R_2) \to \operatorname{coker} \ell$ by setting $q(eR_2) = (e_2+e_3)\otimes\alpha$; p is then "inverse" to q and is determined by

$$p((e_2+e_3)\otimes\alpha) = eR_2 \qquad (12.7)$$

From (12.6) and (12.1) it follows that the restriction $h|M_4$ fits into an exact sequence:

$$0 \to M_4 \xrightarrow{h|M_4} S^{(3,1)} \otimes S^{(2,2)} \otimes S^{(2,2)} \xrightarrow{p\otimes id} F_2(\Sigma_4/R_2) \otimes S^{(2,2)} \to 0 \qquad (12.8)$$

Hence, to prove that h of (8.13) is onto, it will suffice to prove that the composition

$$F_2(\Sigma_4/\Sigma_2) \xrightarrow{h|F_2(\Sigma_4/\Sigma_2)} S^{(3,1)} \otimes S^{(2,2)} \otimes S^{(2,2)} \xrightarrow{p\otimes id} F_2(\Sigma_4/R_2) \otimes S^{(2,2)}$$

is onto. But this composition carries $e\Sigma_2$ to $(132)R_2\otimes\beta$, and this latter element generates $F_2(\Sigma_4/R_2)\otimes S^{(2,2)}$ as a module over $F_2(\Sigma_4)$. So the proof of Lemma 23 is complete.

<u>Lemma 24</u>. (8.13) is exact.

<u>Proof</u>. This now follows from (12.5), Lemma 23, and a dimension count.

Now we turn to the construction of (8.14). Our strategy is first to construct a version of (8.14) from which $S^{(2,2)}$ has been "factored out". We begin by observing that if $\Omega \subset \Sigma_4$ is the subgroup $\Omega = \{e, (12)\}$, then the Σ_4-linear map

$$F_2(\Sigma_4) \xrightarrow{\cong} F_2(\Sigma_4/\Omega) \otimes S^{(2,2)} \qquad (12.9)$$

determined by $e \to e\Omega\otimes\alpha$ is an isomorphism. Then considering (12.9) and (8.7) together, we see that one way to obtain the desired short-exact sequence (8.14) might be to construct first a short-exact sequence:

$$0 \to S^{(3,1)} \otimes S^{(2,2)} \xrightarrow{k'} F_2(\Sigma_4/\Omega) \xrightarrow{g'} V_4 \oplus S^{(2,2)} \to 0 \qquad (12.10)$$

and then tensor it with $S^{(2,2)}$. In fact, a suitable map g' is obtaine by setting

$$g'(e\Omega) = (e_3, \beta) \tag{12.11}$$

This is quickly verified to be onto. To construct k', we first observe that there is a unique Σ_4-linear map $k'': V_4 \otimes S^{(2,2)} \to F_2(\Sigma_4/\Omega)$ satisfying

$$\begin{aligned} k''(e_1 \otimes \alpha) &= (13)\Omega + (24)(13)\Omega \\ k''(e_1 \otimes \beta) &= (13)\Omega + (143)\Omega \end{aligned} \tag{12.12}$$

One checks easily that the diagram commutes:

$$\begin{array}{ccc} V_4 \otimes S^{(2,2)} & \xrightarrow{k''} & F_2(\Sigma_4/\Omega) \\ {\scriptstyle \varepsilon \otimes S^{(2,2)}} \downarrow & & \downarrow {\scriptstyle g'} \\ S^{(2,2)} & \longrightarrow & V_4 \oplus S^{(2,2)} \end{array} \tag{12.13}$$

Here $\varepsilon: V_4 \to F_2$ is as in Section 8, and the bottom arrow is the standard inclusion of summand into sum. But $S^{(3,1)} = \ker \varepsilon$, so if k'' is restricted to $S^{(3,1)} \otimes S^{(2,2)}$, it follows from (12.13) that it maps into $\ker g'$. We define

$$k' = k'' | S^{(3,1)} \otimes S^{(2,2)} \tag{12.14}$$

One checks easily that k' is monic; hence onto $\ker g'$ by a dimension count. So (12.10) has been constructed, and it is exact. Now we tensor the modules in (12.10) with $S^{(2,2)}$, and define maps k and g by $k = k' \otimes (\mathrm{id})$, $g = g' \otimes (\mathrm{id})$. There results an exact sequence of the form

$$\begin{array}{ccccccc} 0 \to S^{(3,1)} \otimes S^{(2,2)} \otimes S^{(2,2)} & \xrightarrow{k} & F_2(\Sigma_4/\Omega) \otimes S^{(2,2)} & \xrightarrow{g} & (V_4 \oplus S^{(2,2)}) \otimes S^{(2,2)} & \to 0 \\ & & \| & & \| & \\ & & F_2(\Sigma_4) & & P \oplus (S^{(2,2)} \otimes S^{(2,2)}) & \end{array} \tag{12.15}$$

where we have used (12.9) and (8.7) as we had planned. We identify (12.15) with (8.14), and so have completed the proof of Proposition 20.

We construct the exact sequence (8.11) by splicing (8.13) and (8.14); we set $f = kh$. The proof of Proposition 19 will be complete if we can demonstrate (8.12). But the functor $F_2 \otimes_{\Sigma_4} (\;\;)$ is right exact. Applying it to (8.11) gives an exact sequence:

$$F_2 \oplus (F_2 \underset{\Sigma_4}{\otimes} M_4) \xrightarrow{F_2 \otimes_{\Sigma_4} f} F_2 \xrightarrow{F_2 \otimes_{\Sigma_4} g} F_2 \underset{\Sigma_4}{\otimes} (S^{(2,2)} \otimes S^{(2,2)}) \to 0 \qquad (12.16)$$

where we have used (8.8). But direct computation using (8.3) shows that $F_2 \otimes_{\Sigma_4} (S^{(2,2)} \otimes S^{(2,2)}) = F_2$. Hence $F_2 \otimes_{\Sigma_4} g$ must be an isomorphism; hence $F_2 \otimes_{\Sigma_4} f = 0$. But this last equation implies (8.12), so the proof of Proposition 19 is complete.

Bibliography

1. A. K. Bousfield, E. B. Curtis, D. M. Kan, D. G. Quillen, D. L. Rector and J. W. Schlesinger, The mod-p lower central series and the Adams spectral sequence, Topology 5 (1966), 331-342.

2. E. B. Curtis, Some relations between homotopy and homology, Annals of Math 83 (1965), 386-413.

3. E. B. Curtis, Simplicial homotopy theory, Aarhus University Lecture Notes (1967).

4. E. B. Curtis, Simplicial homotopy theory, Advances in Math 6 (1971), 107-209.

5. A. Dold, Homology of symmetric products and other functors of complexes, Annals of Math 68 (1958), 54-80.

6. M. Hall, A basis for free Lie rings and higher commutators in free groups, Proc. A.M.S. 1 (1950), 575-581.

7. D. M. Kan, Semisimplicial spectra, Ill. Jour. Math. 7 (1963), 463-478.

8. D. M. Kan, Functors involving c.s.s. complexes, Trans. A.M.S. 87 (1958), 330-346.

9. D. M. Kan, A combinatorial definition of homotopy groups, Annals of Math 67 (1958), 282-312.

10. D. M. Kan and G. W. Whitehead, The reduced join of two spectra, Topology 3, Suppl. 2 (1965), 239-261.

11. J. C. Moore, Seminar on algebraic homotopy theory, Princeton University Lecture Notes (1956).

12. D. L. Rector, An unstable Adams spectral sequence, Topology 5 (1966), 343-346.

13. R. Steinberg, Prime power representations of finite linear groups, Canadian Jour. Math. 8 (1956), 580-591.

Vol. 1117: D.J. Aldous, J.A. Ibragimov, J. Jacod, Ecole d'Été de Probabilités de Saint-Flour XIII – 1983. Édité par P.L. Hennequin. IX, 409 pages. 1985.

Vol. 1118: Grossissements de filtrations: exemples et applications. Seminaire, 1982/83. Edité par Th. Jeulin et M. Yor. V, 315 pages. 1985.

Vol. 1119: Recent Mathematical Methods in Dynamic Programming. Proceedings, 1984. Edited by I. Capuzzo Dolcetta, W.H. Fleming and T. Zolezzi. VI, 202 pages. 1985.

Vol. 1120: K. Jarosz, Perturbations of Banach Algebras. V, 118 pages. 1985.

Vol. 1121: Singularities and Constructive Methods for Their Treatment. Proceedings, 1983. Edited by P. Grisvard, W. Wendland and J.R. Whiteman. IX, 346 pages. 1985.

Vol. 1122: Number Theory. Proceedings, 1984. Edited by K. Alladi. VII, 217 pages. 1985.

Vol. 1123: Séminaire de Probabilités XIX 1983/84. Proceedings. Edité par J. Azéma et M. Yor. IV, 504 pages. 1985.

Vol. 1124: Algebraic Geometry, Sitges (Barcelona) 1983. Proceedings. Edited by E. Casas-Alvero, G.E. Welters and S. Xambó-Descamps. XI, 416 pages. 1985.

Vol. 1125: Dynamical Systems and Bifurcations. Proceedings, 1984. Edited by B.L.J. Braaksma, H.W. Broer and F. Takens. V, 129 pages. 1985.

Vol. 1126: Algebraic and Geometric Topology. Proceedings, 1983. Edited by A. Ranicki, N. Levitt and F. Quinn. V, 423 pages. 1985.

Vol. 1127: Numerical Methods in Fluid Dynamics. Seminar. Edited by F. Brezzi, VII, 333 pages. 1985.

Vol. 1128: J. Elschner, Singular Ordinary Differential Operators and Pseudodifferential Equations. 200 pages. 1985.

Vol. 1129: Numerical Analysis, Lancaster 1984. Proceedings. Edited by P.R. Turner. XIV, 179 pages. 1985.

Vol. 1130: Methods in Mathematical Logic. Proceedings, 1983. Edited by C.A. Di Prisco. VII, 407 pages. 1985.

Vol. 1131: K. Sundaresan, S. Swaminathan, Geometry and Nonlinear Analysis in Banach Spaces. III, 116 pages. 1985.

Vol. 1132: Operator Algebras and their Connections with Topology and Ergodic Theory. Proceedings, 1983. Edited by H. Araki, C.C. Moore, Ş. Strătilă and C. Voiculescu. VI, 594 pages. 1985.

Vol. 1133: K.C. Kiwiel, Methods of Descent for Nondifferentiable Optimization. VI, 362 pages. 1985.

Vol. 1134: G.P. Galdi, S. Rionero, Weighted Energy Methods in Fluid Dynamics and Elasticity. VII, 126 pages. 1985.

Vol. 1135: Number Theory, New York 1983–84. Seminar. Edited by D.V. Chudnovsky, G.V. Chudnovsky, H. Cohn and M.B. Nathanson. V, 283 pages. 1985.

Vol. 1136: Quantum Probability and Applications II. Proceedings, 1984. Edited by L. Accardi and W. von Waldenfels. VI, 534 pages. 1985.

Vol. 1137: Xiao G., Surfaces fibrées en courbes de genre deux. IX, 103 pages. 1985.

Vol. 1138: A. Ocneanu, Actions of Discrete Amenable Groups on von Neumann Algebras. V, 115 pages. 1985.

Vol. 1139: Differential Geometric Methods in Mathematical Physics. Proceedings, 1983. Edited by H. D. Doebner and J. D. Hennig. VI, 337 pages. 1985.

Vol. 1140: S. Donkin, Rational Representations of Algebraic Groups. VII, 254 pages. 1985.

Vol. 1141: Recursion Theory Week. Proceedings, 1984. Edited by H.-D. Ebbinghaus, G.H. Müller and G.E. Sacks. IX, 418 pages. 1985.

Vol. 1142: Orders and their Applications. Proceedings, 1984. Edited by I. Reiner and K. W. Roggenkamp. X, 306 pages. 1985.

Vol. 1143: A. Krieg, Modular Forms on Half-Spaces of Quaternions. XIII, 203 pages. 1985.

Vol. 1144: Knot Theory and Manifolds. Proceedings, 1983. Edited by D. Rolfsen. V, 163 pages. 1985.

Vol. 1145: G. Winkler, Choquet Order and Simplices. VI, 143 pages. 1985.

Vol. 1146: Séminaire d'Algèbre Paul Dubreil et Marie-Paule Malliavin. Proceedings, 1983–1984. Edité par M.-P. Malliavin. IV, 420 pages. 1985.

Vol. 1147: M. Wschebor, Surfaces Aléatoires. VII, 111 pages. 1985.

Vol. 1148: Mark A. Kon, Probability Distributions in Quantum Statistical Mechanics. V, 121 pages. 1985.

Vol. 1149: Universal Algebra and Lattice Theory. Proceedings, 1984. Edited by S. D. Comer. VI, 282 pages. 1985.

Vol. 1150: B. Kawohl, Rearrangements and Convexity of Level Sets in PDE. V, 136 pages. 1985.

Vol 1151: Ordinary and Partial Differential Equations. Proceedings, 1984. Edited by B.D. Sleeman and R.J. Jarvis. XIV, 357 pages. 1985.

Vol. 1152: H. Widom, Asymptotic Expansions for Pseudodifferential Operators on Bounded Domains. V, 150 pages. 1985.

Vol. 1153: Probability in Banach Spaces V. Proceedings, 1984. Edited by A. Beck, R. Dudley, M. Hahn, J. Kuelbs and M. Marcus. VI, 457 pages. 1985.

Vol. 1154: D.S. Naidu, A.K. Rao, Singular Pertubation Analysis of Discrete Control Systems. IX, 195 pages. 1985.

Vol. 1155: Stability Problems for Stochastic Models. Proceedings, 1984. Edited by V.V. Kalashnikov and V.M. Zolotarev. VI, 447 pages. 1985.

Vol. 1156: Global Differential Geometry and Global Analysis 1984. Proceedings, 1984. Edited by D. Ferus, R.B. Gardner, S. Helgason and U. Simon. V, 339 pages. 1985.

Vol. 1157: H. Levine, Classifying Immersions into $\mathbb{R}^4$ over Stable Maps of 3-Manifolds into $\mathbb{R}^2$. V, 163 pages. 1985.

Vol. 1158: Stochastic Processes – Mathematics and Physics. Proceedings, 1984. Edited by S. Albeverio, Ph. Blanchard and L. Streit. VI, 230 pages. 1986.

Vol. 1159: Schrödinger Operators, Como 1984. Seminar. Edited by S. Graffi. VIII, 272 pages. 1986.

Vol. 1160: J.-C. van der Meer, The Hamiltonian Hopf Bifurcation. VI, 115 pages. 1985.

Vol. 1161: Harmonic Mappings and Minimal Immersions, Montecatini 1984. Seminar. Edited by E. Giusti. VII, 285 pages. 1985.

Vol. 1162: S.J.L. van Eijndhoven, J. de Graaf, Trajectory Spaces, Generalized Functions and Unbounded Operators. IV, 272 pages. 1985.

Vol. 1163: Iteration Theory and its Functional Equations. Proceedings, 1984. Edited by R. Liedl, L. Reich and Gy. Targonski. VIII, 231 pages. 1985.

Vol. 1164: M. Meschiari, J.H. Rawnsley, S. Salamon, Geometry Seminar "Luigi Bianchi" II – 1984. Edited by E. Vesentini. VI, 224 pages. 1985.

Vol. 1165: Seminar on Deformations. Proceedings, 1982/84. Edited by J. Ławrynowicz. IX, 331 pages. 1985.

Vol. 1166: Banach Spaces. Proceedings, 1984. Edited by N. Kalton and E. Saab. VI, 199 pages. 1985.

Vol. 1167: Geometry and Topology. Proceedings, 1983–84. Edited by J. Alexander and J. Harer. VI, 292 pages. 1985.

Vol. 1168: S.S. Agaian, Hadamard Matrices and their Applications. III, 227 pages. 1985.

Vol. 1169: W.A. Light, E.W. Cheney, Approximation Theory in Tensor Product Spaces. VII, 157 pages. 1985.

Vol. 1170: B.S. Thomson, Real Functions. VII, 229 pages. 1985.

Vol. 1171: Polynômes Orthogonaux et Applications. Proceedings, 1984. Edité par C. Brezinski, A. Draux, A.P. Magnus, P. Maroni et A. Ronveaux. XXXVII, 584 pages. 1985.

Vol. 1172: Algebraic Topology, Göttingen 1984. Proceedings. Edited by L. Smith. VI, 209 pages. 1985.

Vol. 1173: H. Delfs, M. Knebusch, Locally Semialgebraic Spaces. XVI, 329 pages. 1985.

Vol. 1174: Categories in Continuum Physics, Buffalo 1982. Seminar. Edited by F.W. Lawvere and S.H. Schanuel. V, 126 pages. 1986.

Vol. 1175: K. Mathiak, Valuations of Skew Fields and Projective Hjelmslev Spaces. VII, 116 pages. 1986.

Vol. 1176: R.R. Bruner, J.P. May, J.E. McClure, M. Steinberger, H_∞ Ring Spectra and their Applications. VII, 388 pages. 1986.

Vol. 1177: Representation Theory I. Finite Dimensional Algebras. Proceedings, 1984. Edited by V. Dlab, P. Gabriel and G. Michler. XV, 340 pages. 1986.

Vol. 1178: Representation Theory II. Groups and Orders. Proceedings, 1984. Edited by V. Dlab, P. Gabriel and G. Michler. XV, 370 pages. 1986.

Vol. 1179: Shi J.-Y. The Kazhdan-Lusztig Cells in Certain Affine Weyl Groups. X, 307 pages. 1986.

Vol. 1180: R. Carmona, H. Kesten, J.B. Walsh, École d'Été de Probabilités de Saint-Flour XIV – 1984. Édité par P.L. Hennequin. X, 438 pages. 1986.

Vol. 1181: Buildings and the Geometry of Diagrams, Como 1984. Seminar. Edited by L. Rosati. VII, 277 pages. 1986.

Vol. 1182: S. Shelah, Around Classification Theory of Models. VII, 279 pages. 1986.

Vol. 1183: Algebra, Algebraic Topology and their Interactions. Proceedings, 1983. Edited by J.-E. Roos. XI, 396 pages. 1986.

Vol. 1184: W. Arendt, A. Grabosch, G. Greiner, U. Groh, H.P. Lotz, U. Moustakas, R. Nagel, F. Neubrander, U. Schlotterbeck, One-parameter Semigroups of Positive Operators. Edited by R. Nagel. X, 460 pages. 1986.

Vol. 1185: Group Theory, Beijing 1984. Proceedings. Edited by Tuan H.F. V, 403 pages. 1986.

Vol. 1186: Lyapunov Exponents. Proceedings, 1984. Edited by L. Arnold and V. Wihstutz. VI, 374 pages. 1986.

Vol. 1187: Y. Diers, Categories of Boolean Sheaves of Simple Algebras. VI, 168 pages. 1986.

Vol. 1188: Fonctions de Plusieurs Variables Complexes V. Séminaire, 1979–85. Edité par François Norguet. VI, 306 pages. 1986.

Vol. 1189: J. Lukeš, J. Malý, L. Zajíček, Fine Topology Methods in Real Analysis and Potential Theory. X, 472 pages. 1986.

Vol. 1190: Optimization and Related Fields. Proceedings, 1984. Edited by R. Conti, E. De Giorgi and F. Giannessi. VIII, 419 pages. 1986.

Vol. 1191: A.R. Its, V.Yu. Novokshenov, The Isomonodromic Deformation Method in the Theory of Painlevé Equations. IV, 313 pages. 1986.

Vol. 1192: Equadiff 6. Proceedings, 1985. Edited by J. Vosmansky and M. Zlámal. XXIII, 404 pages. 1986.

Vol. 1193: Geometrical and Statistical Aspects of Probability in Banach Spaces. Proceedings, 1985. Edited by X. Fernique, B. Heinkel, M.B. Marcus and P.A. Meyer. IV, 128 pages. 1986.

Vol. 1194: Complex Analysis and Algebraic Geometry. Proceedings, 1985. Edited by H. Grauert. VI, 235 pages. 1986.

Vol. 1195: J.M. Barbosa, A.G. Colares, Minimal Surfaces in $\mathbb{R}^3$. X, 124 pages. 1986.

Vol. 1196: E. Casas-Alvero, S. Xambó-Descamps, The Enumerative Theory of Conics after Halphen. IX, 130 pages. 1986.

Vol. 1197: Ring Theory. Proceedings, 1985. Edited by F.M.J. van Oystaeyen. V, 231 pages. 1986.

Vol. 1198: Séminaire d'Analyse, P. Lelong – P. Dolbeault – H. Skoda. Seminar 1983/84. X, 260 pages. 1986.

Vol. 1199: Analytic Theory of Continued Fractions II. Proceedings, 1985. Edited by W.J. Thron. VI, 299 pages. 1986.

Vol. 1200: V.D. Milman, G. Schechtman, Asymptotic Theory of Finite Dimensional Normed Spaces. With an Appendix by M. Gromov. VIII, 156 pages. 1986.

Vol. 1201: Curvature and Topology of Riemannian Manifolds. Proceedings, 1985. Edited by K. Shiohama, T. Sakai and T. Sunada. VII, 336 pages. 1986.

Vol. 1202: A. Dür, Möbius Functions, Incidence Algebras and Power Series Representations. XI, 134 pages. 1986.

Vol. 1203: Stochastic Processes and Their Applications. Proceedings, 1985. Edited by K. Itô and T. Hida. VI, 222 pages. 1986.

Vol. 1204: Séminaire de Probabilités XX, 1984/85. Proceedings. Edité par J. Azéma et M. Yor. V, 639 pages. 1986.

Vol. 1205: B.Z. Moroz, Analytic Arithmetic in Algebraic Number Fields. VII, 177 pages. 1986.

Vol. 1206: Probability and Analysis, Varenna (Como) 1985. Seminar. Edited by G. Letta and M. Pratelli. VIII, 280 pages. 1986.

Vol. 1207: P.H. Bérard, Spectral Geometry: Direct and Inverse Problems. With an Appendix by G. Besson. XIII, 272 pages. 1986.

Vol. 1208: S. Kaijser, J.W. Pelletier, Interpolation Functors and Duality. IV, 167 pages. 1986.

Vol. 1209: Differential Geometry, Peñíscola 1985. Proceedings. Edited by A.M. Naveira, A. Ferrández and F. Mascaró. VIII, 306 pages. 1986.

Vol. 1210: Probability Measures on Groups VIII. Proceedings, 1985. Edited by H. Heyer. X, 386 pages. 1986.

Vol. 1211: M.B. Sevryuk, Reversible Systems. V, 319 pages. 1986.

Vol. 1212: Stochastic Spatial Processes. Proceedings, 1984. Edited by P. Tautu. VIII, 311 pages. 1986.

Vol. 1213: L.G. Lewis, Jr., J.P. May, M. Steinberger, Equivariant Stable Homotopy Theory. IX, 538 pages. 1986.

Vol. 1214: Global Analysis – Studies and Applications II. Edited by Yu.G. Borisovich and Yu.E. Gliklikh. V, 275 pages. 1986.

Vol. 1215: Lectures in Probability and Statistics. Edited by G. del Pino and R. Rebolledo. V, 491 pages. 1986.

Vol. 1216: J. Kogan, Bifurcation of Extremals in Optimal Control. VIII, 106 pages. 1986.

Vol. 1217: Transformation Groups. Proceedings, 1985. Edited by S. Jackowski and K. Pawalowski. X, 396 pages. 1986.

Vol. 1218: Schrödinger Operators, Aarhus 1985. Seminar. Edited by E. Balslev. V, 222 pages. 1986.

Vol. 1219: R. Weissauer, Stabile Modulformen und Eisensteinreihen. III, 147 Seiten. 1986.

Vol. 1220: Séminaire d'Algèbre Paul Dubreil et Marie-Paule Malliavin. Proceedings, 1985. Edité par M.-P. Malliavin. IV, 200 pages. 1986.

Vol. 1221: Probability and Banach Spaces. Proceedings, 1985. Edited by J. Bastero and M. San Miguel. XI, 222 pages. 1986.

Vol. 1222: A. Katok, J.-M. Strelcyn, with the collaboration of F. Ledrappier and F. Przytycki, Invariant Manifolds, Entropy and Billiards; Smooth Maps with Singularities. VIII, 283 pages. 1986.

Vol. 1223: Differential Equations in Banach Spaces. Proceedings, 1985. Edited by A. Favini and E. Obrecht. VIII, 299 pages. 1986.

Vol. 1224: Nonlinear Diffusion Problems, Montecatini Terme 1985. Seminar. Edited by A. Fasano and M. Primicerio. VIII, 188 pages. 1986.

Vol. 1225: Inverse Problems, Montecatini Terme 1986. Seminar. Edited by G. Talenti. VIII, 204 pages. 1986.

Vol. 1226: A. Buium, Differential Function Fields and Moduli of Algebraic Varieties. IX, 146 pages. 1986.

Vol. 1227: H. Helson, The Spectral Theorem. VI, 104 pages. 1986.

Vol. 1228: Multigrid Methods II. Proceedings, 1985. Edited by W. Hackbusch and U. Trottenberg. VI, 336 pages. 1986.

Vol. 1229: O. Bratteli, Derivations, Dissipations and Group Actions on C*-algebras. IV, 277 pages. 1986.

Vol. 1230: Numerical Analysis. Proceedings, 1984. Edited by J.-P. Hennart. X, 234 pages. 1986.

Vol. 1231: E.-U. Gekeler, Drinfeld Modular Curves. XIV, 107 pages. 1986.